Martian Geometry Book 1

Preface

This preface refers to twelve new books of Martian anomalies. Each book is approximately 250-270 pages in length, they also have the same introduction which is about 70 pages long. There are about ten more books partially completed to be published, the books cover anomalies all over Mars and have about 3000 images in total. If you like these books, and would like to support this work, then you can buy the books on Amazon. You can search for "Greg Orme" and "Martian Hypotheses" there. You can also support this work at Patreon at this link: https://www.patreon.com/ultor. If you enjoy the books you can also help with reviewing them at Amazon.

The aim is to raise money with these books to fund an institute to study these formations. If these are artificial then they will need to be studied by scientists from many fields such as biology (examining the faces, their bodies, and fish sculptures), geology (analysing the materials used in their construction), anthropology (why repeated faces with crowns were constructed, perhaps gods or rulers), mathematics (for geometric formations), sociology (how these societies worked), economists (working out how the society functioned, for example with farming, fishing, working together for large scale constructions), engineering (how these formations were constructed), and archaeology (examining ruins). How this would be done is not clear, but this institute would try to make a start on understanding these formations. No one really knows how to study an extinct alien civilization, if this is one. Most likely, if they are real, then a more professional organization would take over this work later. The intention then is to bridge the gap between amateur analysis of these formation to a much better funded organization, perhaps at the government level. The evidence gives a reasonable case for artificiality, but much study needs to be done to determine how plausible this is.

The introduction is repeated at the start of each book. If you have read it you might skip forward to the new images. However it may be valuable to read it more than once, to see how the images you see are connecting into these classifications. Often the images have a lot of details, each time they are examined more of these can be seen. They might also inspire you to see other connections, for example one image might be similar to another in a different part of Mars. This is likely to happen, even with so many images the surface of this hypothesis is barely being scratched. Mars has an area similar to the land area of Earth, this is because much of Earth is covered in oceans. For this much land then 3000 images is likely to have missed many important discoveries.

You can also use the indexes in each book, they refer to many similar formations throughout them. For example, if you are looking at hypothetical road formations then roads in many different areas can be found in the indexes. It would be possible then to quickly see all the different kinds of hypothetical roads in all 10 books. The idea behind the introduction is to give an outline to the global hypothesis, how these different formations connect together into a hypothetical Martian civilization. It's important then to get an intuition of how these formations connect together globally.

Some areas for example might have hypothetical roads for transport, other might have hypothetical tubes like a covered road. Different terrain, available materials, and climate might have led to one being used over the other. It may be as Mars cooled it became necessary to travel under cover because of the cold. Another possibility is predators or meteors made traveling on roads too dangerous. Also there are many hypothetical dam formations, but the construction techniques vary between areas. Some are formed with dam walls attached to the crater, when they break some show a cavity under them and others do not. This would indicate the dam wall was dug into this cavity to keep it from sliding down the crater wall. In other areas this was not necessary, it may be that there the crater wall was harder rock which the dam wall could be cemented to. Some show columns and layers in them but others have evenly spaced vertical grooves on the dam walls. Some dams are excavated out of the crater wall or the material at the bottom of the crater, these may depend on the rock type in the crater. For example, if the crater wall is too easily broken then an excavated dam might have been the best engineering solution. Some areas have hollow hills, these are where a hollow habitat may have been built on an existing hill or the whole hill was constructed. In some areas these have layers similar to a Cobler Dome, this is where bricks form the dome in decreasing circles as the dome is built up. These are called amphitheatres as a friendly name, the first amphitheatre formation looked more like seating around an amphitheatre. Other hypothetical buildings have no layers in their roofs. This may have depended on the materials available. Many appear to have a smooth skin like cement which has broken up in some parts of the roof, and is intact in others. In many areas this is more intact on the southern side, as the skin breaks off the softer inner parts of the roof appear to have eroded faster and collapse. The one sided erosion may imply a prevailing wind, or as the oceans and air froze at the pole this created the erosion.

There are also large areas of walls and room like shapes, these are hypothetical cities. Other areas connect these hollow hills together with tubes or roads as another kind of hypothetical city. Still others seem to be made of tubes that connect together in intersections called a tube nexus. This may have been because of the climate further from the equator, for example tubes might have been used to travel through in colder areas. The Martian Faces are mainly discussed in books 11 and 12, a reprint of published peer reviewed papers. These differ according to where they are. The Cydonia Face, Nefertiti, and King Face all fall on a great circle, this is hypothesized to have been an old equator that lines up with a known previous pole position west of Hellas Crater. The newly discovered Queen Face is in Cydonia but not near the old equator. If the faces were used to mark latitudes and longitudes then the overall system remains obscure. For example there is a large hyperbola shown close to the old equator. Another is far from this equator, but drawing a line from it to Nefertiti gives a right angle to this old equator. Joining these two hyperbolas and the King Face gives an Isosceles Triangle. The hypothesis of these mapping system is highly speculative at this stage.

Canals, lakes, and water channels also vary in different areas. West of Cydonia there is an extensive array of hypothetical canals, also east and west of Elysium Mons. Some of these connect to larger lakes which may be artificial. Some hypothetical dams have water channels to direct water into a dam, and to collect an overflow to another dam.

There are also darker areas often bounded by walls or geometric shapes. These may have been farms, why they appear in some areas like around Cydonia and in Isidis remains unanswered. Other areas contain hypothetical artefacts but no farm formations, so these creatures would have used a different way of collecting food.

The idea of these books then is not just to prove artificiality, but to try to prove a global hypothesis of how the whole civilization functioned. Once the evidence becomes plausible enough, and the shock wears off, this larger question is much more interesting. Each section is labelled with the title hypothesis to make clear these notions are being proposed along with the evidence there. The sections all have many keywords connecting to the index. If you see a connection to a kind of formation then it is easy to find similar formations. In seeing the global hypothesis the different pieces of the puzzle are more likely to come together, for example the hypothesis of dams sounds less plausible if it is not connected to the hypothesis of buildings and farms. Together they give the ideas of habitation, food, and water. The conclusions can be controversial. However there is so much evidence it was better to put it all together into a more comprehensive hypothesis. Otherwise people are looking at isolated formations like faces without seeing the overall context in which they appear.

Introduction

Many people have seen, or heard of, the discovery of faces on Mars. Often they are sceptical about this. One common objection is the faces look too much like us to be an alien race, so researchers are recognizing faces in the terrain that aren't there. This has also been an objection to possible discoveries of bones, statues, even small animals. The mainstream view is that these are the products of people's imaginations, often this is a fair comment. Historically though, people have believed in a Martian civilization, whether still existing or extinct. This was explored in many science fiction books from Edgar Rice Burroughs and Arthur C. Clarke to Robert Heinlein. Many expected Mars to be habitable, or even inhabited, when the Mariner 6 and 7 spacecraft went to Mars in 1969. What was found instead was a near airless world devoid of water. The conventional wisdom was turned on its head, that Mars had never been inhabited and probably never had any life at all.

From this time forward the mainstream scientific opinion was that Mars had always been devoid of life much like our own Moon, so anything that looked artificial was just people seeing things. This is called Pareidolia, seeing illusory faces and animals often in clouds and random patterns. The problem in overcoming these legitimate objections was that spacecraft imagery was low resolution, it could only map the surface of Mars very slowly. So if signs of an extinct Martian civilization did get imaged then they would likely be ambiguous in this low resolution, and be dismissed as fringe science and illusions. But these anomalies have kept turning up as the spacecraft imagery became higher in resolution, more able to see signs of this civilization if they existed. Mars is now largely mapped to a fairly high resolution, called the HiRise and CTX images, so many unusual formations have been found. The situation has also continued to be toxic for mainstream science, some use their imaginations too much and see things that really are not there. This tends to scare away mainstream researchers, they are rightfully concerned that too much speculation can damage their careers. But other formations are not so easily dismissed.

Another complication is that this hypothetical Martian civilization would have died out perhaps billions of years ago. This is because Mars had a warm climate and oceans long ago according to NASA, but being further from the sun it cooled with the atmosphere and oceans freezing at the poles. With billions of years of erosion many possibly artificial formations look more natural over time. The evidence has then been ambiguous and highly eroded, but with thousands of possible artefacts being found.

One problem for mainstream science was in understanding what was actually being claimed by researchers. Mixing more plausible artefacts with illusions also makes the claims less logical. For example finding skulls and boats runs into the objection of bone and wood quickly eroding under the surface conditions. They might also give the impression that boats may have been used in an area that had no oceans or rivers. Separating the more plausible artefacts then improves the quality of these hypotheses. This may help to answer the questions of who constructed them, where they lived, how they created these formations and why. If hypothetical aliens came to Mars, then why would they build faces and not another kind of formation. Some might have preferred finding large geometric shapes or perhaps a representation of an equation. These have been found as well. But the problem then was not just what was found made little sense, but that it did not fit into the preconceptions of mainstream science of what they should find.

It became necessary to try to connect these ambiguous formations together into a global hypothesis. In that case mainstream scientists and others could see all the evidence and how it connected together. As will be shown, the evidence looks like a civilization but one profoundly alien in some ways. It likely covered most of Mars, life tends to extend to wherever it can survive. So, to understand this global hypothesis, images from all over the globe of this evidence need to be viewed and seen holistically. Sentient creatures should have learned to tame the climate and can live in wider temperature ranges, also where water is plentiful or scarce. We should expect a hypothetical Martian civilization to do the same. In different areas the evidence should point to different adaptations.

Methodology

The main methods used with these hypotheses are falsification, the law of large numbers, and the reduction to the absurd. Falsification means that the null hypothesis, that these formations are random geology, cannot be true. This is because geology perhaps could not create structures like this. The other method is the law of large numbers. That there are too many of these structures to be from the occasional coincidence. For example the parabola appears to have been used extensively in these formations, it has been used on Earth in many dams because of its load bearing properties. It is also used in parabolic domes. In these Martian formations there are 945 parabolas which are shown and outlined. These outlines are from geometric parabolic shapes, in some cases they might be widened or narrowed. This does not affect their load bearing properties, they are still described by a simple mathematical formula $y=ax^2$ where a is a variable. This is a large number, there are formations like dams in many craters and most of them are parabolas as will be shown. It would seem highly unlikely that they eroded into parabolic shapes as these dams are formed in many different ways. Parabolas are not known to be associated naturally with formations like these. In some cases a reduction to the absurd might be applicable. This might be hard to define scientifically but it may be apparent to some readers that a natural explanation is absurd. This should be used with some caution as some patterns can form by random chance or be illusions. However the human eye is good at seeing real patterns and is not so easily fooled.

A basic global hypothesis

The next section goes through a number of different types of hypothetical artefacts. These should be looked at as a whole, how each connects to the others. They can be regarded as components of a viable civilization such as buildings, water supplies, farms, roads, artistic works, etc. The significance of a hypothetical road then is also what possible buildings it connects to. A farm is significant in the context of possible buildings near it. Possibly artificial canals and lakes are significant in terms of their proximity to ancient oceans, also to dams in craters collecting groundwater.

Faces

The Queen Face

One of the most controversial problems with the evidence accumulated has been the discovery of Martian Faces. That they appear to look like us raises the suspicion of Pareidolia, like seeing faces in clouds. However Mars and Earth would have had their ecosystems connected by panspermia, this is where life can be transferred from one planet to another by meteors. We may then have had a similar genetic background, and so plants and animals may have evolved to look similar on both planets. Panspermia is a just a hypothesis, but we don't know whether DNA from Mars might have caused us to evolve later looking similar to Martian life. The Queen Face was discovered by the author recently, it is close to the Cydonia Face which was the first Martian Face discovered in 1976. There are about 30 Martian faces of varying degrees of plausibility. Some might see these reducing to the absurd, that the idea these could all form naturally as absurd in a way that is hard to define. Others might see the number of faces as statistically significant, a product of the law of large numbers. Still other might be unconvinced or believe they are random or illusory. Some find them quite shocking with the impression of artificiality they give.

This shows two versions of the Queen Face from different CTX images. It appears to have hat like a crown, like most of the other Martian faces.

The High Face

Most of the Martian faces are found in a small valley in Libya Montes, near the better known Crowned or King Face. This is often referred to as the King's Valley, a similar name to the Valley of the Kings in Egypt. The High Face is named because it is high on a cliff overlooking the valley. The faces are discussed in two papers in Martian Hypotheses Volume 11. A statistical argument can be made, as to why so many faces would be found next to each other or to be on a great circle bisecting Mars.

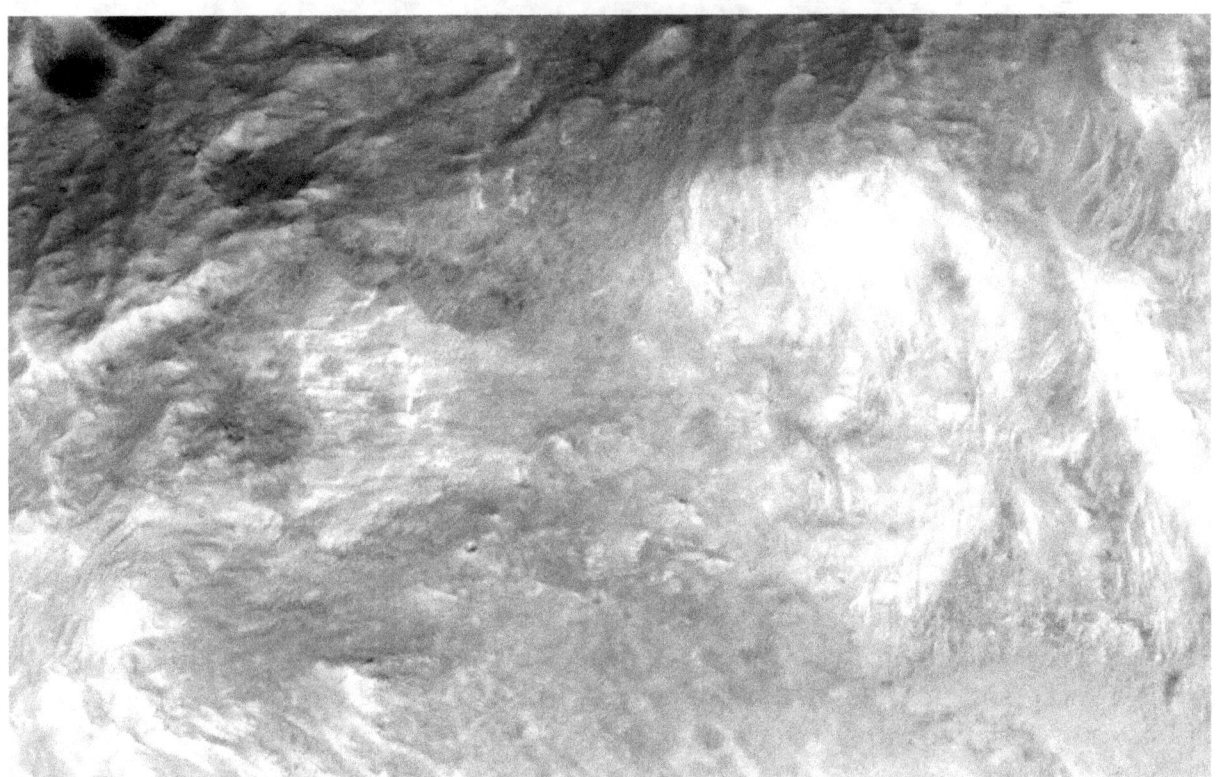

The Meridiani Face

This face was discovered in a Viking image by a Martian researcher Terry James. It is also discussed in Volume 11.

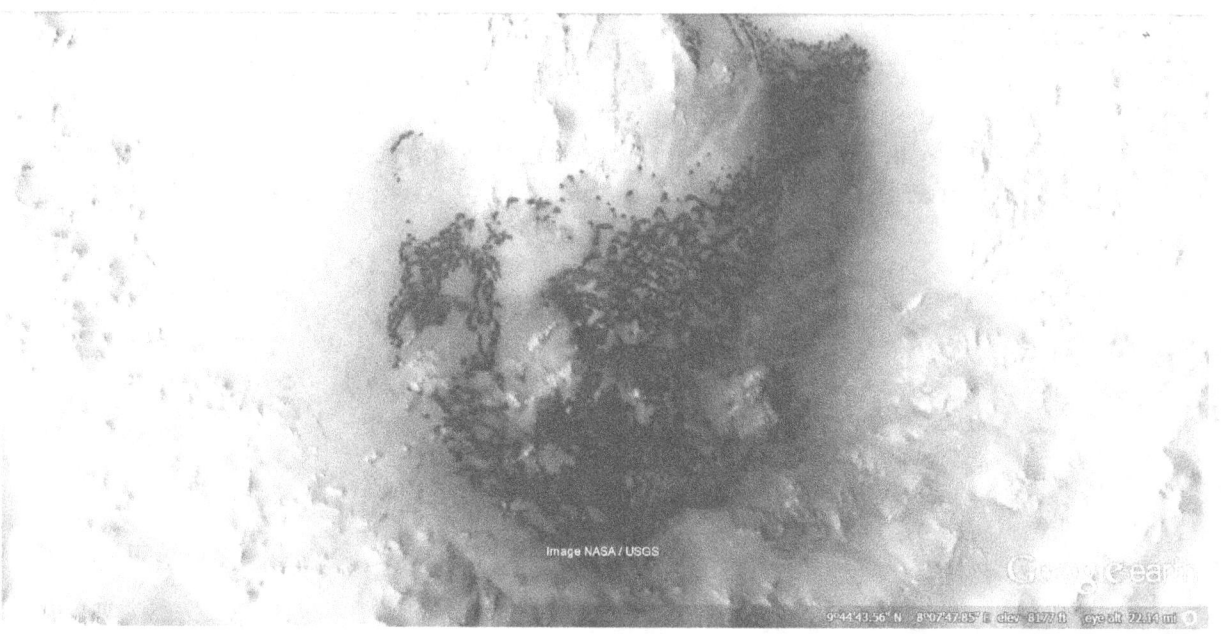

Image NASA / USGS

Nefertiti

This face was discovered by JP Levasseur, it is discussed in Volume 11. The two inserts are from higher resolution images that were recently taken by the HiRise orbiter, they were added by the author. It missed the whole face but shows some of the hat and face. It represents a successful prediction, that higher resolution imagery would make these formations more face like rather than appearing more natural.

The King Face

The King Face was discovered by the author in June 2000. It has been called the Crowned Face, however with the discovery of the feminine looking Queen Face the name King Face may be more appropriate. Whether they had sexes or if we could tell the difference is another hypothesis.

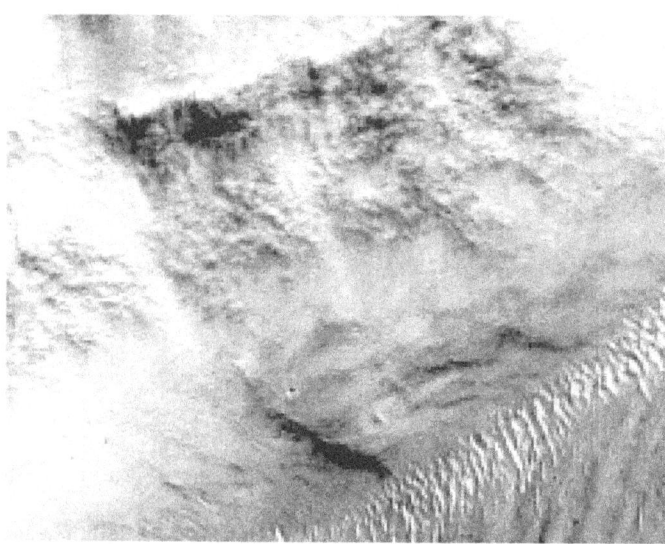

Dams

In many craters there are formations that look like dams, these seem to follow an old Martian equator implying that water may have been liquid in an equatorial zone. This old equator hypothesis is discussed more in Volumes 11 and 12. Most of these dams are parabolic in shape, the hypothesis is that parabolas are well suited for load bearing in dams. From here the analysis from the book is included with each example image.

Cymd259c

Hypothesis

These dams are in the same crater, A which appears parabolic and B have smooth walls with a few cracks as shown. B at 4 o'clock has a sharp edge to the dam wall in good condition. C at 4 and 6 o'clock show a secondary dam perhaps to catch the overflow, the second line at 6 o'clock shows the base of this wall. D shows another section, perhaps parabolic, with a cracked wall at 5 o'clock. C at 10 o'clock shows a probable parabolic arch. There appear to be faint vertical ridges on the upper part of the dam walls as seen in other dams, these may be for strengthening the wall such as there being pillars inside.

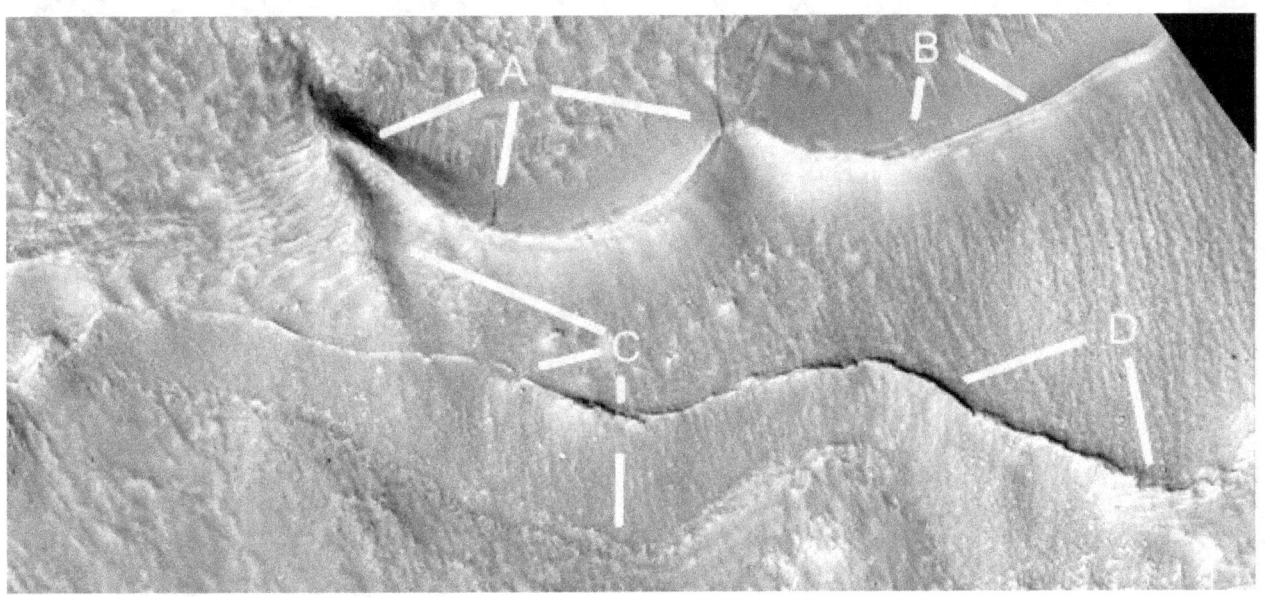

Cymd259c2

Hypothesis

A parabola is shown.

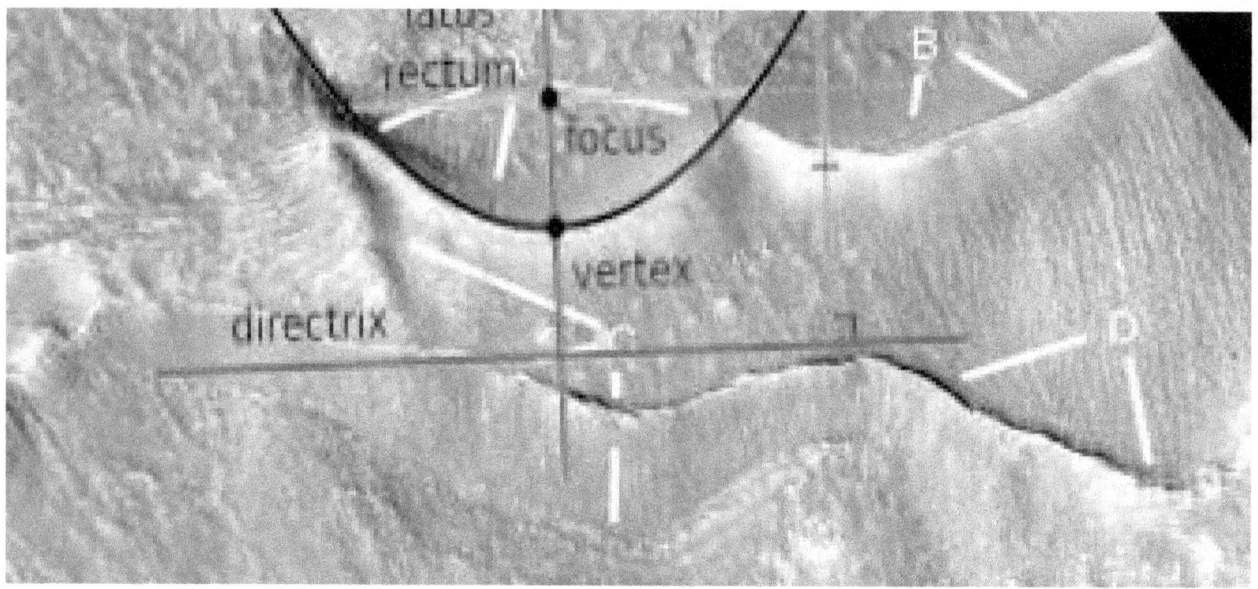

Cymd280a

Hypothesis

A shows how the skin on the dam wall is peeled off, at 3 o'clock is has many pits like on the skin of hollow hills. At 4 o'clock this rough interior is exposed but just below it the skin is smooth. At 6 o'clock is another edge of the smooth skin. B shows at 8 o'clock. How it is peeling off, at 5 o'clock it is more stable. At 10 o'clock there are many pits as it degrades, at 2 o'clock it shows the lip of the dam has broken off. C shows a smooth area that goes up to the broken lip of the dam wall like an external layer, perhaps a patch.

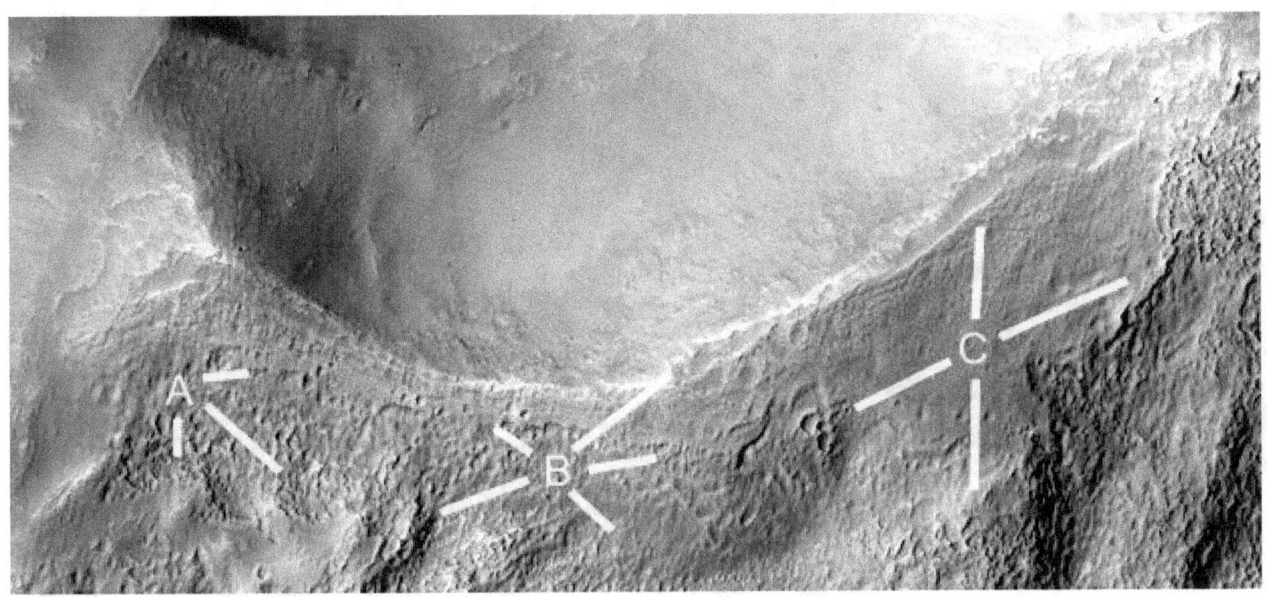

Cymd280a2

Hypothesis

A parabola is shown.

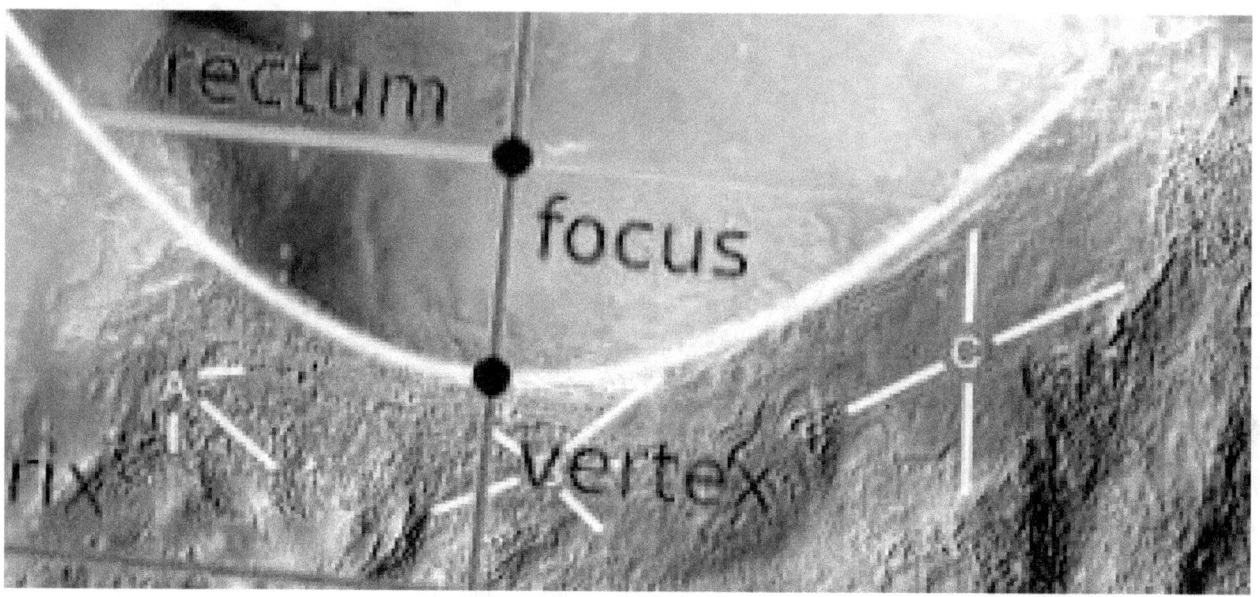

Cymd280i

Hypothesis

Engineers might examine how this wall is fracturing at A to D, Also D at 2 o'clock shows the thicker base holding the dam wall in place. Above C the dam floor is smooth like cement, higher up and outside the dam the terrain is much rougher.

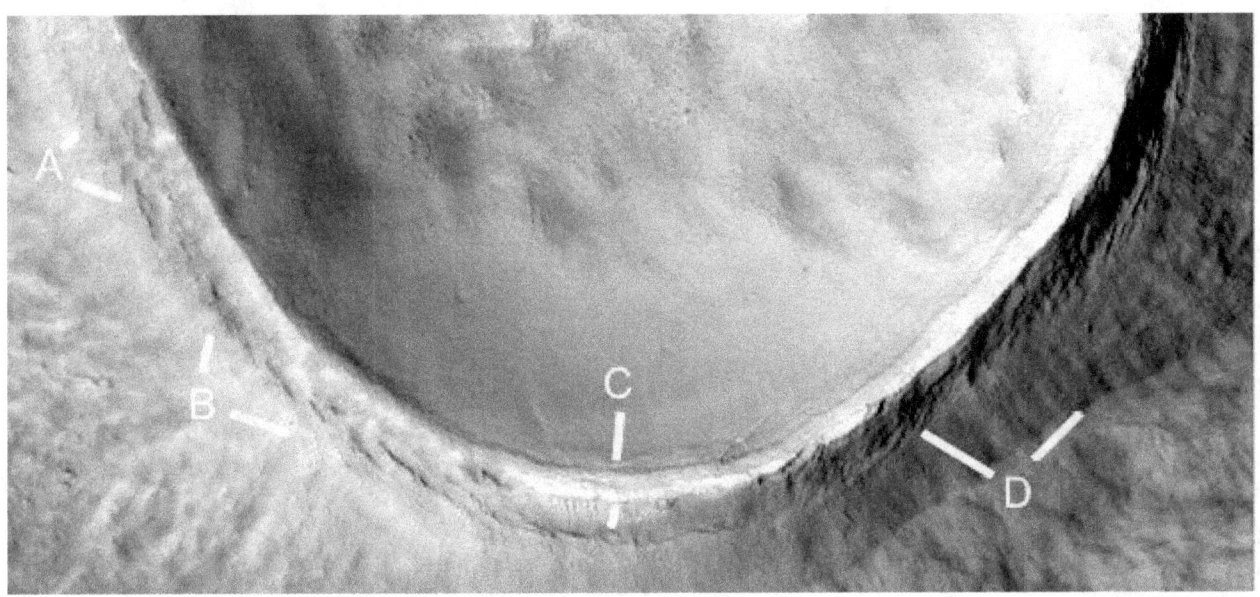

Cymd280i2

Hypothesis

A parabola is shown.

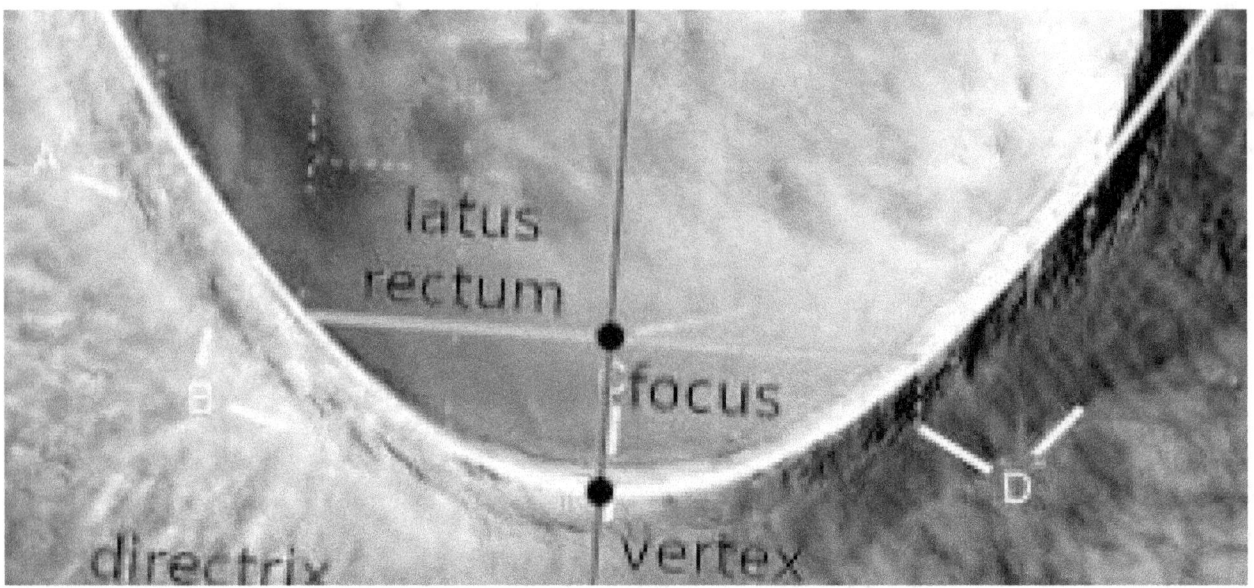

Cymd408a

Hypothesis

An unusual shape pointing up the crater wall, A is one dam, B may show some creep or cold flow in the dam, this where over time rock might slowly flow like a viscous liquid. C shows a smooth dam floor like cement, different to the terrain outside the dams. D at 7 o'clock also shows the smooth dam floor compared to the ground above it. At 2 o'clock the wall is eroded or breaking.

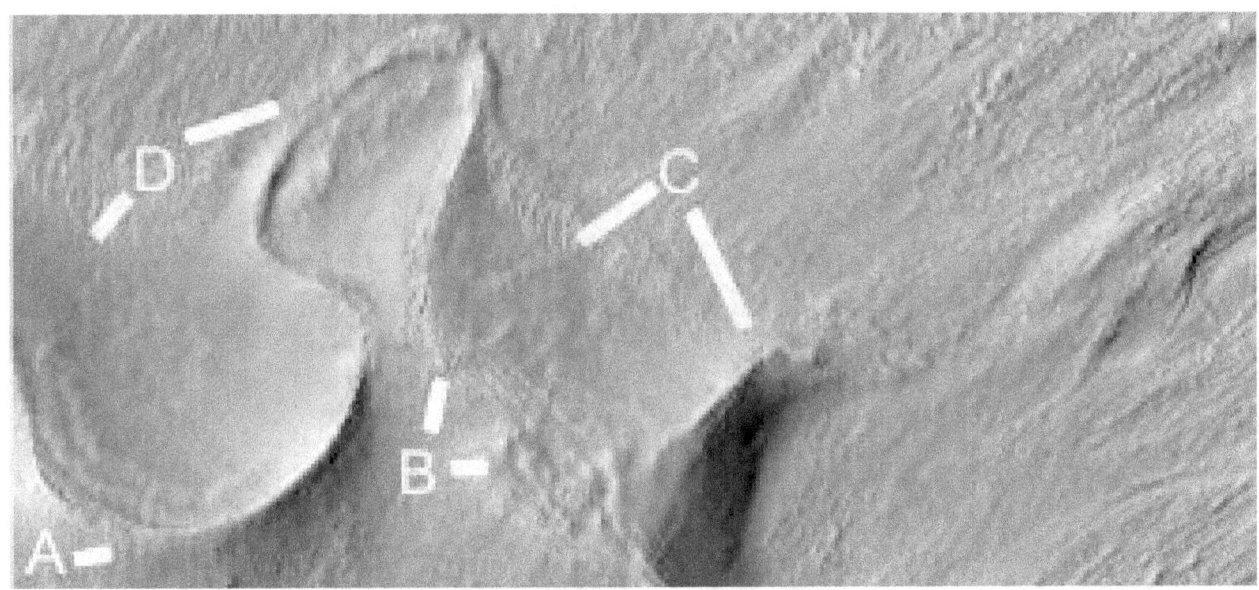

Cymd408a2

Hypothesis

This shows 4 parabolas making up the formation. These would have used the load bearing properties of the parabola to resist erosion. The straight dam at B may have broken because it did not use a parabola.

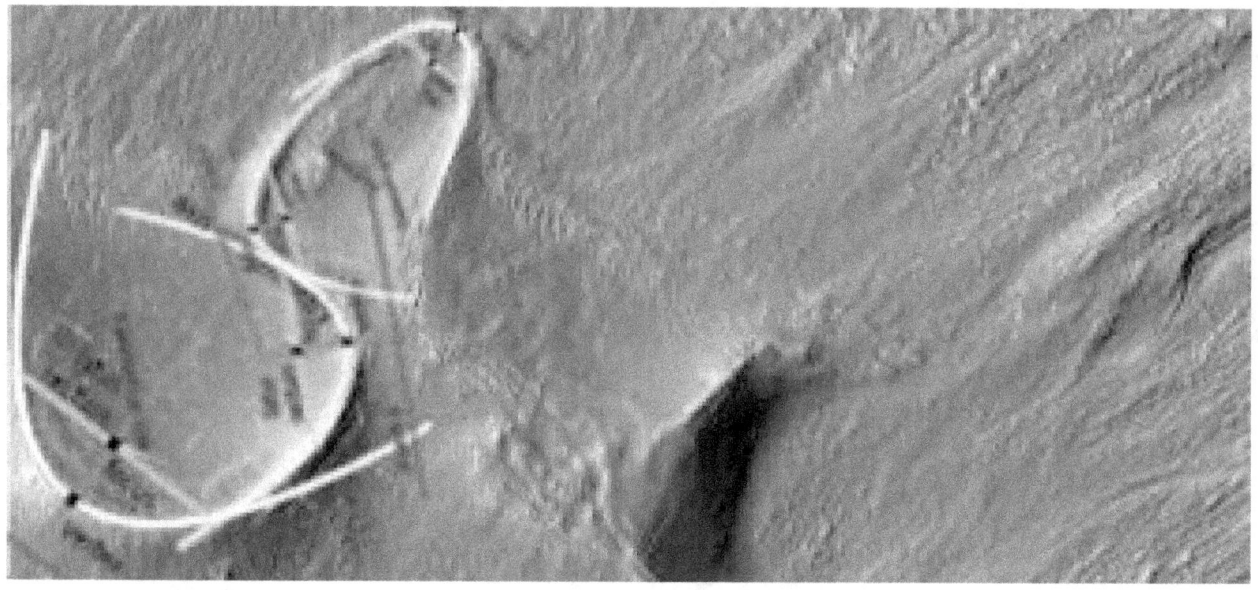

Argd1444a

Hypothesis

Eighteen parabolic dams are shown. A few others are too eroded to determine their shape. It would seem impossible for eighteen mud slumps to happen to form perfect parabolas, above them the materials look highly random by contrast.

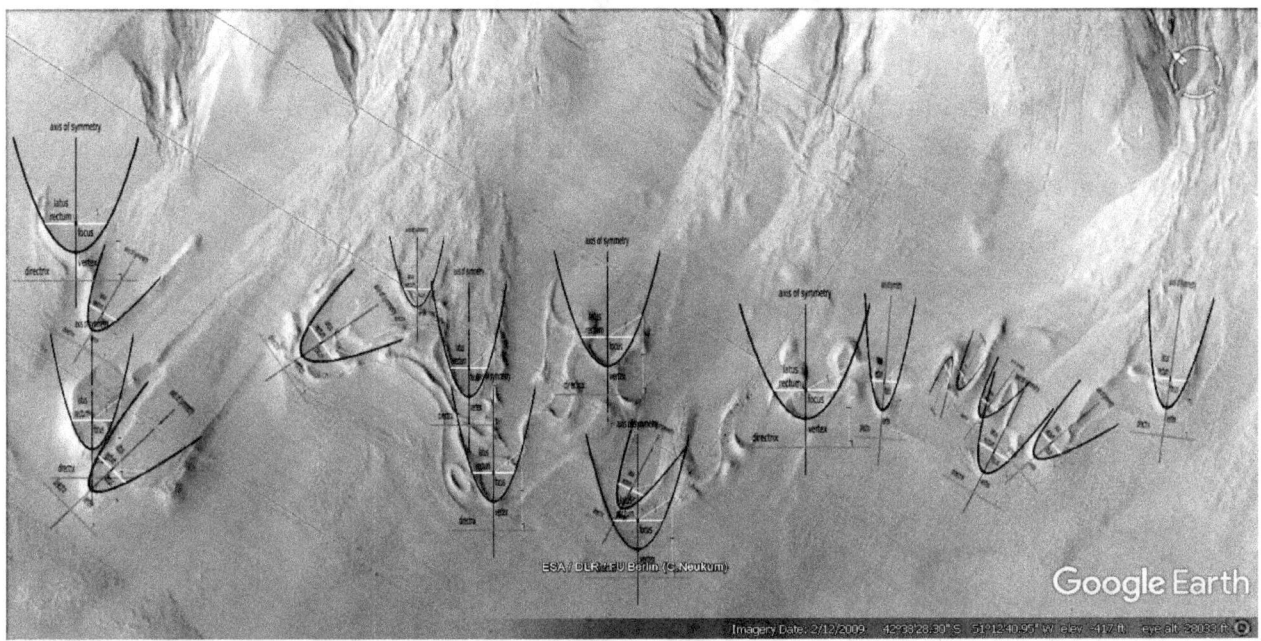

Canals

Some areas near hypothetical Martian buildings and dams have these canal like formations. The hypothesis is that water was important in this civilization, they used dams in craters to collect water often associated with water channels and perhaps pipes. In other areas canals may have brought water from the lakes and oceans, perhaps irrigating farming and residential areas or even for transport using boats. This is what we use canals for on Earth.

Prca480

Hypothesis

More of these tube shapes, A shows dark spots along it like it is breaking up. B at 9 o'clock is like a hollow hill as seen in many other areas, the dark patch on top may be the roof. B at 5 o'clock shows more collapsed areas. C at 7 o'clock shows the bank is well defined, at 4 and 8 o'clock the tube shape changes from dark to pale. At 10 and 4 o'clock the bank is also well defined.

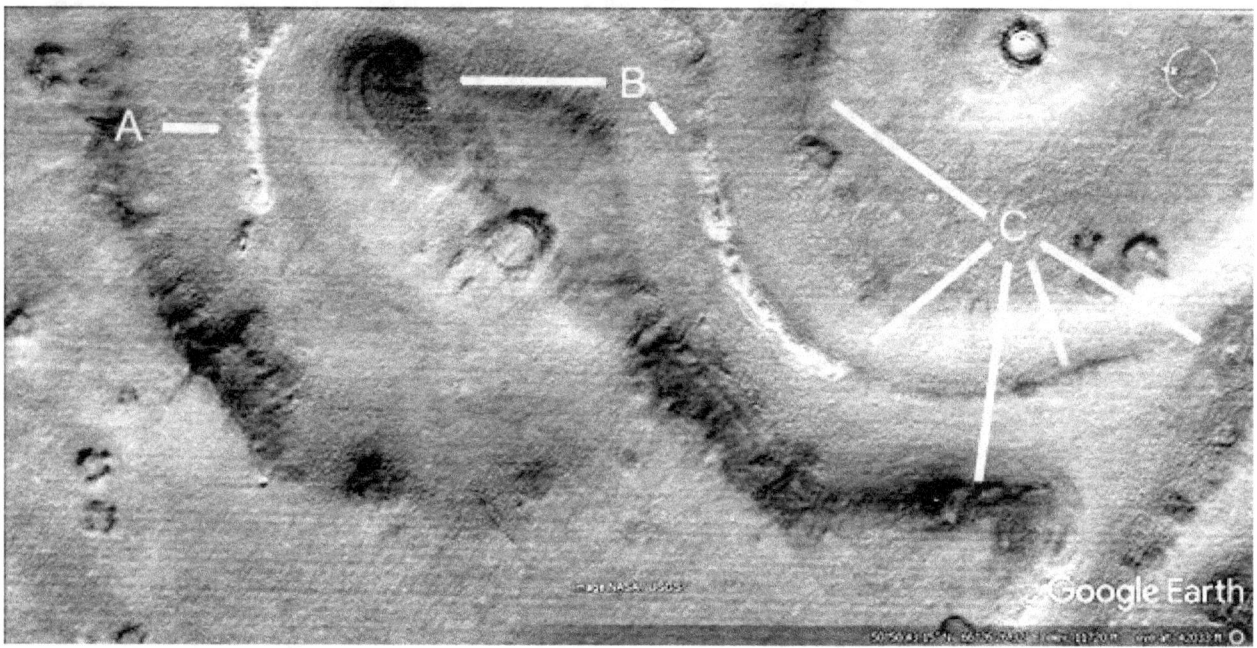

Prca480a

Hypothesis

This part of the tube shape is a near perfect parabola as shown, unlikely to occur by chance. The tube shape is also about the same height and width wherever seen, it does not vary much randomly like a natural formation from weather erosion. Also parabolas are shown in canals as well as dams, a natural hypothesis would need to explain how geological processes formed parabolas in each. They also appear in hypothetical buildings and as walls around possible farms.

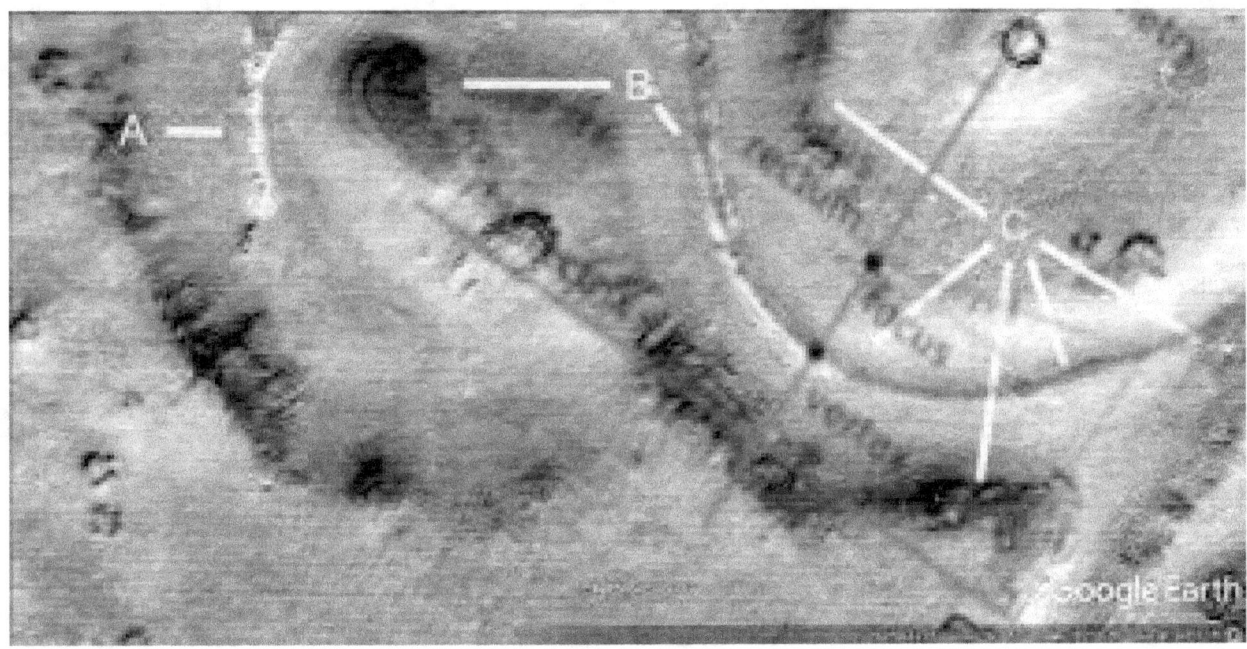

Ect1619

Hypothesis

A shows a much thicker wall with a line running along it as a peak, from 4 o'clock to B at 5 o'clock, up to E. This may have been a habitat connected by hollow walls. At 2 and 6 o'clock A shows a clean edge like cement to the dam floor. B at 9 o'clock shows a double wall like a collapsed tube. At 3 o'clock B shows a small hill or dark area. C may be a collapsed hollow hill, the ridge shown may have been an interior support and part of the larger hollow wall. D shows a darker line perhaps a collapsed wall, also a narrow wall like those in Hellas at 1 o'clock second leg.

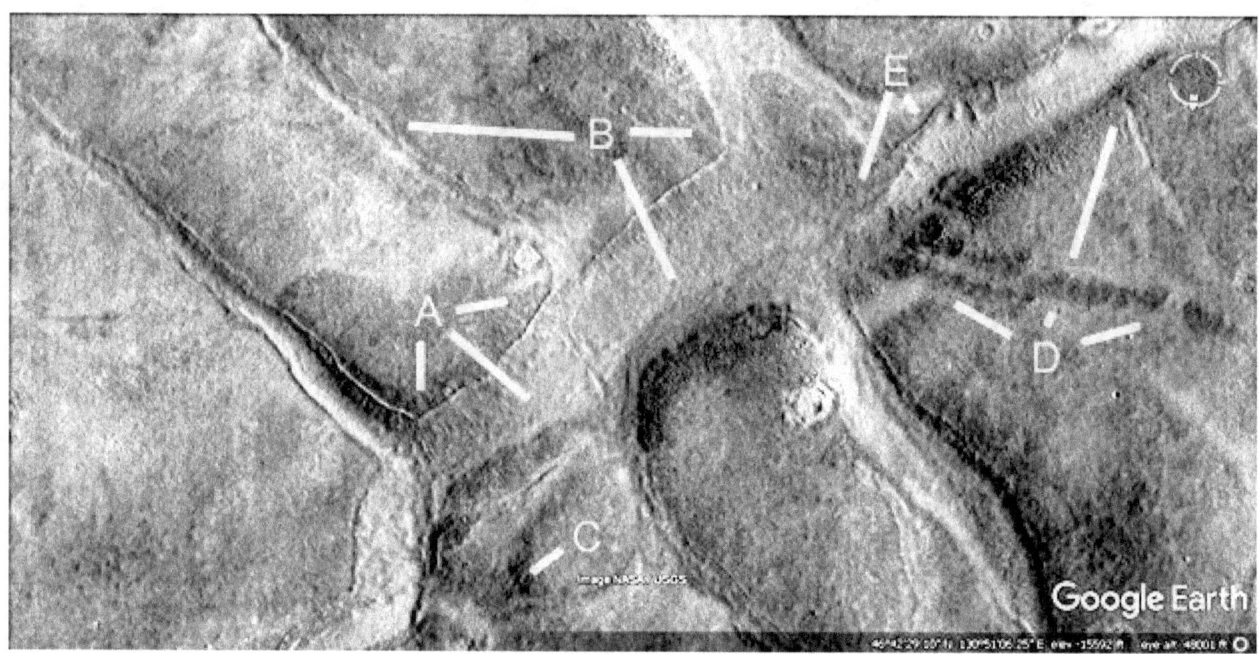

Ect1619a

Hypothesis

Four parabolas are shown.

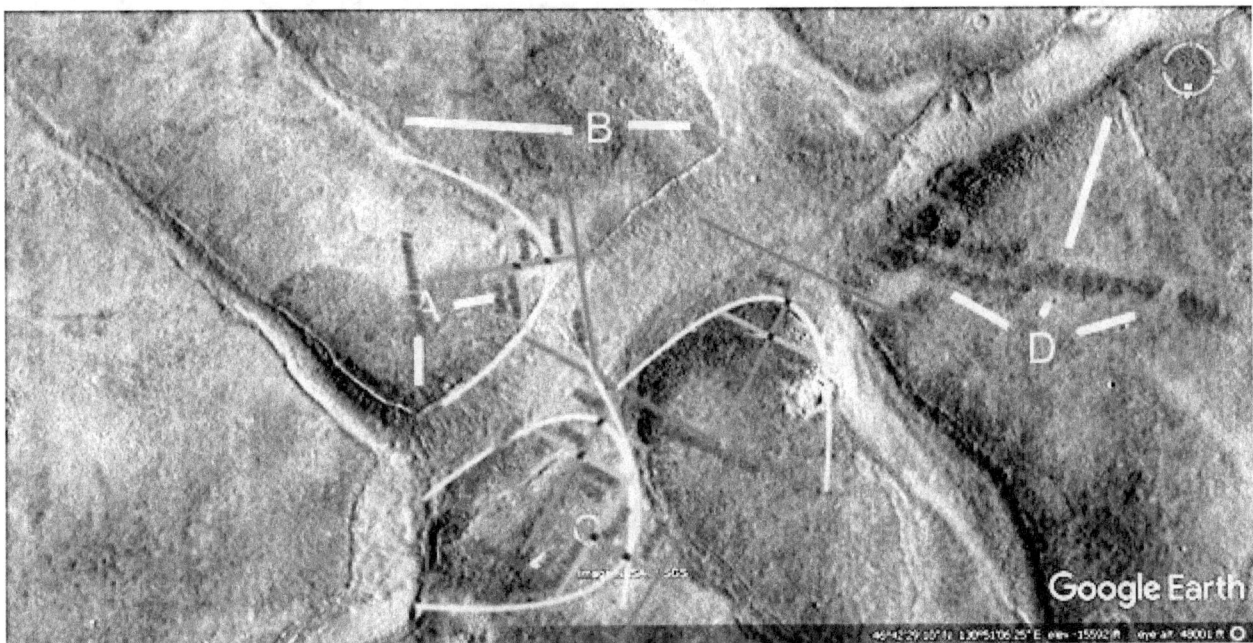

Ect1643

Hypothesis

A shows more ridges like grout, these connect into the canal wall at B but do not extend into the canal embankment. C shows regular spacing like tiles at 11 o'clock, squarish tiles at 3 o'clock, and a collapsed tile segment at 6 o'clock. D shows a gap growing between the bank and the wall, also with regular tile spacings. At 6 o'clock second leg there is a ridge like grout. E shows more grout connecting to the canal wall like a single segment. This cannot be cracks then because it must be the same material as the wall, probably cement. F shows more tiles.

Ect1643a

Hypothesis

A parabola is shown.

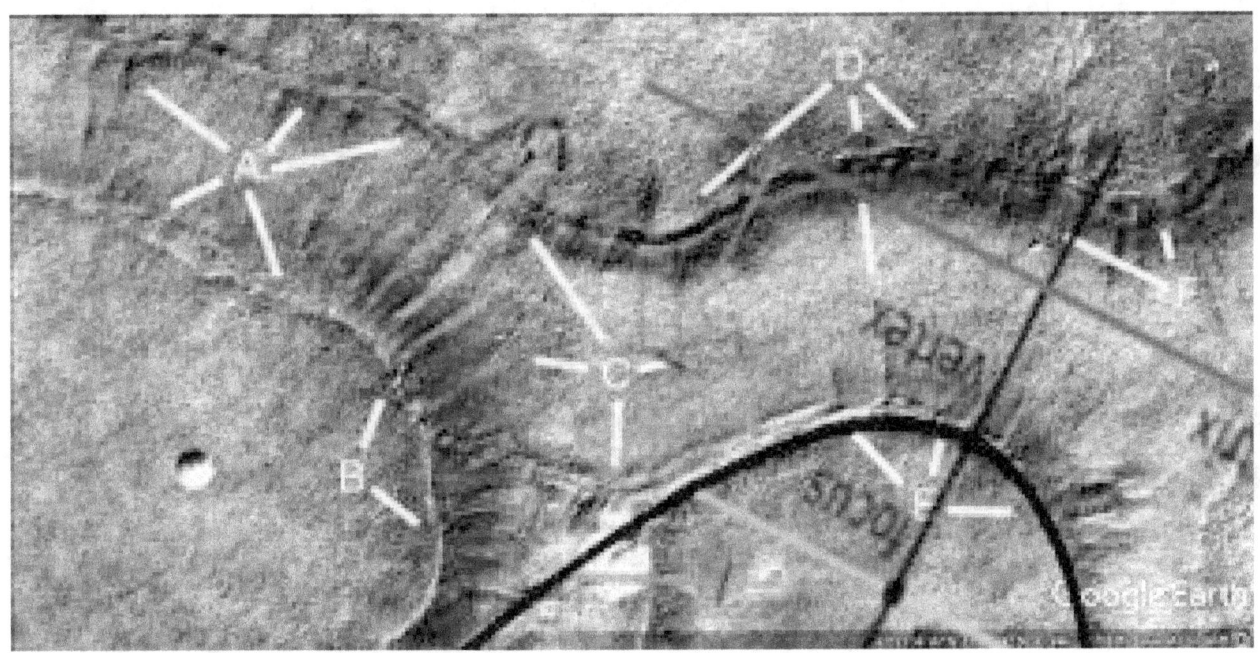

Water channels

Water channels can encompass the conduits feeding dams in crater, they can extend up to the hypothesis of large scale canals. They would have been important, to direct water into dams instead of being dissipated into the ground. Also there are overflow water channels which appear to direct water from an overflowing dam to another so as not to waste water.

Prd965c

Hypothesis

These may have been canals or pit dams, they are highly geometric in shape. A shows a dam for water at 12 o'clock, another wall for a dam and channel at 3 to 5 o'clock. B shows a wall for a canal from 2 to 7 o'clock, it has a groove running along the top like a double wall.

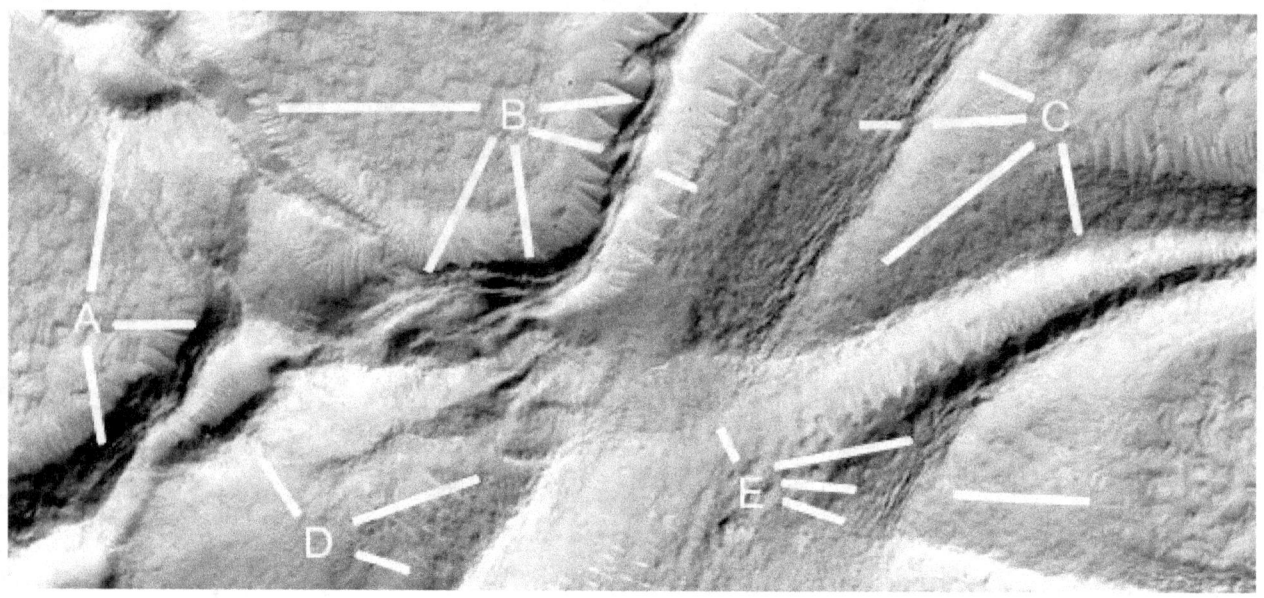

Prd965c2

Hypothesis

Part of a parabola is shown. The lines show how straight parts of the formation are.

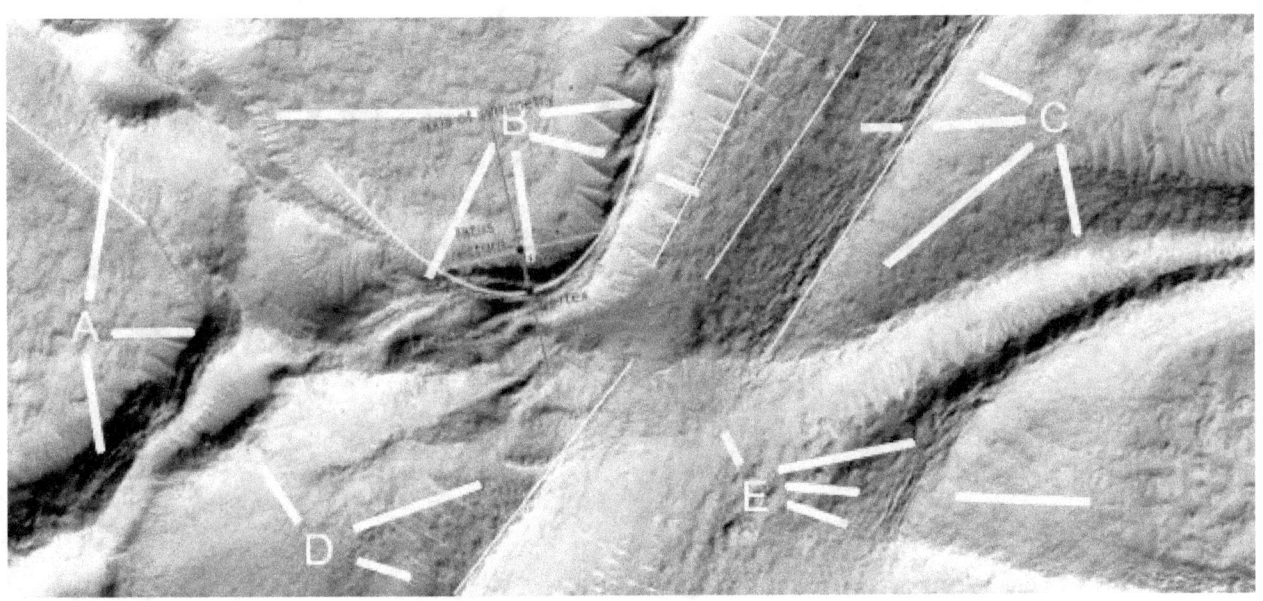

Cymd454h

Hypothesis

A and B show the sides of a water channel, water would have flowed across this at C to another dam. The shape appears so artificial that a natural explanation is hard to sustain.

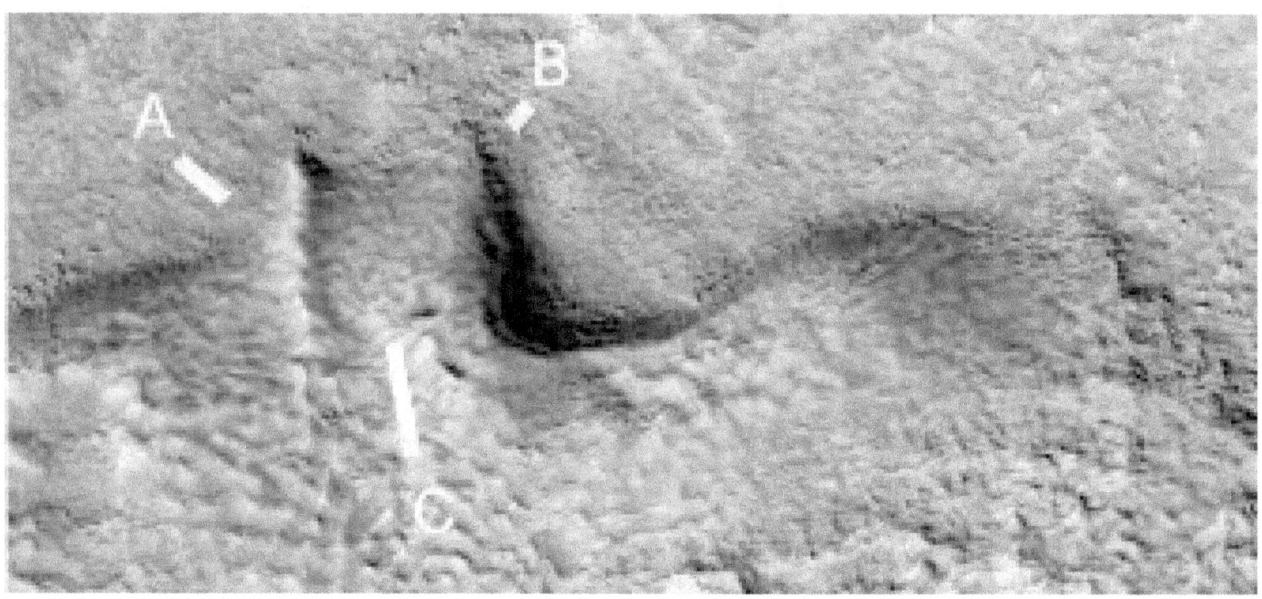

Cymd454h2

A parabola is shown.

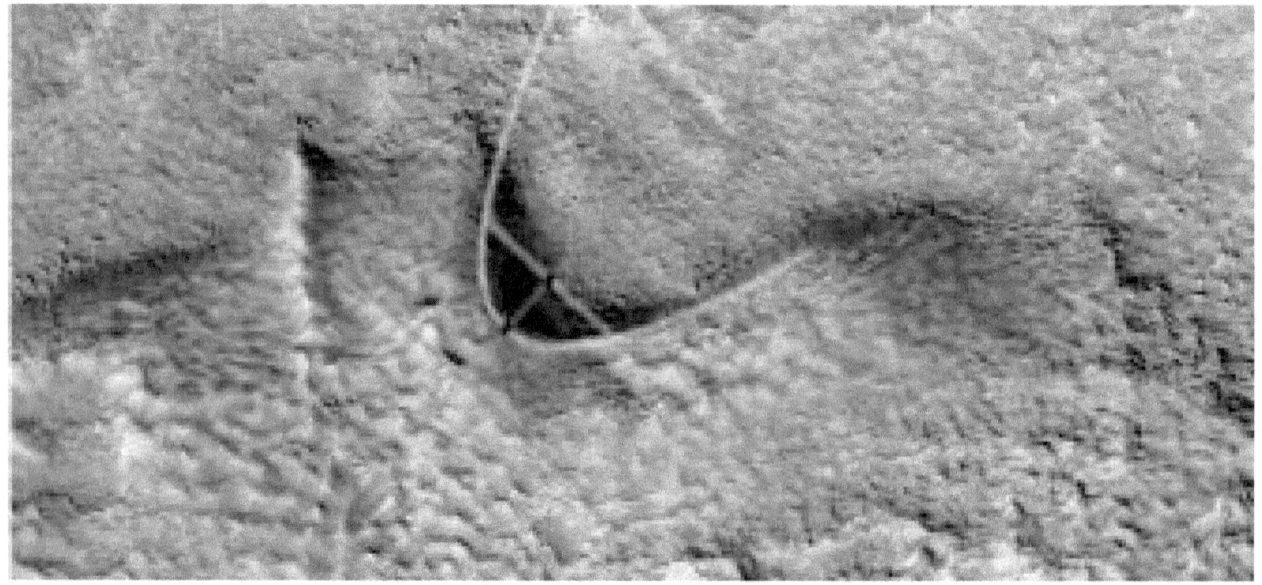

Held1095f

Hypothesis

A shows more dams, turned on its side to fit into the page. B shows a dam wall in good condition at 11 and 3 o'clock, one with cracks at 5 o'clock. C shows more cracks at 5 and 6 o'clock, in good condition at 7 o'clock. D and E also show walls in good condition. F shows more cracks developing.

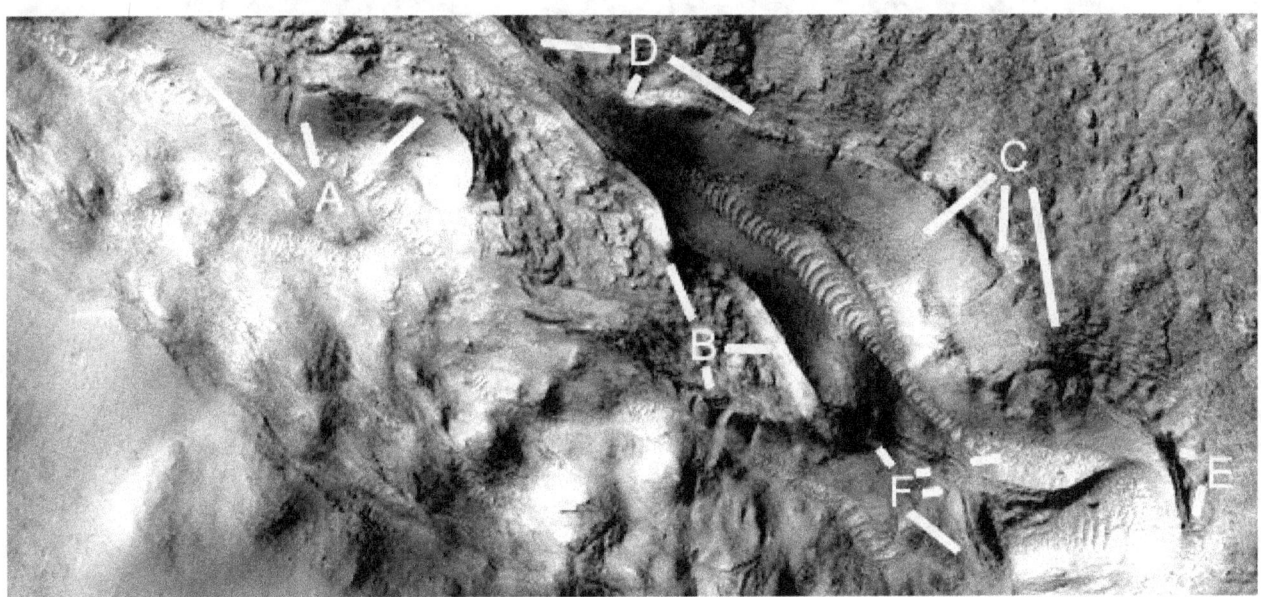

Held1095f2

Hypothesis

At least 5 parabolas occur in the formation.

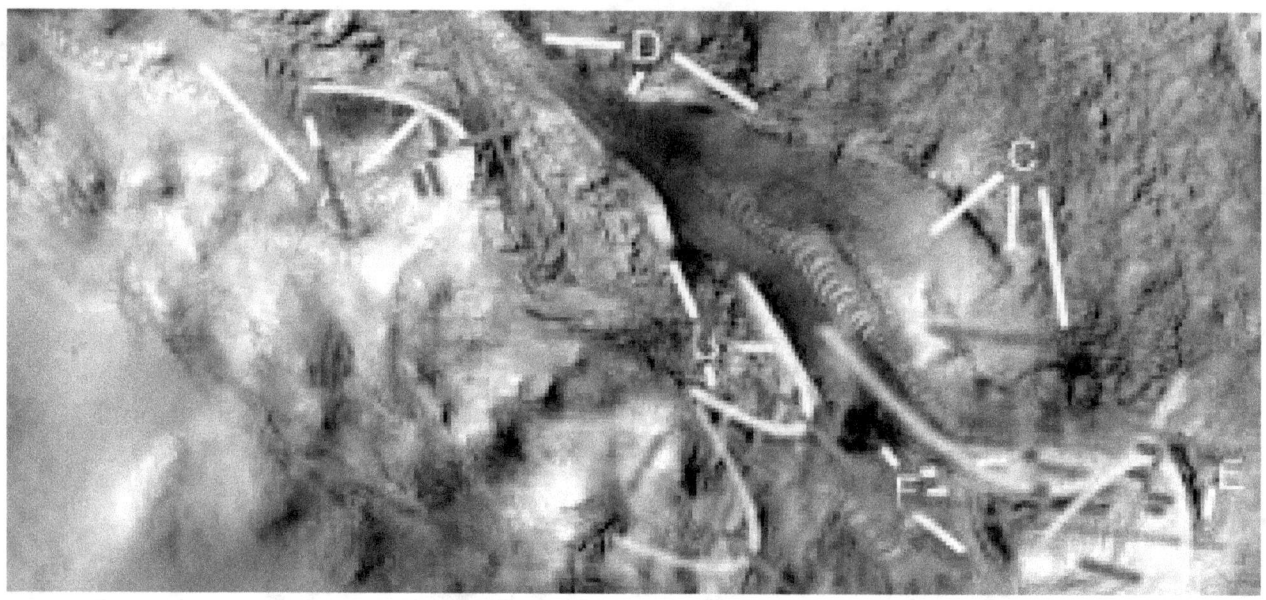

Ect1731k

Hypothesis

A shows a water channel going into a pit dam, B shows another water channel coming from this from 10 to 4 o'clock, also another water channel at 7 o'clock second leg. C shows a water channel coming from the other side of the pit dam to B. D shows a small water channel connecting two pit dams.

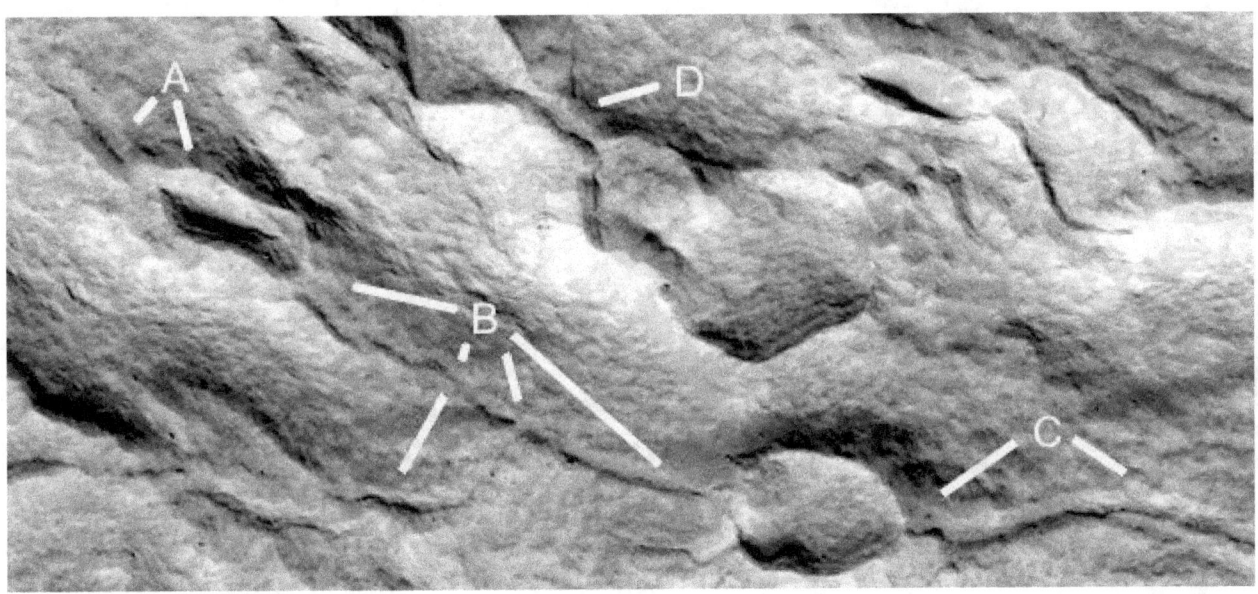

Ect1731k2

Hypothesis

Eight parabolas are shown, though there would also be some smaller ones and the water channel at C.

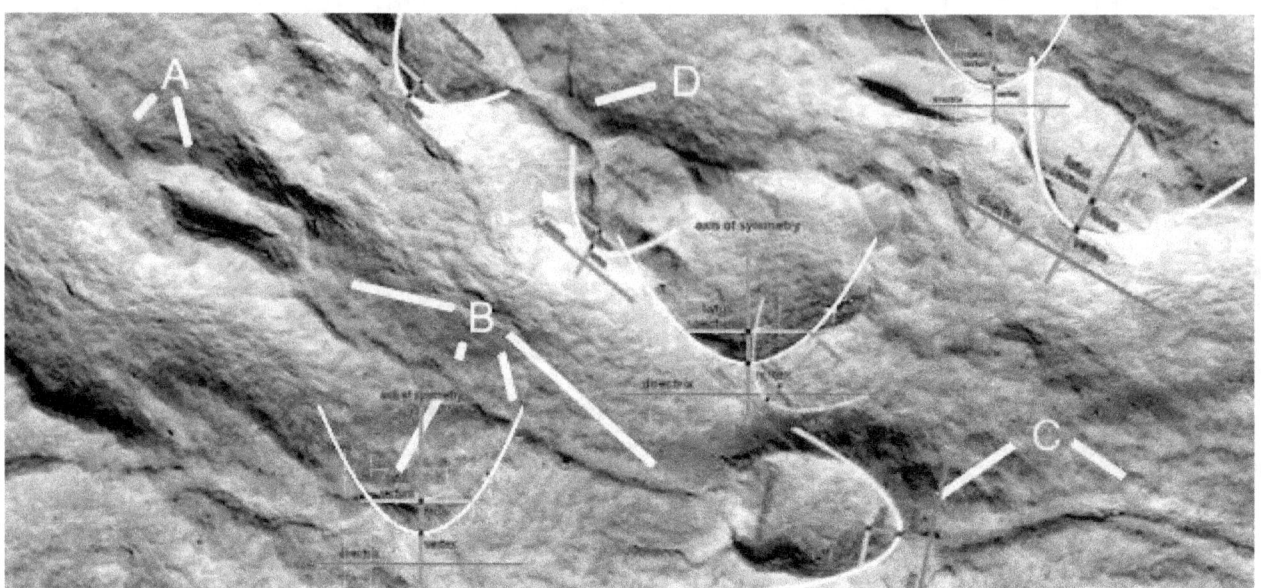

Cities

There are formations that look like cities, these are also clustered around this old Martian equator. Some are also clustered around large extinct volcanoes like Olympus Mons. It adds to the global hypothesis, that these creatures lived together in these buildings in warmer areas.

Cymhh209o

Hypothesis

A shows many rooms, also the walls here appear to be doubled or are collapsed tubes. This is important for the room hypothesis, if someone could go to each room in these tubes then each is accessible. If not then how many could be used is problematic. The thicker ridges also appear hollow at some points elsewhere, B shows a main tube that has some collapsed areas along it. C shows an area that may have eroded to the bare ground, there are faint walls here the same as in the other parts. C at 11 o'clock has very high walls as see from the shadows. Engineers could calculate the height of these walls from the shadow knowing the sun angle from HiRise. The higher the wall the longer the shadow would be inside the room. At C at 8 o'clock the walls are lower as if eroding. D at 5 o'clock shows a rounded formation of rooms like a nexus, at 8 o'clock the walls have collapsed apparently leaving some pillars standing in some cases. E shows a zig zag in this wall or tube, as if the access to it gives straight sections for the entrances. F shows areas where the ceiling appears to have either fallen onto the walls or is still secured above them in parts.

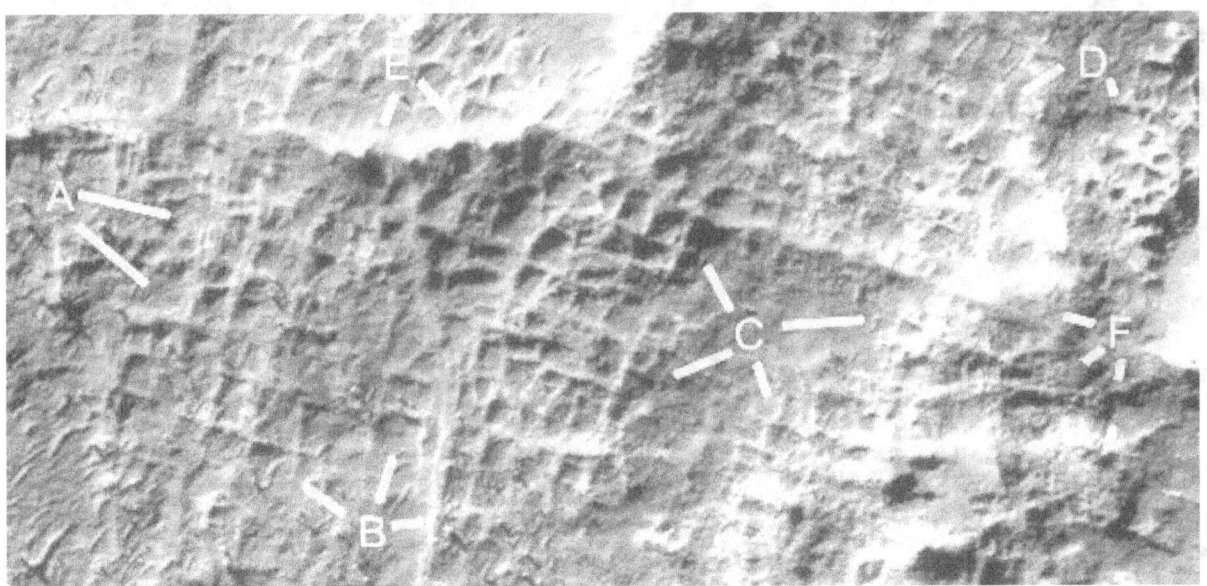

Cymhh361i

Hypothesis

The three dimensional impression is even stronger here, A shows rooms appearing under the smooth ceiling material. B may also be tubes or suspended roads as there is an impression of empty space under them. C at 9 o'clock shows rooms with no ceilings, at 4 o'clock there is still some ceiling or they are full of soil. D at 9 o'clock is like a hill of rooms, at 1 and 2 o'clock there is a road like formation that goes on to 12 and 2 o'clock. The letter E is in a depression surrounded by higher rooms like at 7 and 8 o'clock. F shows more variations in the elevations of the rooms from the shadow. G has many straight walls and may have right angles from directly above it. The rooms at H appear to be partially eroded.

Cymhh469g

Hypothesis

A at 10 o'clock shows a hill with room like shapes on its lower side, at 3 and 5 o'clock are more rooms. B and C show many walled rooms. D shows rooms that may be partially buried by the dark soil, or they ended in this open area. E shows more degraded rooms, F at 10 o'clock shows a nexus where many walls converge to it. At 3 and 4 o'clock there are perhaps rooms under the dark soil. G at 10, 12, and 1 o'clock as well as H at 12 o'clock follow this edge of the rooms, this section may be an intact ceiling with rooms under it.

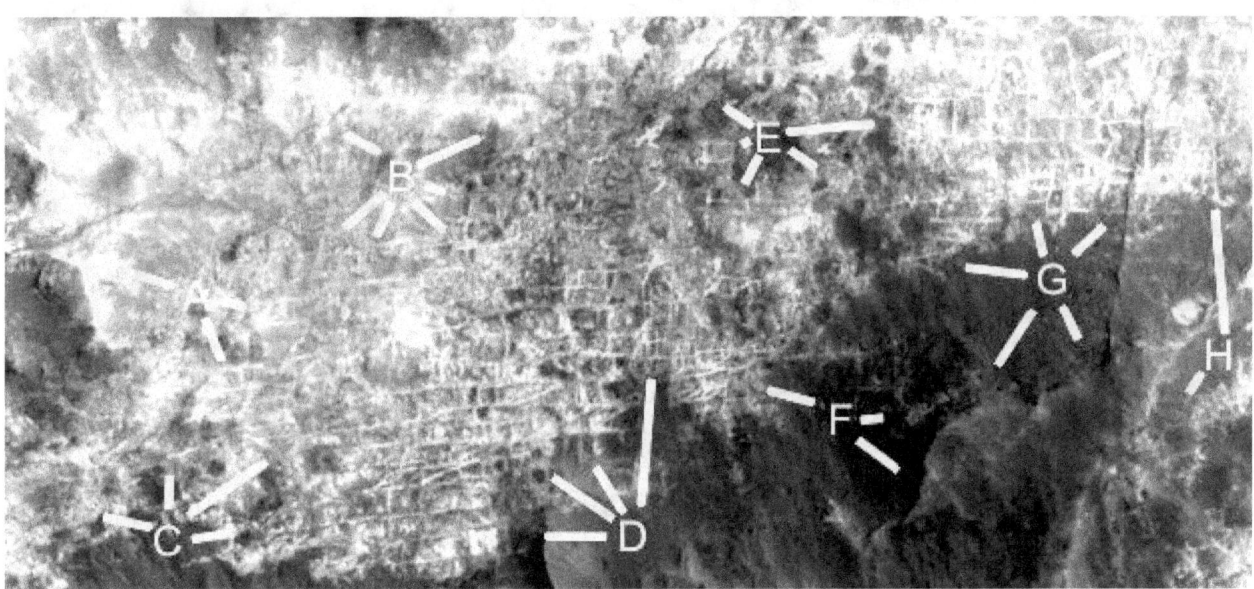

Cymhh469g2

Hypothesis

There are many lines here showing how straight the walls are, but many more could have been drawn as well.

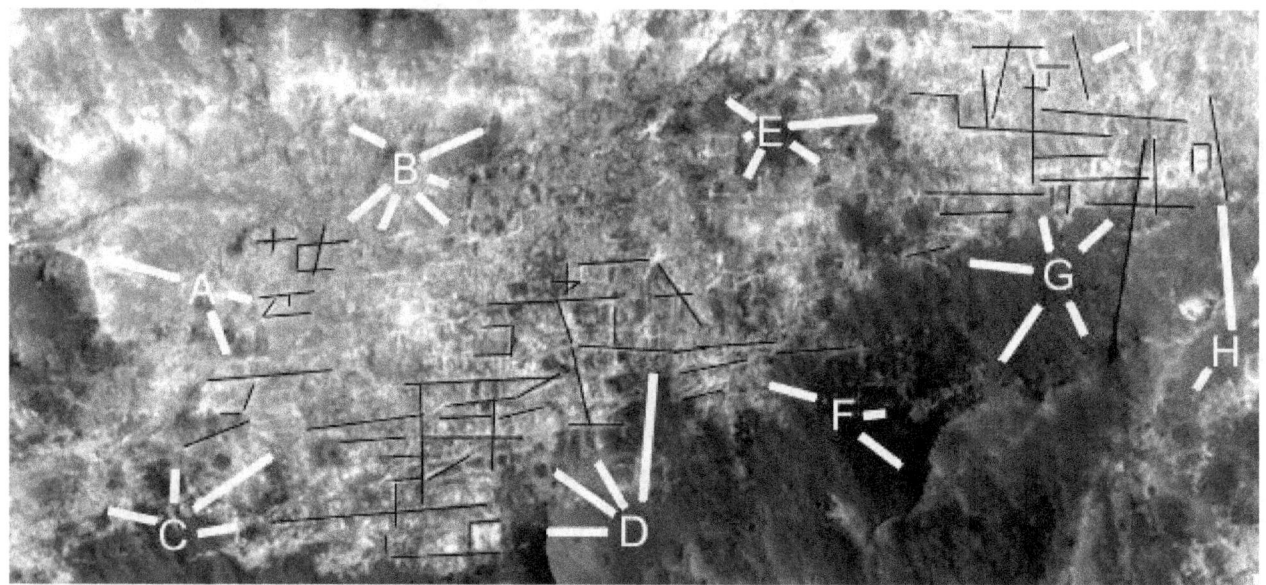

Buildings

Some individual formations look like large buildings, sometimes incorporating parabolas.

Cymhh467

Hypothesis

A may show some collapsed hollow hills. B shows some straight ridges, perhaps interior supports of this larger formation. From C to D is a curved interior support. E may be a collapsed section, F shows some tubes or walls.

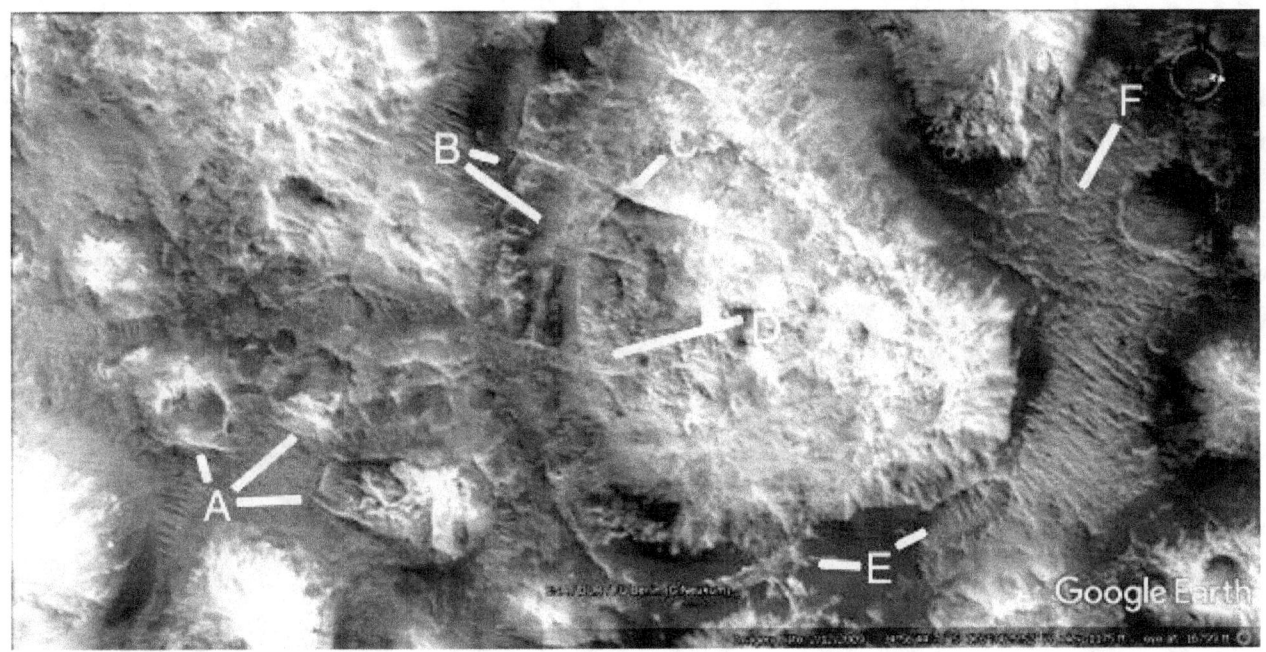

Cymhh467a

Hypothesis

There are two parabolas in this formation, as well as the straight walls.

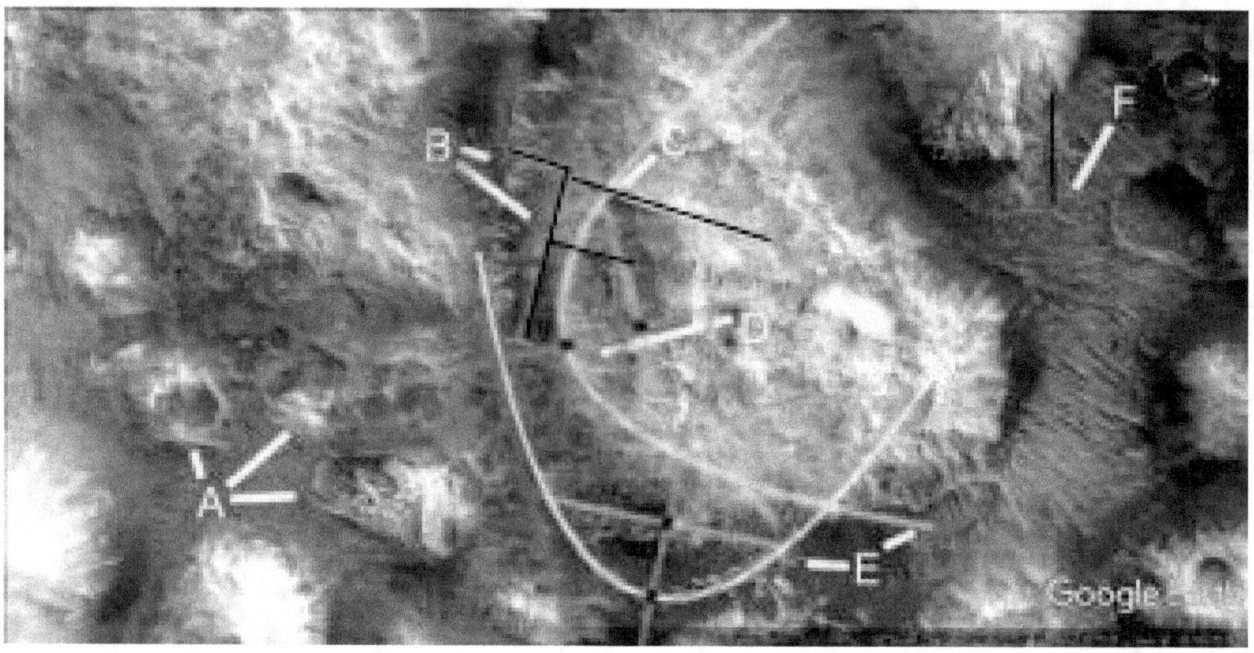

The hills often shows collapsed segments on their roofs so being hollow is implied. That adds to the hypothesis, that they lived in these hollow hills, and travelled between them on these roads.

Prhh944c

Hypothesis

The top of the layer here is shown at A at 12 o'clock, at 10 o'clock is a tube. B shows multiple layers under it, this may be the construction technique. C shows a broken wall segment at 8 o'clock second leg, this may be two thinner layers broken together. At the first leg is a tube. At 9 o'clock second leg is another broken layer. At 6 o'clock the tube appears to come from here, this has a collapsed side and a gap between it and 8 o'clock first leg. At 12 o'clock the texture of the roof is different to the wall layers.

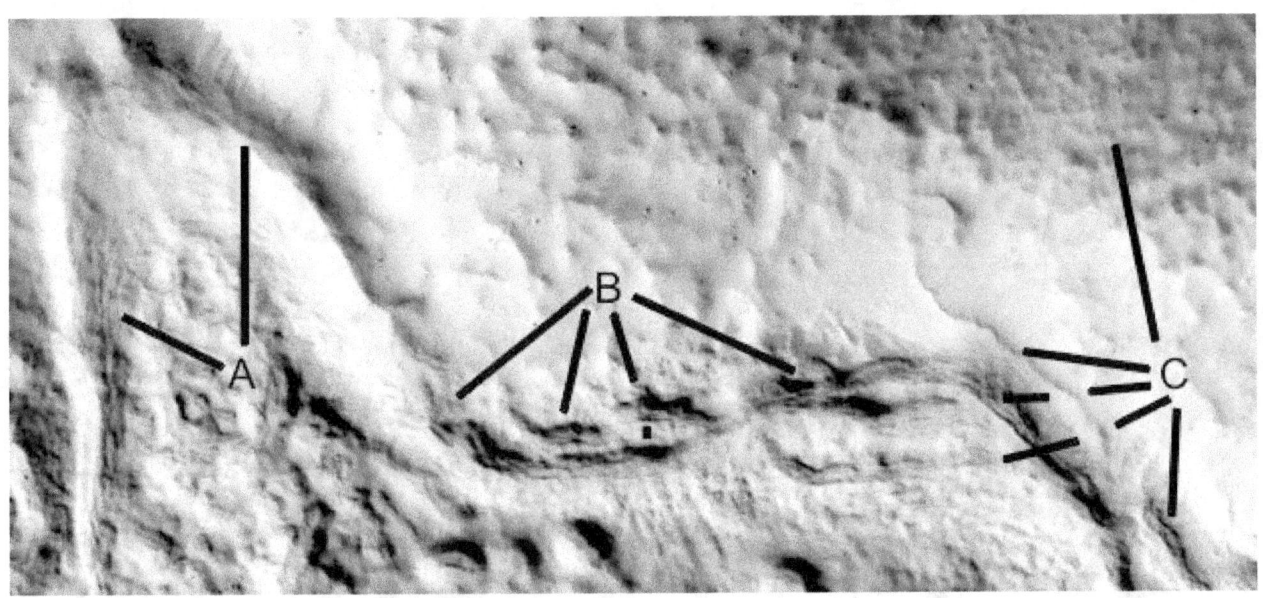

Prhh944c2

Hypothesis

Three parabolas are shown, like a parabolic wave. This can be an approximation to ocean waves which are elliptical.

Prhh944f

Hypothesis

A shows tubes or eroded segments on the roof. B shows contours which may have been used for strengthening the roof. C shows a settled area. D shows many parabolic arcs to strengthen the roof at 9 and 10 o'clock, at 2 o'clock there is an exposed grid perhaps used for reinforcing the roof.

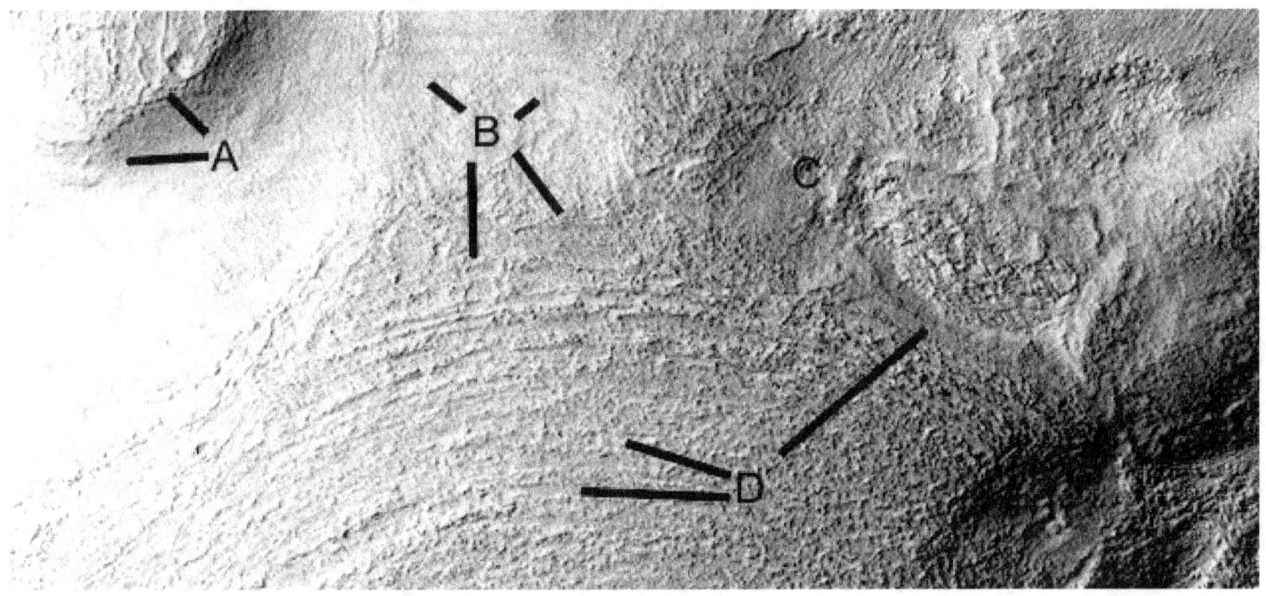

Prhh944f2

Hypothesis

Three parabolas are shown, there are several more but these are the clearest. The axis of symmetry of each is closely aligned but each parabola is smaller than the one surrounding it.

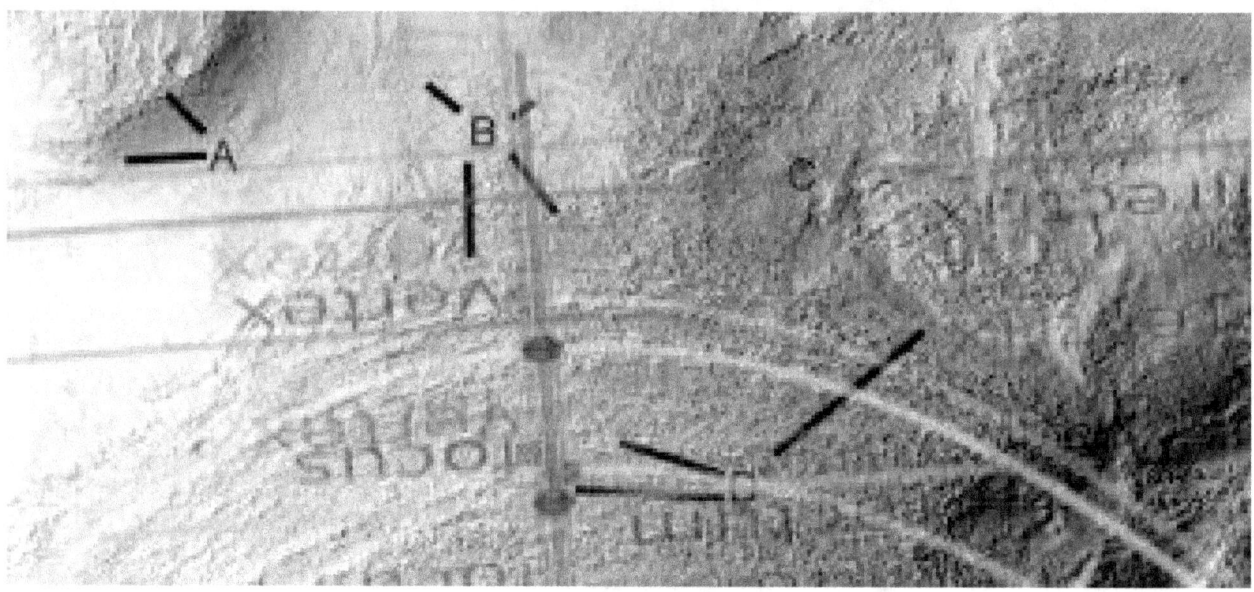

Prhh944j

Hypothesis

This may be a Cobler Dome where the parabolic layers of bricks are exposed. They are less visible at A at 10 o'clock, at 4 o'clock the top of the hill may be peeling off. B shows a smooth skin like cement that may have broken off on the upper side exposing the layers. C shows the parabolic layers, D shows two skins that have eroded away exposing the arcs.

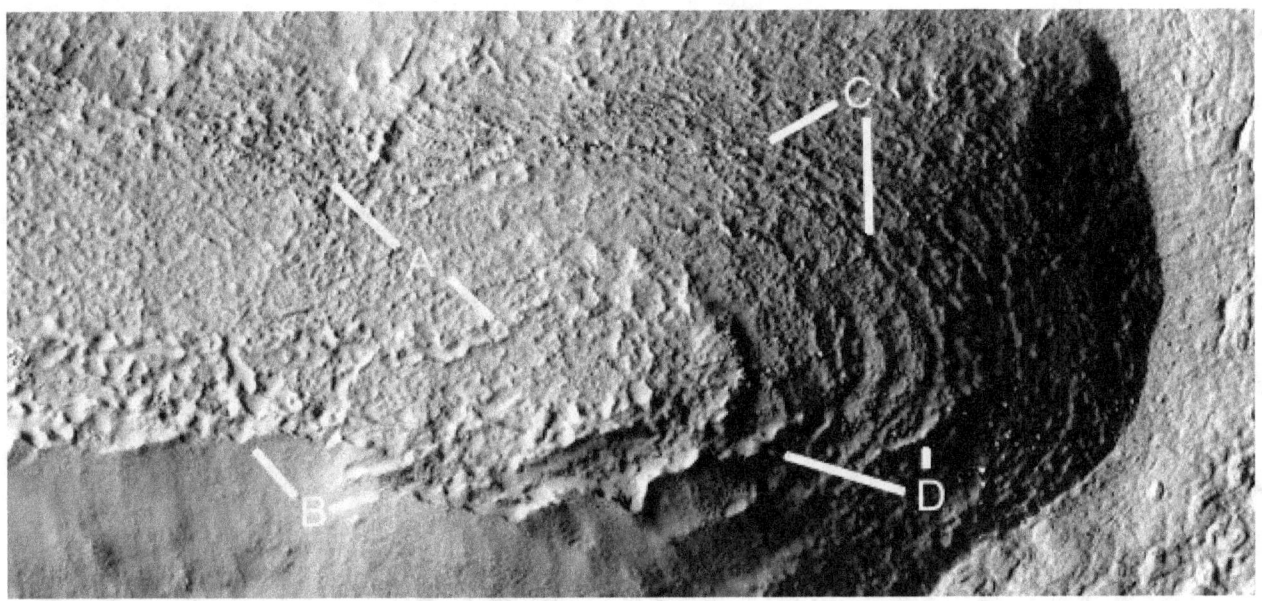

Prhh944j2

Hypothesis

Three parabolas are shown, there are several more which are too faint. Straight ridges are also overlaid by lines.

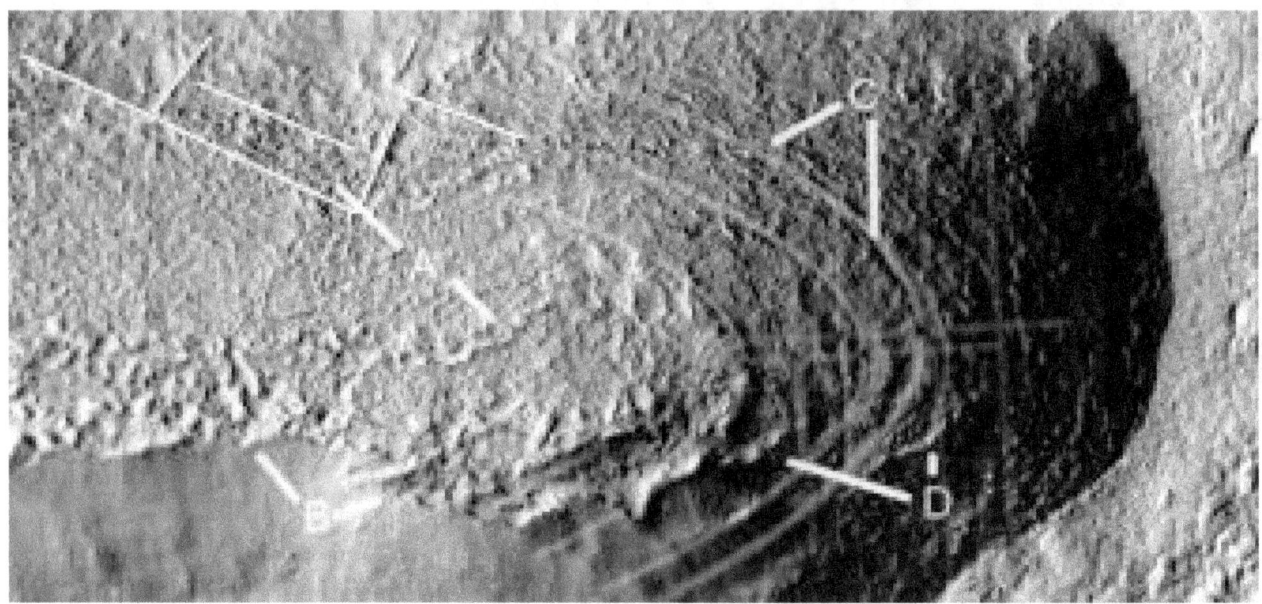

Helhh1117

Hypothesis

A shows the curved segments of the hollow hill roof. B may be a collapsed segment of the roof. C at 2 and 4 o'clock may be a tube, at 5 o'clock an interior support with some settled segments of the roof around it. D at 1 o'clock may show a tunnel going into the hill continuing on at 4 o'clock perhaps as a collapsed tube.

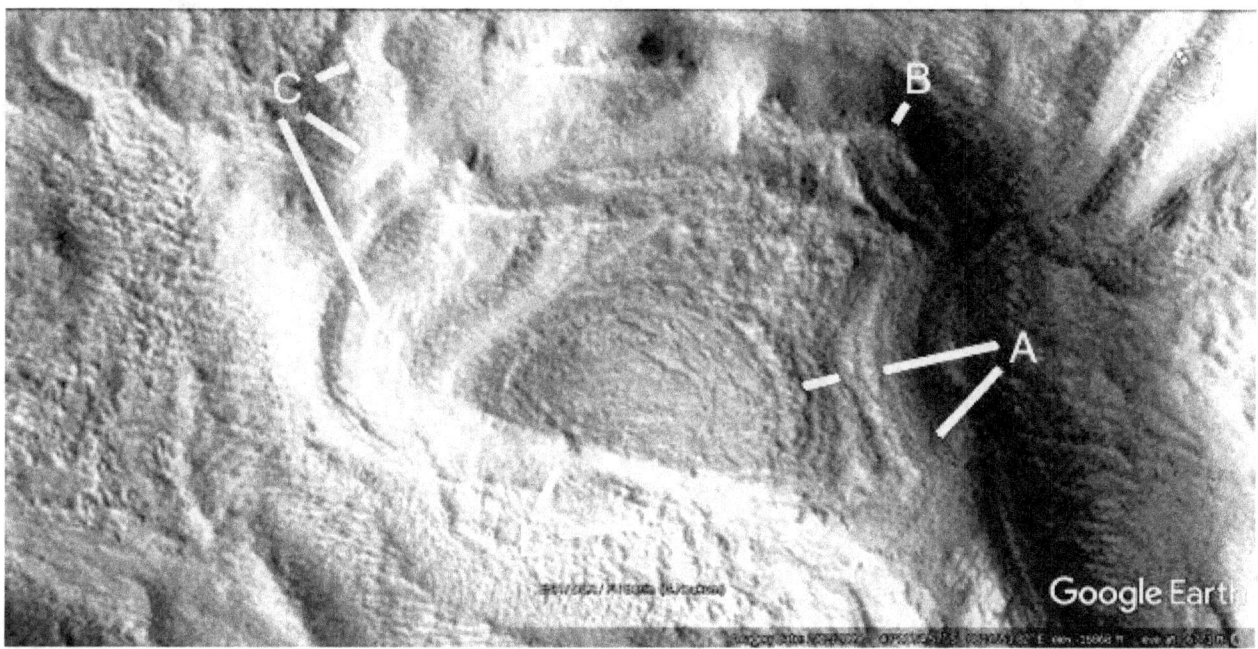

Held117a

Hypothesis

The edge of the rounded segment of the roof forms a parabola, the flat side lines up well with the latis rectum, the name for the line through the focus. The ends of a parabolic formation often deviate from the perfect parabola, shown at E. This may be because the parabola was not used to be a geometric statement to be viewed. Instead it was hypothetically used to make the formations stronger. These edge at E would serve no purpose to continue here as a parabola. This corner may also have been a small parabola to make it stronger.

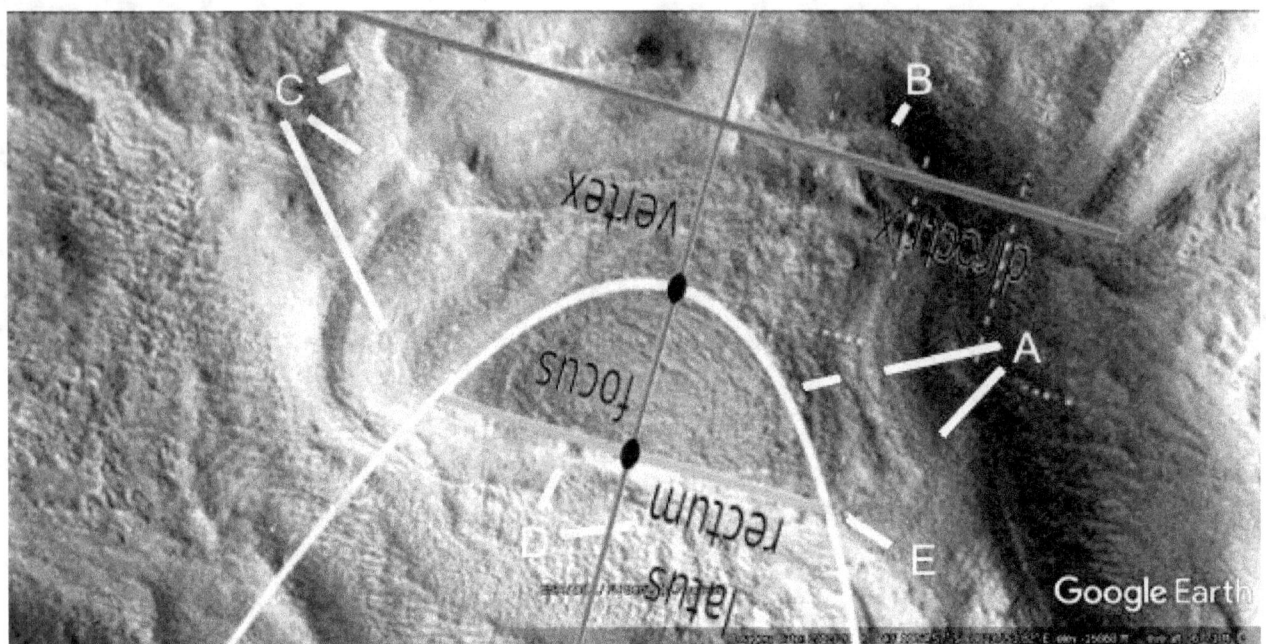

Walled fields

The hypothesis is that these may have been used for farming, or for pools of water containing fish.

Held1186

Hypothesis

These walls are much straighter and with more right angles between them.

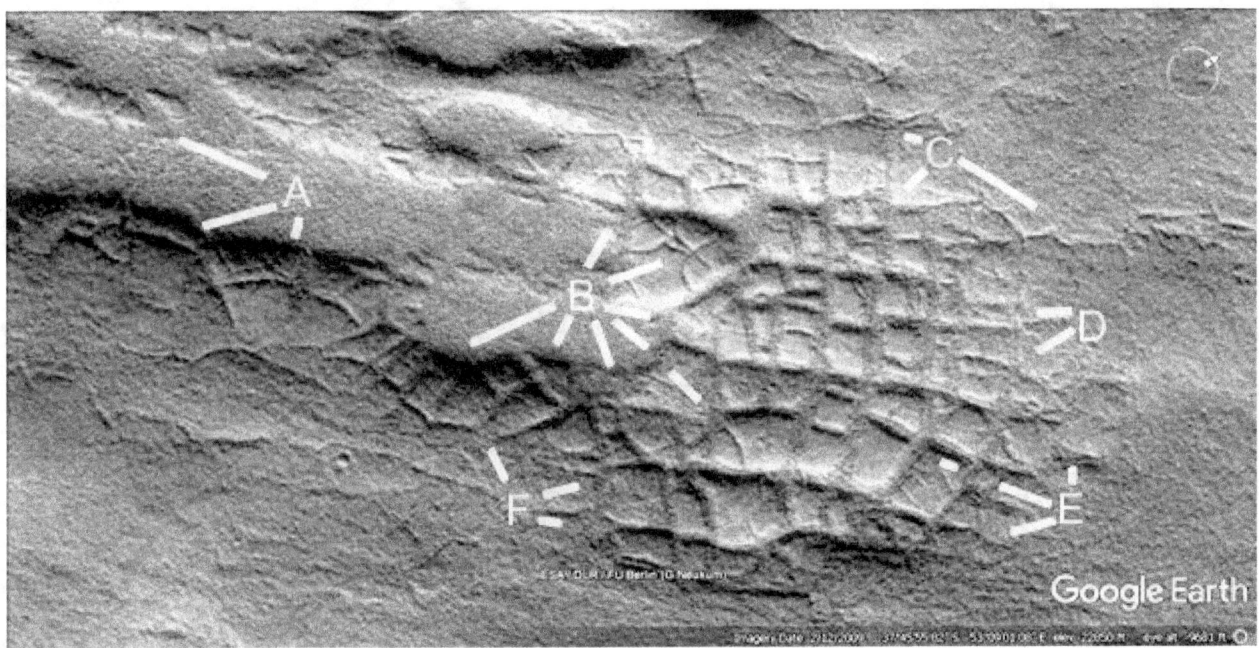

Held1222c

Hypothesis

The walled fields are in better condition here, without gaps. A shows some joins with little erosion, at 8 o'clock however is a much more eroded wall.
B shows an eroded wall at 10 o'clock and where one wall passes over another at 7 o'clock.
C shows a much thicker wall between 6 and 10 o'clock, this extends under a wall to a thin wall between 1 and 4 o'clock at D. E shows some wall erosion at 3 and 9 o'clock.

Held1222c2

Hypothesis

The lines indicate how straight the walls are.

Held1222e

Hypothesis

This shows how many walls are hollow. The wall at A at 6 and 7 o'clock has collapsed indicating it was a tube. At 4 and 8 o'clock the walls are intact, it implies these tubes would give a passage in and out of the hills. B shows more collapsed walls, at 3 o'clock one goes into a small hill perhaps a habitat. Above C at 10 o'clock the tube has partially collapsed, the wall forms a side of this hill. At 5, 7, and 8 o'clock the walls have collapsed, at 4 o'clock the wall goes into another hill which may be a habitat. D, F, and G shows more collapsed walls. E shows more narrow walls going through a possible habitat at 2 o'clock.

Held1244

Hypothesis

A shows a possible habitat at 4 o'clock, B shows two others at 8 and 11 o'clock. These may be like the typical hill in this area when the outer skin erodes away. A at 6 o'clock shows many fine walls or tubes going into a nexus at B at 4 o'clock, also with a circle of walls around it. This would be similar to Earth roads where a central meeting place might be bypassed with this ring road. C shows more walls, D shows how they go into a hill at 6 and 9 o'clock. This hill is much flatter, it connects the hypothesis of the other hills in the image being like for example Held1232. It appears as if the roof has collapsed onto the ground. E shows a wider wall coming out of the hill at A.

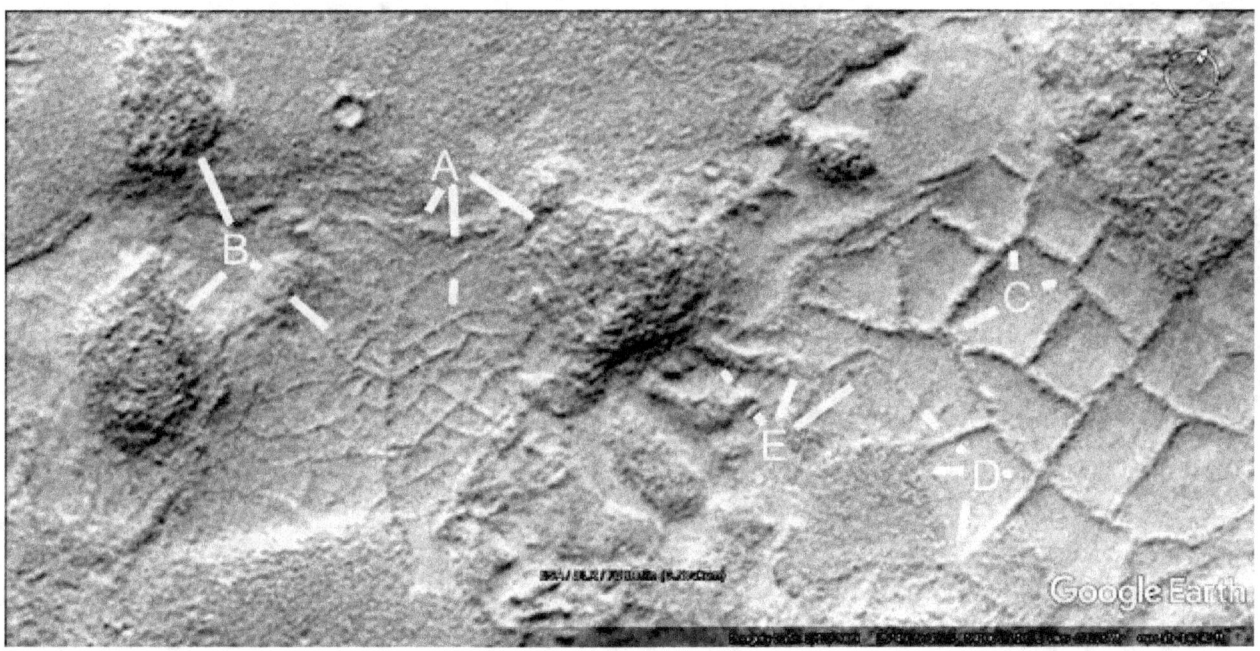

Held1258

Hypothesis

A also implies the hill is artificial, it is approximately parallel to the Latis Rectum of the parabolic wall. B is probably a collapsed hill at 8 o'clock, a wall comes out of it at 7 o'clock. C also shows a network of walls coming out of a hill. The walls at D appear more eroded.

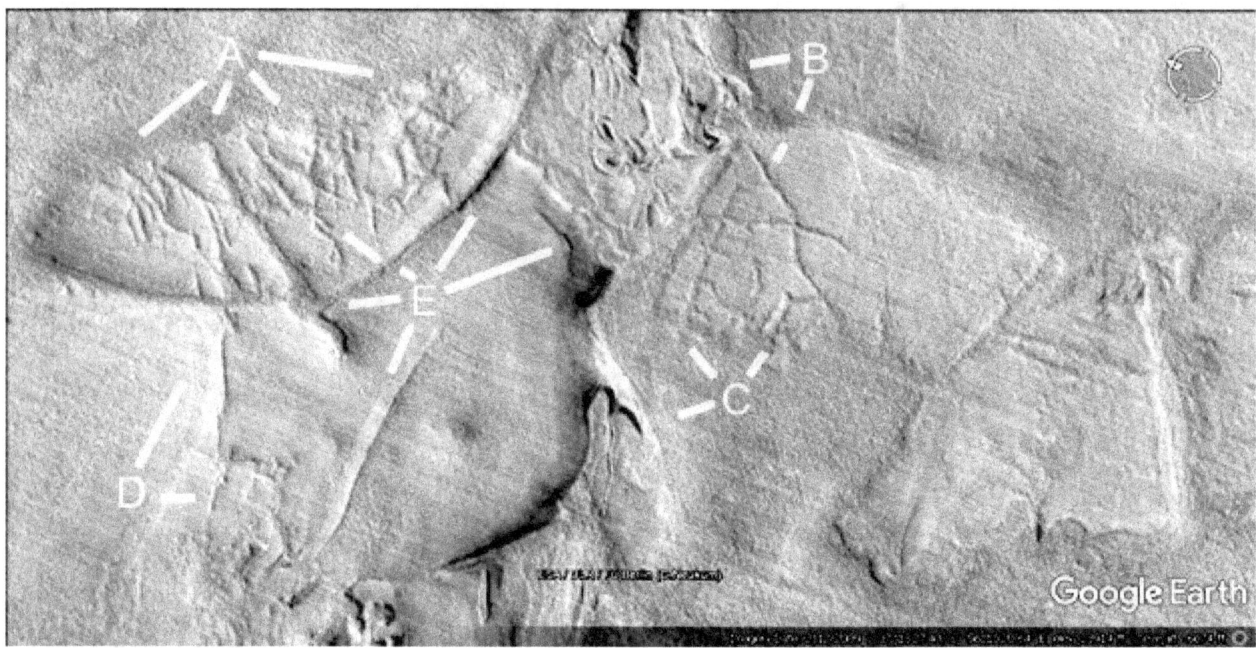

Held1258b

Hypothesis

A parabola is shown, also the lines indicate how straight the walls are.

Held1295b

Hypothesis

A appear to show a water channel or perhaps roadway, perhaps water could come through here and fill some of the walled areas. B shows some of these walls, C shows a parabola. D shows another curved wall, probably a parabola but not long enough to check. Shows many walled fields with smaller walls subdividing them.

Held1295b2

Hypothesis

A parabola is shown, also the lines show how straight the walls are.

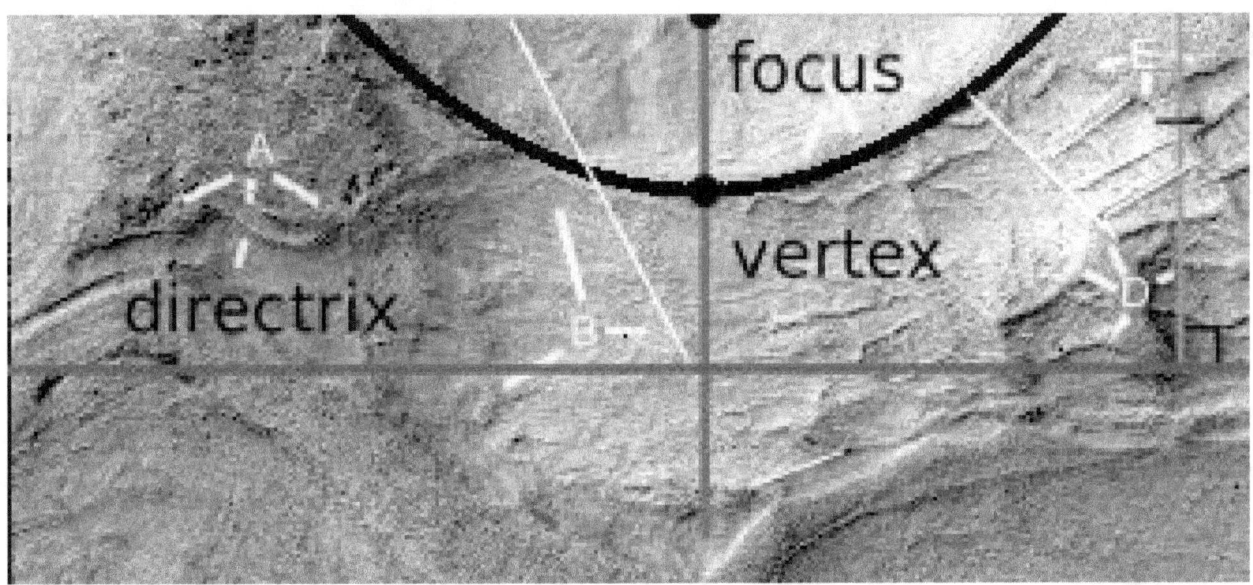

Roads

Some formations also look like roads, they often appear between hills that are hollow. The hypothesis these hills are buildings, either completely constructed or adapted from geological formations. It further ads to the global hypothesis, we use roads and so we might expect Martians to have built them to travel between buildings and cities.

Prhh498

Hypothesis

The hollow hill has collapsed at A, B shows a straight wall still standing. C shows another road going into the hill perhaps with two lanes, this extends to D at 10 and 1 o'clock. There may be another road at 7 o'clock.

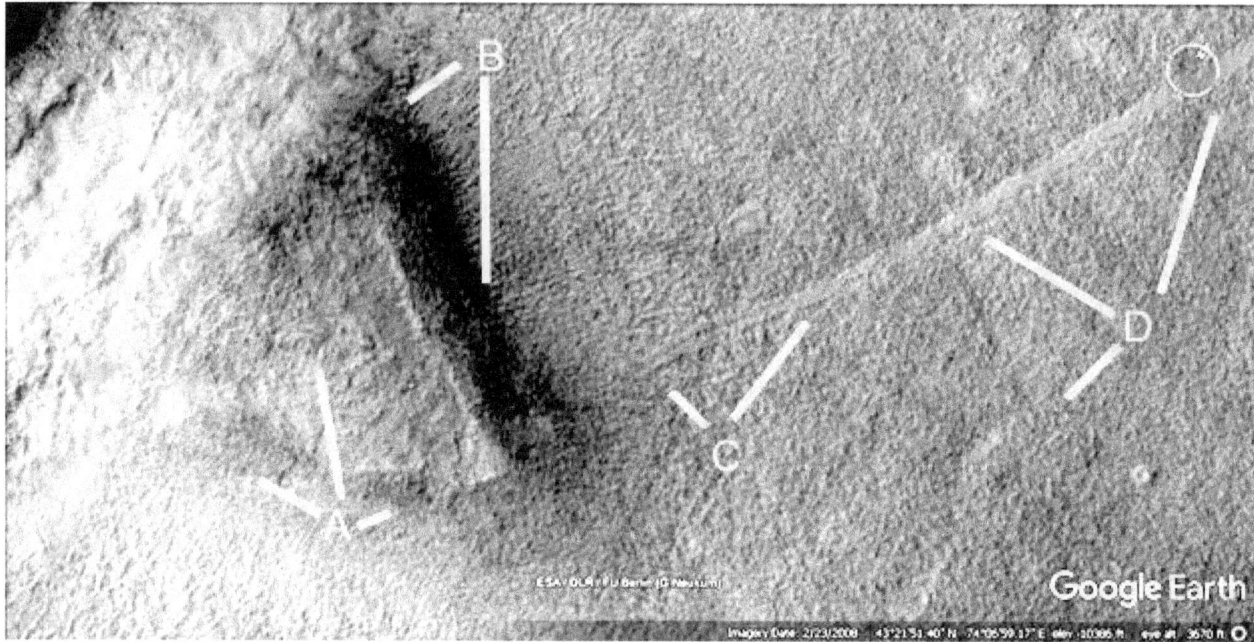

Prr499

Hypothesis

This is a closeup of a road, much smoother than the surrounding terrain like cement. It extends past A to B where a tube or raised road intersects it. C shows this tube going down from 10 o'clock, then possibly at 6 and 7 o'clock into the crater.

Prr508

Hypothesis

A shows the road continuing on over the pale material, B and C also show pits like altered craters perhaps with the same road material to act as dams.

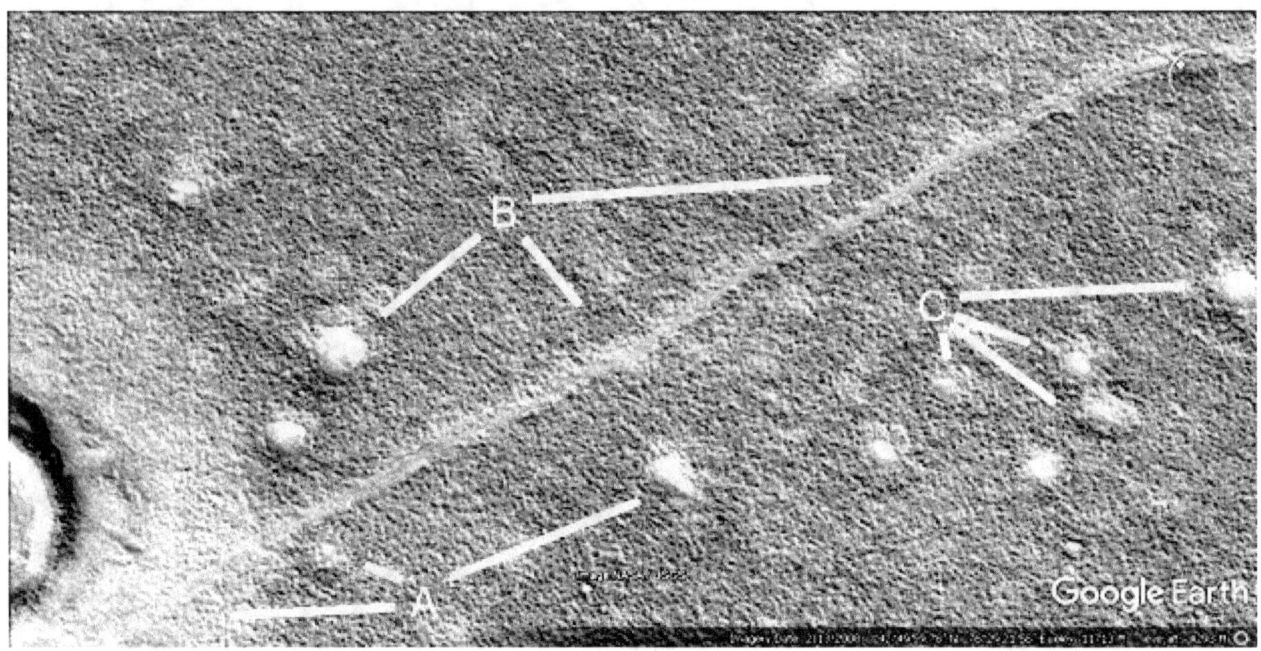

Prr533a

Hypothesis

This closeup of the road shows right angled shapes in it, perhaps like bricks or tiles. This impression continues along the road where it seems to vary in an angular rather than a smooth way. The center is very smooth compared to the surrounding terrain as shown by comparing A at 1 and 5 o'clock. B shows a shape like a gutter along the road's side. C shows a small pit at 10 o'clock that appears to be connected to the road, perhaps a former hollow hill, at 2 o'clock is an angular section on the side of the road.

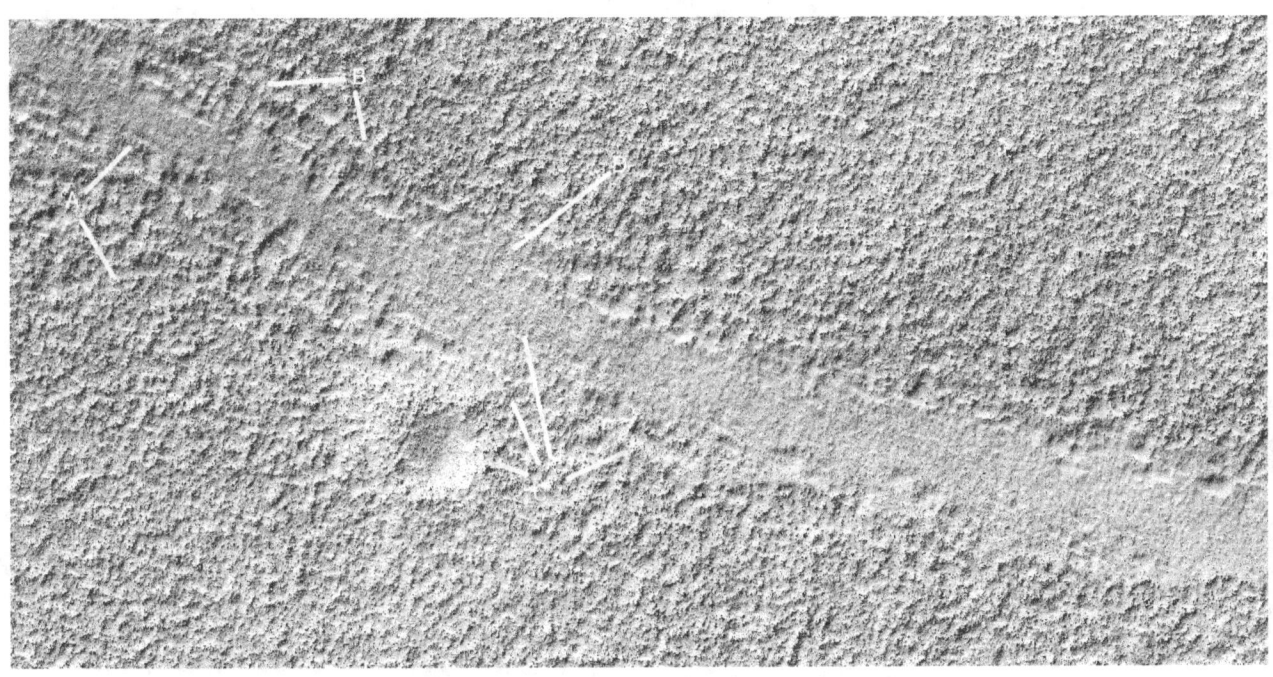

Prhh1821

Hypothesis

A shows more roads, they connect to a crater at 5 o'clock. B shows a road at 6 o'clock going into a small hollow hill, another at 4 o'clock going into a hollow hill. C shows a road connecting to a complex of hollow hills. D and E show many more roads connecting to hollow hills. F and G show roads connecting to the large crater. H shows a major intersection going up the image.

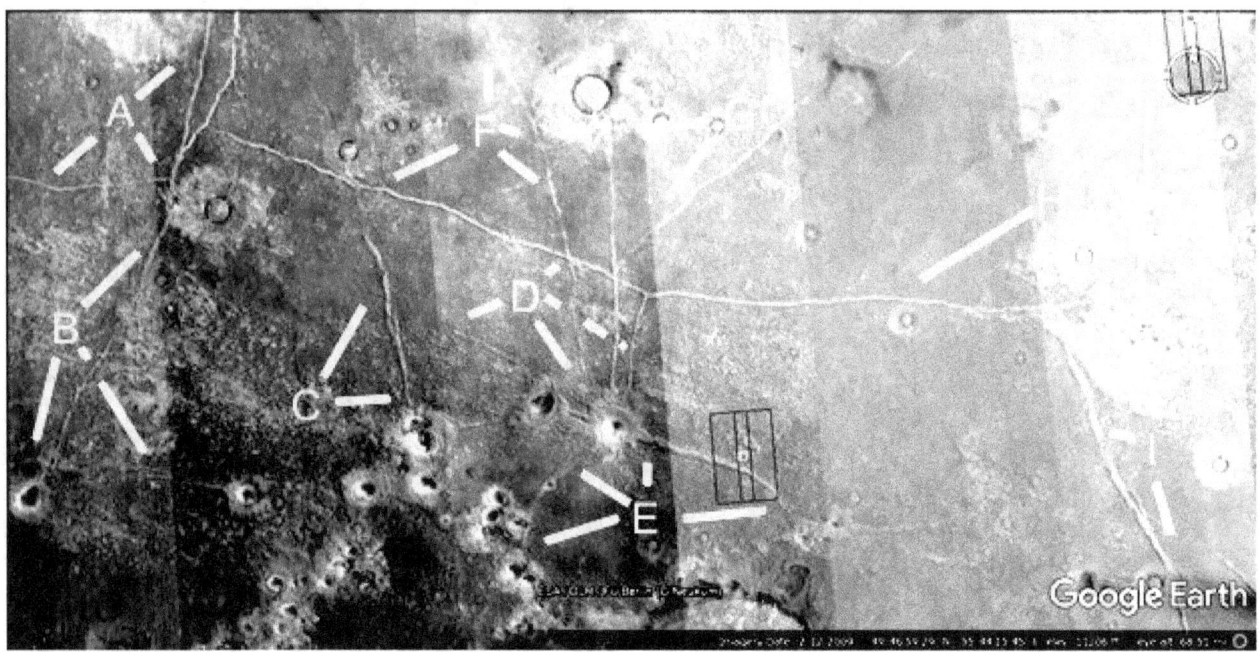

Tubes

A further hypothesis is that some roads were enclosed like tubes. These hypothetical Martians then could have travelled through them to avoid the cold, predators, meteors, etc. Some may also have been raised roads, for example the ground may have been swampy or covered in water. So, much as we do on Earth, they may have built roads raised above this ground to travel on.

Prt641

Hypothesis

A shows a curved tube going from the walled hill at 4 and 5 o'clock to the small crater at 1 o'clock. B at 8'clock shows the walls of the hill, at 7 o'clock a tube comes out of the hill, at 1 and 4 o'clock are two more hollow hills. D shows the curved tube, it connects to another tube shown by B at 8 o'clock. At 9 o'clock is a small tube from the larger one, at 10 o'clock the smaller hill appears to have collapsed. This main tube continues up through E to the right.

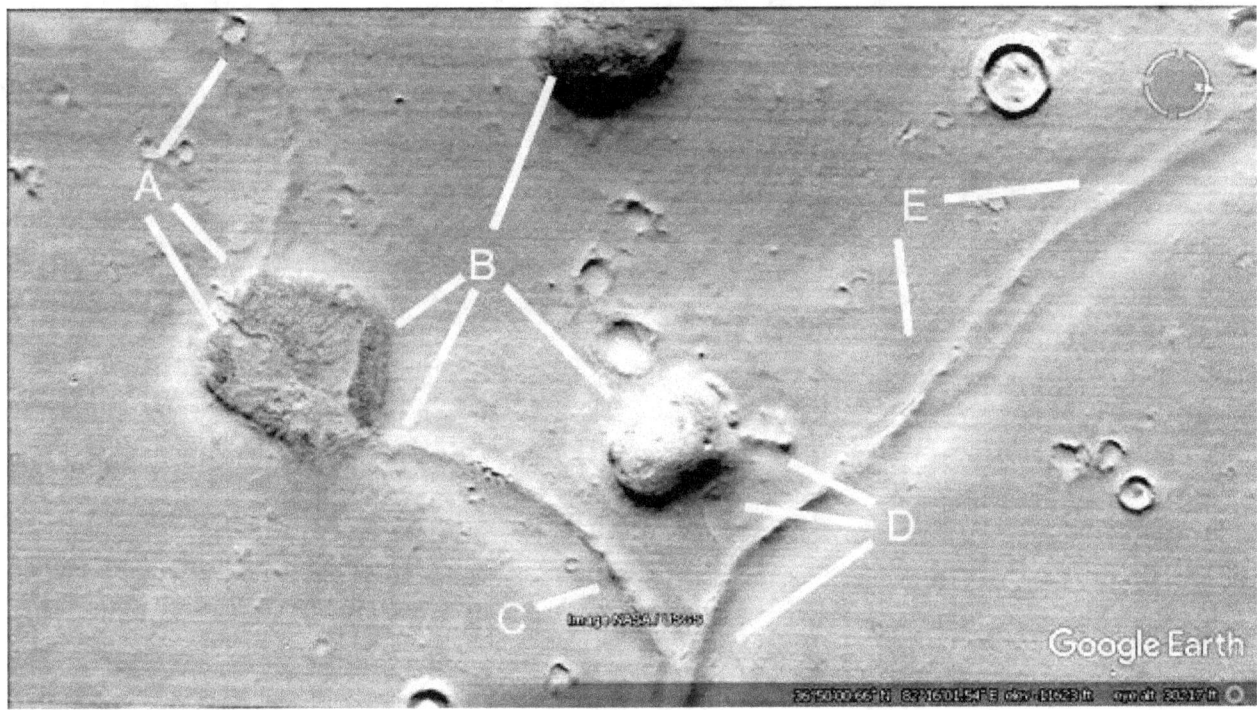

Prt641a

Hypothesis

Two parabolas are shown.

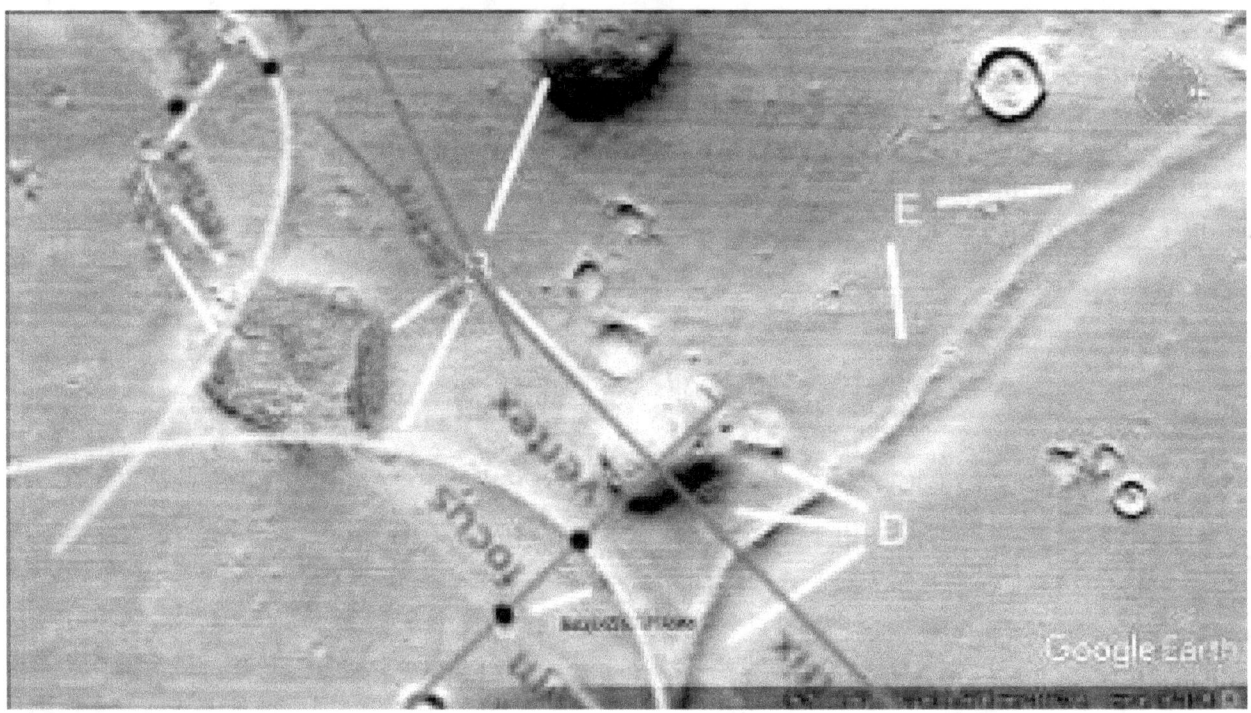

Prt798

Hypothesis

A shows a hollow hill with cavities in the roof, it connects to a wider part of the hill at 6 o'clock. This has a twisted shape like a rope, it continues on through the twisted tube at B to connect to a collapsing hill at 2 o'clock. At 8 o'clock there is another tube. At 3 o'clock the roof has collapsed. D shows another tube going into the hill at 8 o'clock, this connects to the tube at 5 o'clock. This in turn connects to the hill above D with tubes at right angles to it. E shows a collapsed roof at 10 o'clock, at 11 o'clock is a tube. Bat E at 12 o'clock up to F at 6 o'clock is a symmetrical wall.

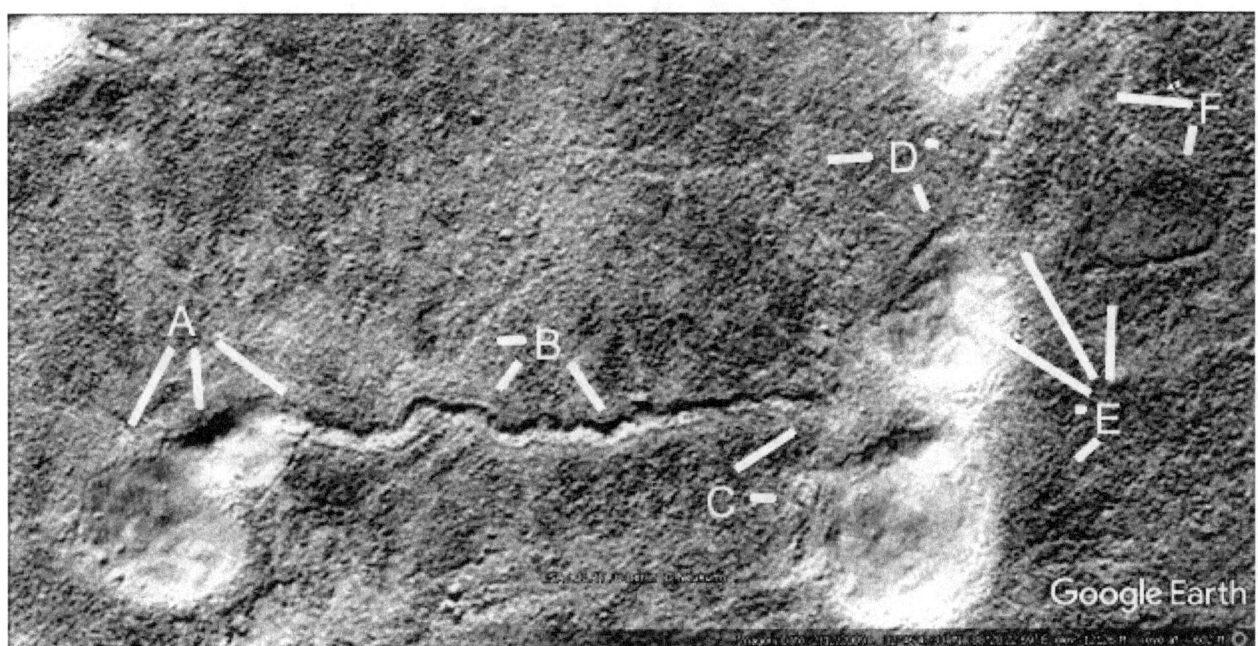

Prt804

Hypothesis

A shows more tubes between collapsed hills. B shows layers in the hill at 2 o'clock like a Cobler Dome. At 11 o'clock the tube from the chain of hills enters the hollow hill. At 3 o'clock is a thicker tube connected to a small hill. C at 8 o'clock shows the circular roof of the hill, it contains two parabolas, at 4 o'clock a tube goes into a small hill with a cavity on the roof. From 11 to 3 o'clock are other tubes. D at 5 o'clock shows the edge of this circular roof, the rest of D shows other tubes. E shows an arc of tubes connected to some collapsing hills.

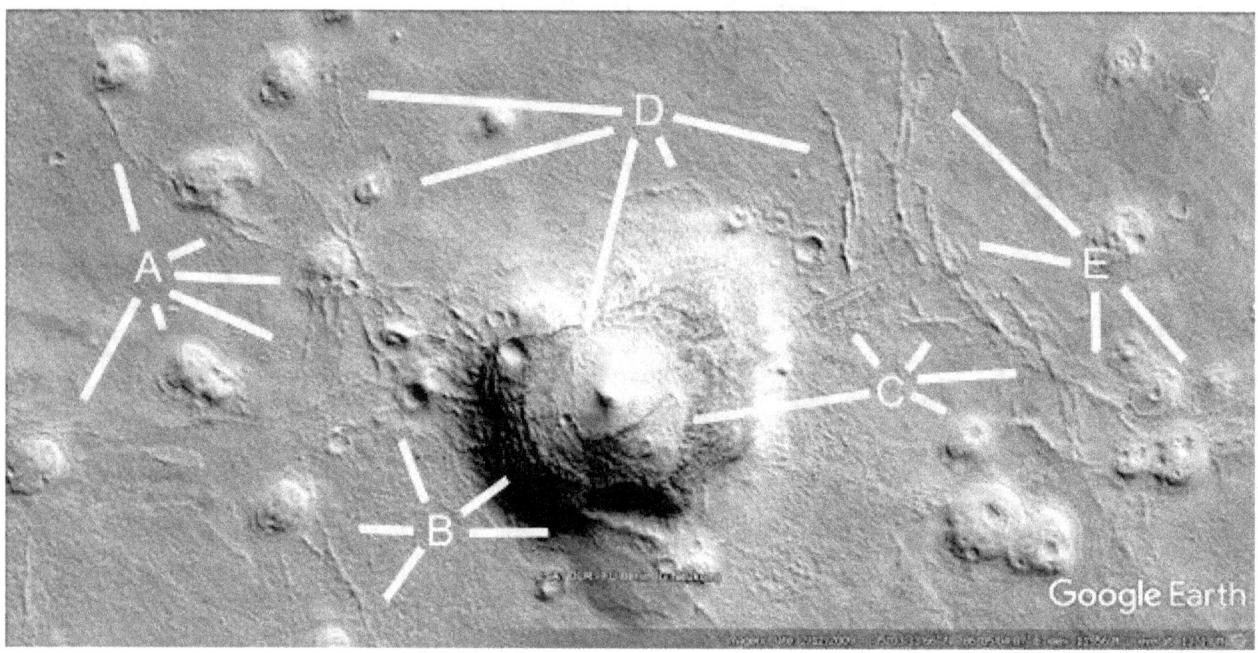

Prt804a

Hypothesis

The roof is close to a circle, here a circle is overlaid onto it. Also two parabolas are drawn onto the dark marks on the roof.

Prt814

Hypothesis

A from 5 to 7 o'clock shows two collapsed hills connected by a tube, the holes in the roof may have been rooms. At 8 o'clock is a tube. B at 10 o'clock shows a collapsed hill connected by a tube to A at 7 o'clock. B from 4 to 7 o'clock shows small hills connected by tubes, also some tubes go to the crater under it. C at 6 o'clock shows many tubes connected to the crater, at 7 o'clock a tube goes through a collapsed hill over to 4 o'clock and then up to the nexus at F at 1 o'clock. At 4 o'clock a forked tube comes out of a collapsed hill. C from 10 to 2 o'clock shows a tube coming out of the collapsed hill continuing over to the nexus. D and E show more tubes connecting to the hills and over to the crater at E at 4 o'clock.

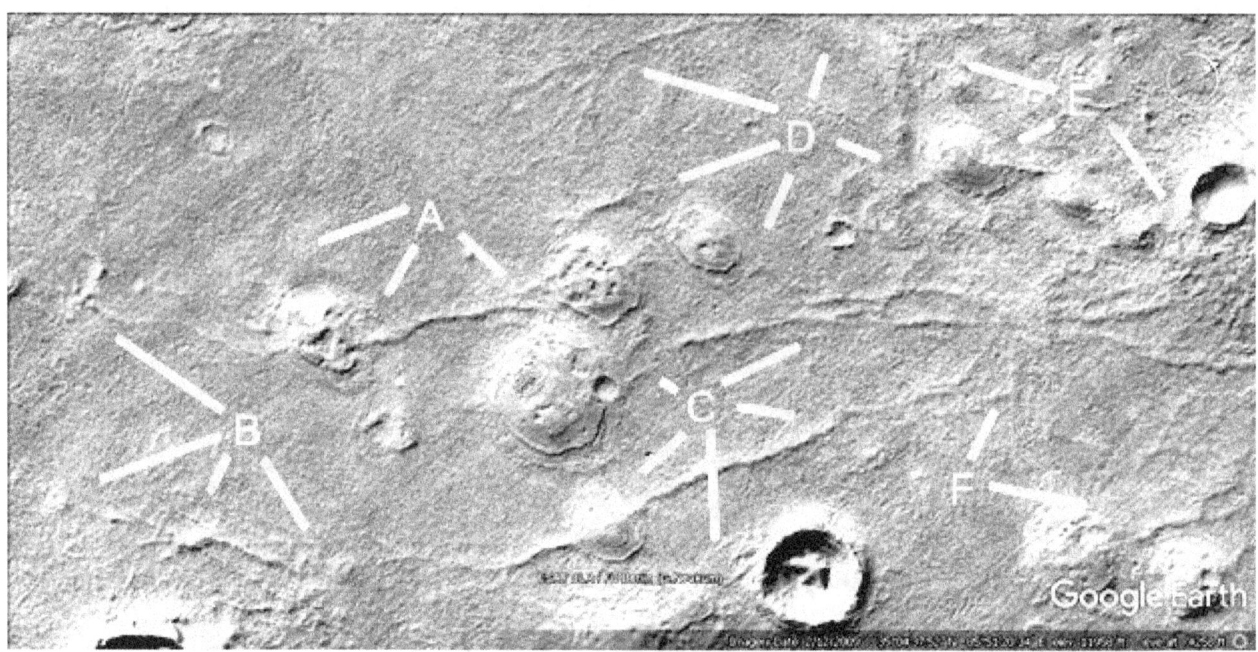

Tube cities

The hypothesis is these large numbers of tubes connected together to form habitats and cities. Some of these may have been underground, others connect to artificial looking hills.

Prt662

Hypothesis

A shows a wavy tube, B shows a clear area surrounded by tubes like a field. C shows tubes going into a crater at 6 and 8 o'clock, at 1 o'clock they go into a rounded area, also shown by F at 10 o'clock, under a nexus. D shows more tubes going into this nexus. E at 6 o'clock shows an intersection of tubes then this goes down, making a right angled turn into a hollow hill at F at 1 o'clock. E at 12 o'clock shows a T intersection, at 4 o'clock there are about four faint parallel tubes going up the image. F at 7 and 8 o'clock shows tubes going into three collapsed hills, also shown by G. H may be a large habitat, at 9 o'clock a tube crosses other tubes at 10 o'clock going up to I at 2,4, and 6 o'clock and a collapsed hill. At 10 and 11 o'clock faint tubes go into the crater. J shows more tubes going into the collapsed hill.

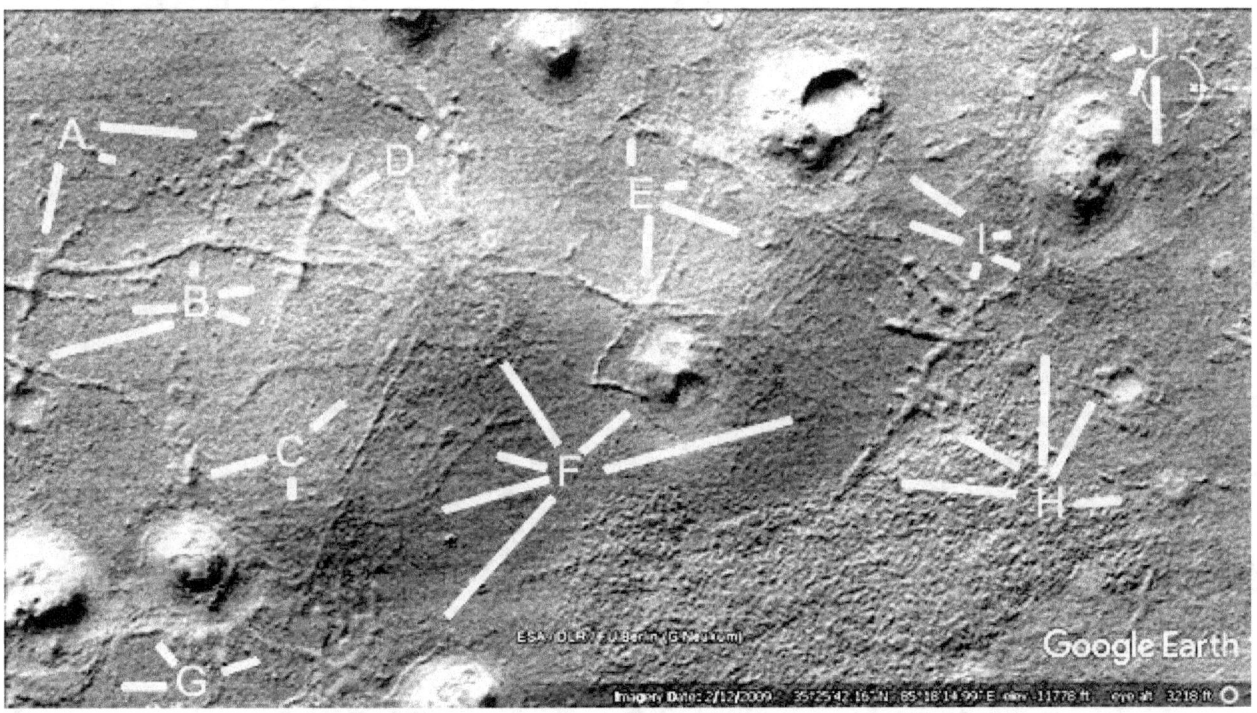

Prt682

Hypothesis

The tubes come together in a large nexus here, there also seems to be flat areas like cement over the tubes. These might act as a roof with rooms under them. A shows a tube crossing another at 2 o'clock, this connects to another tube at 10 o'clock. At 6 o'clock is the edge of the outer circular shape of the nexus. This may have allowed movement around the nexus without going into the centre, like an Earth ring road in many cities. B shows a continuation of the ring road at 3 o'clock, a forked tube at 10 o'clock and at 9 o'clock, and a narrow fork at 8 o'clock. C shows a larger tube at 10 o'clock where it appears to end on top of a small platform. At 1 o'clock the tube is hollow like the roof collapsed. D shows a tube ending at 11 o'clock, some tubes crossing at right angles in a mesh at 2 o'clock. E shows two tubes parallel to each other, further along one tube crosses over the other like a knot. F shows a small hill connecting to the tube at 3 o'clock, a loop of a tube at 5 o'clock with a central tube. From 8 to 10 o'clock is the flattened part of the nexus, whether from erosion or a roof. G shows a small nexus.

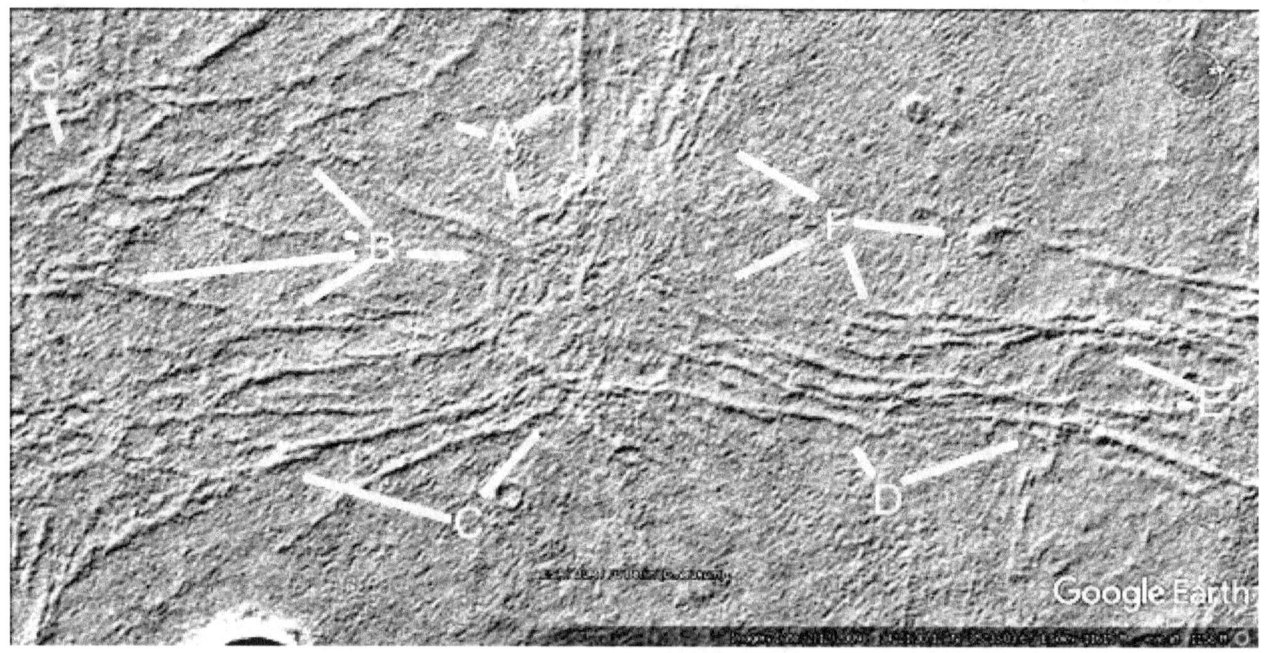

Prt714

Hypothesis

A shows a large nexus at 4 o'clock, it appears to have flat sheets of cement over it so some segments might be rooms. At 1 and 2 o'clock parallel tubes go to the nexus. B shows a squarish area surrounded by tubes, at 7 o'clock there are more like squarish walled segments. At 1 o'clock the crater appears to have been overed over on the right side or this can be an exposed room in the nexus. A wider tube is at 5 o'clock. C shows a T intersection of tubes at 1 o'clock, the tube goes down crossing a long hill at 5 o'clock going into a crater. Another tube crosses the hill from 6 to 7 o'clock. D shows another nexus at 2 o'clock again with flattened segments of a roof. At 4 o'clock this connects to a hill collapsing in many areas. Parallel tubes are shown at 1 o'clock. E shows more tubes, some going into a crater at 4 o'clock. F shows an arc of parallel tubes. G shows tubes exiting under the collapsing hill.

Prt753

Hypothesis

A shows many parallel tubes going through the long hill, continuing as E and E to the large nexus between E and F. This is a flat sheet like a roof in many areas. A at 5 o'clock and D at 7 o'clock show tubes crossing the parallel tubes so someone could have moved from one to another more easily. Above I there are nine parallel tubes going to the nexus, B shows about eight more parallel tubes. Under this is H with a grid or mesh of tubes, this continues on through C with more meshed tubes to the nexus. F shows about six more parallel tubes from 8 to 11 going to the nexus, between E and F there are about twelve more tubes going into the nexus. Between F and G there are about seven more tubes going to the nexus, many more of these form a tube mesh as well.

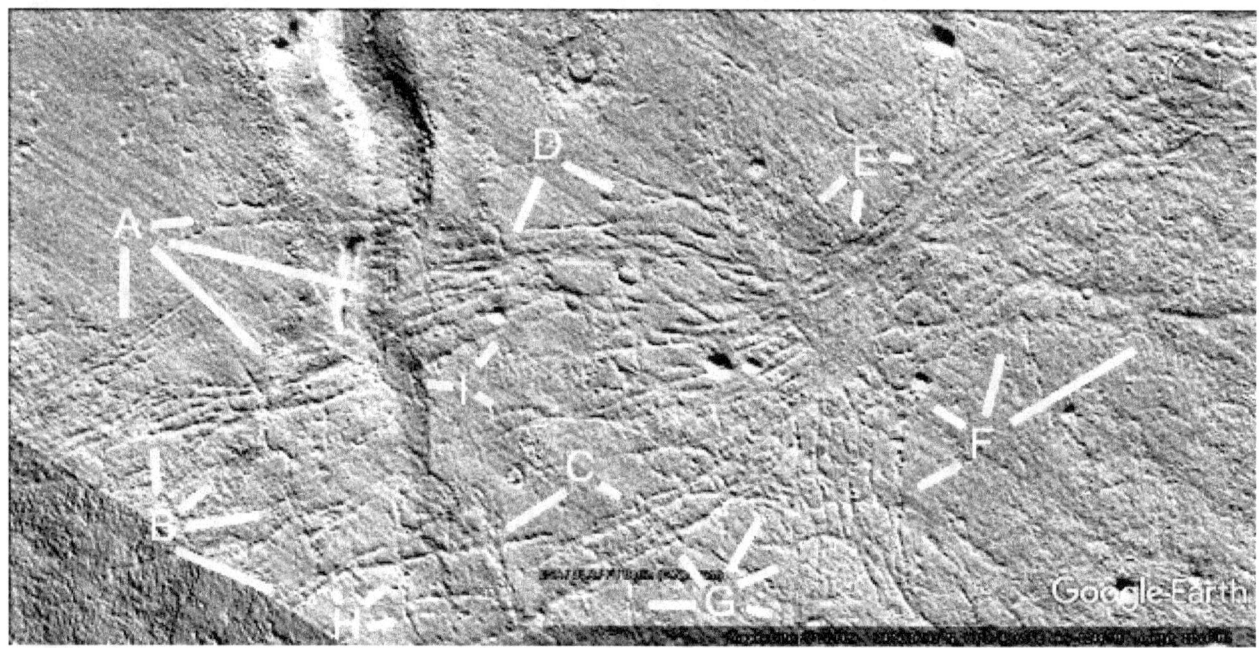

Some areas appear to be bounded, the hypothesis is they were farmlands or walled off for some reason. Often they have a parabolic boundary.

Farms

The hypothesis is that these large areas were farms, they are often bounded by parabolas with walls. We have something similar on Earth, we build walled fields and larger farms.

Prt857

Hypothesis

A, B, and C show many parallel tubes inside this farming area. Some connect to the craters at A at 7 o'clock. Between A and B there are about six parallel tubes, between B and C there are about four. B from 2 to 4 o'clock shows a tube going into the crater. D shows where many of these tubes converge, there may have been a hollow hill here. E at 7 o'clock shows a small hill and a straight tube extends up the image.

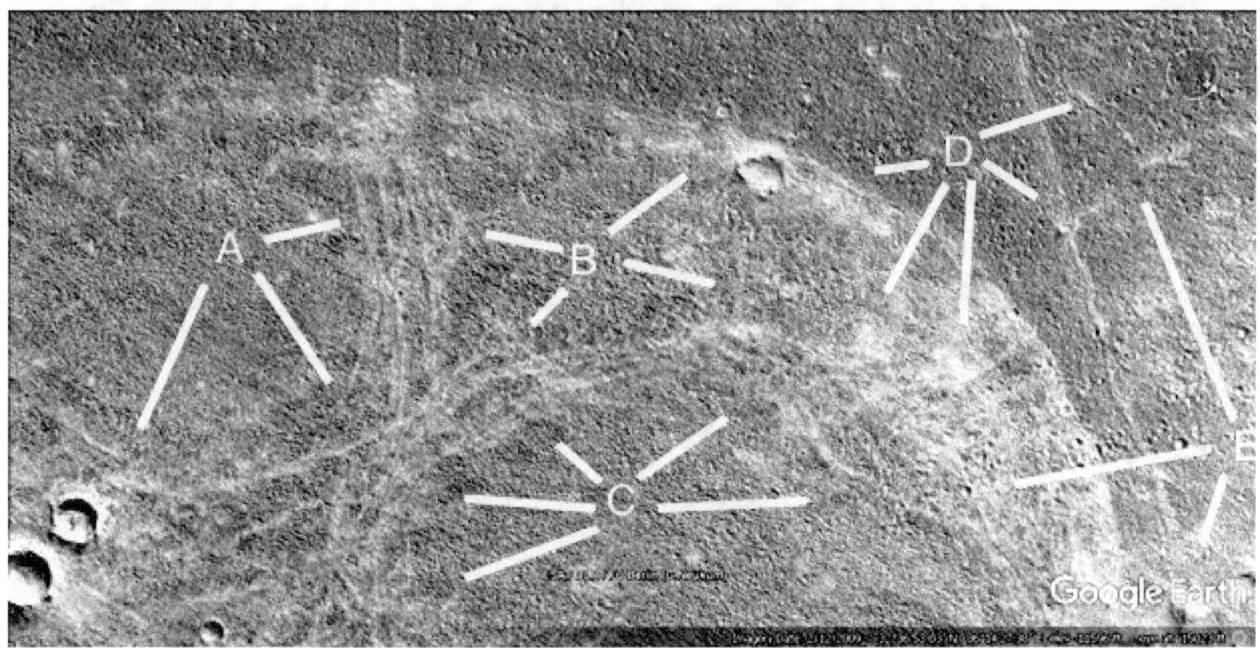

Prt857a

Hypothesis

A parabola is shown. Also the line shows how straight the long tube is.

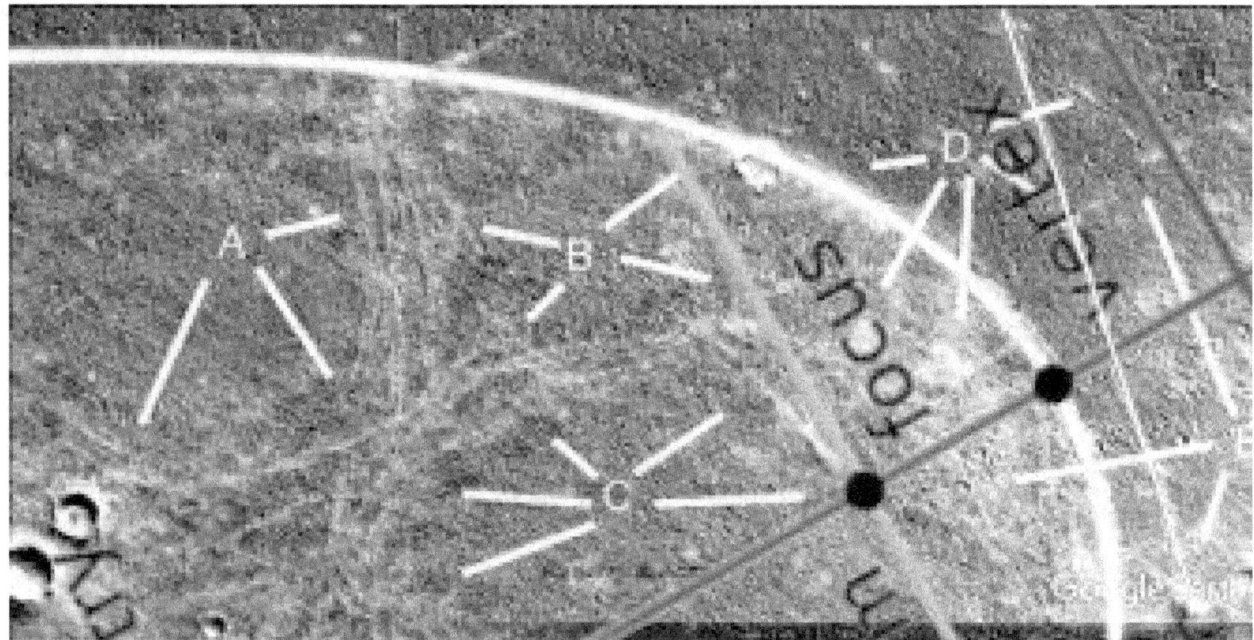

Ecydhh1941

Hypothesis

These curved shapes may have been used for agriculture. Found in many areas of Mars the boundaries are often parabolas. A shows a road or tube going into a crater, B shows the other side of this road and one of the curved pale areas. C shows more of these often shaped as parabolas. At 4 o'clock there is a wall or tube according to the shadows. D shows another tube at 12 o'clock, at 2 o'clock is the other side of the hollow hill. At 7 o'clock is a paler segment of the field. E shows more curved fields and a tube at 3 o'clock going down to a hollow hill at 6 o'clock. F shows another segment of the tube. G shows a tube going to the large crater at 7 o'clock. H, I, J, and K show more tubes and hollow hills.

Ecydhh1941a

Hypothesis

Three parabolas are shown, however the pale curves may all have been parabolas.

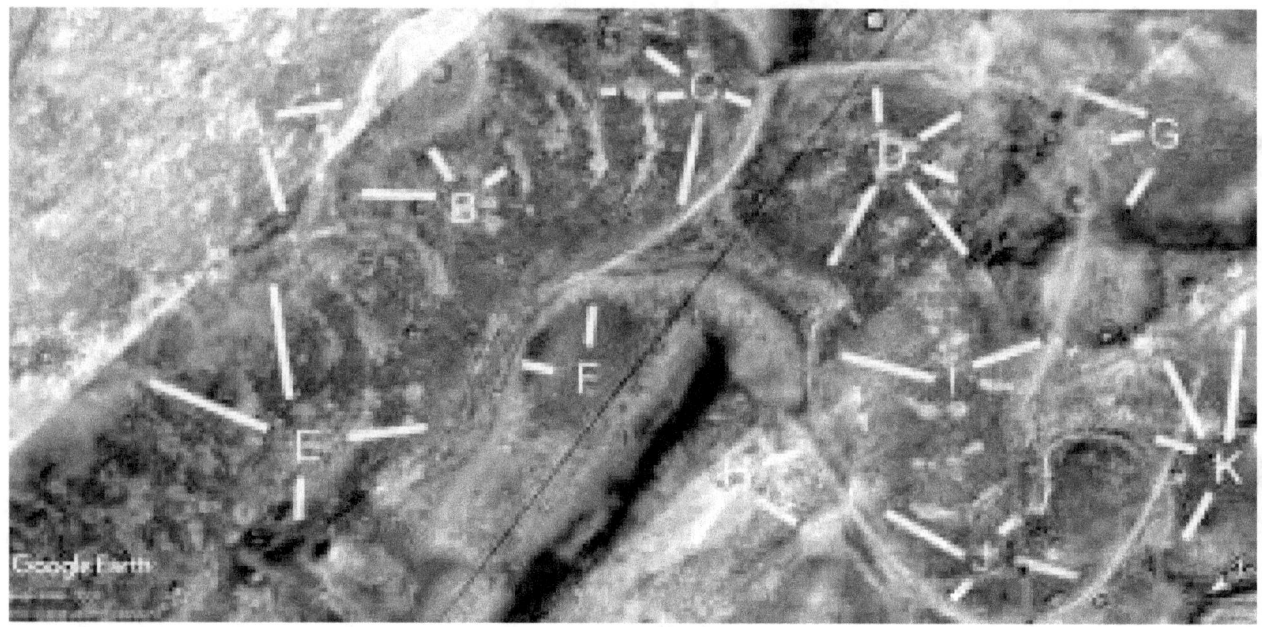

Ecydt1974

Hypothesis

Many walls and pale fields are shown, these may also have been farms.

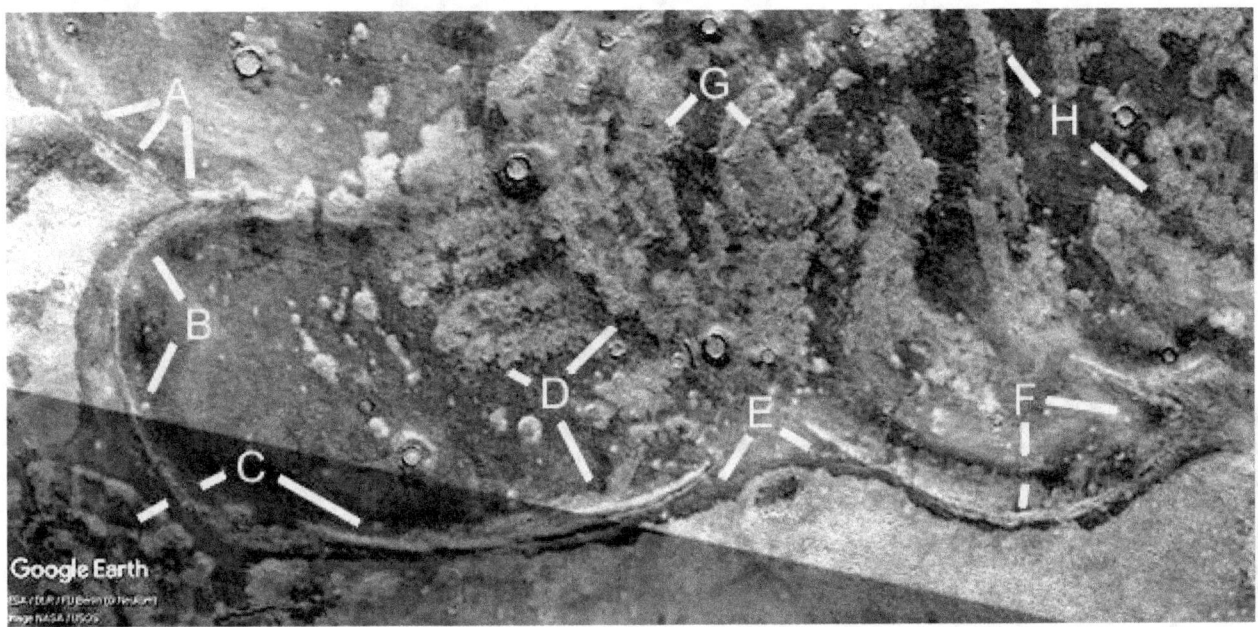

Ecydt1974a

Hypothesis

Three parabolas are shown.

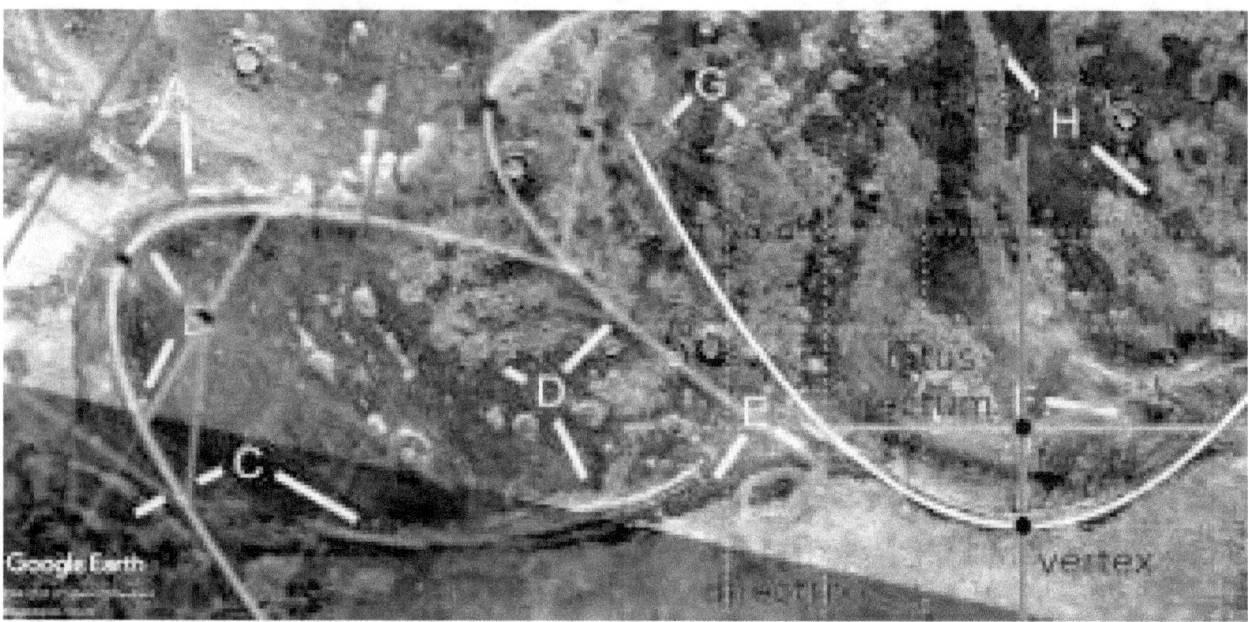

Ishh2306

Hypothesis

These may have been walled fields as often seen near Cydonia. B shows two collapsed hills from 5 to 7 o'clock, C may show tubes or roads in the field. D shows a tube between two craters at 12 o'clock. At 3 and 4 o'clock is a hill connected to a crater.

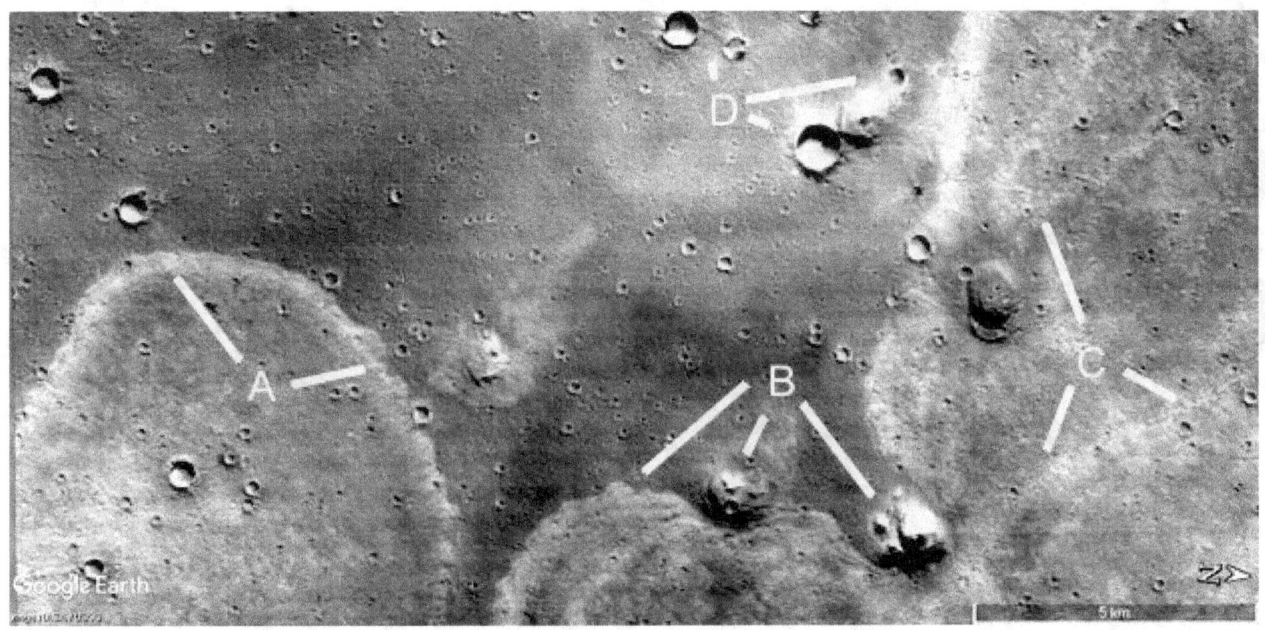

Ishh2306a

Hypothesis

Five parabolas are shown.

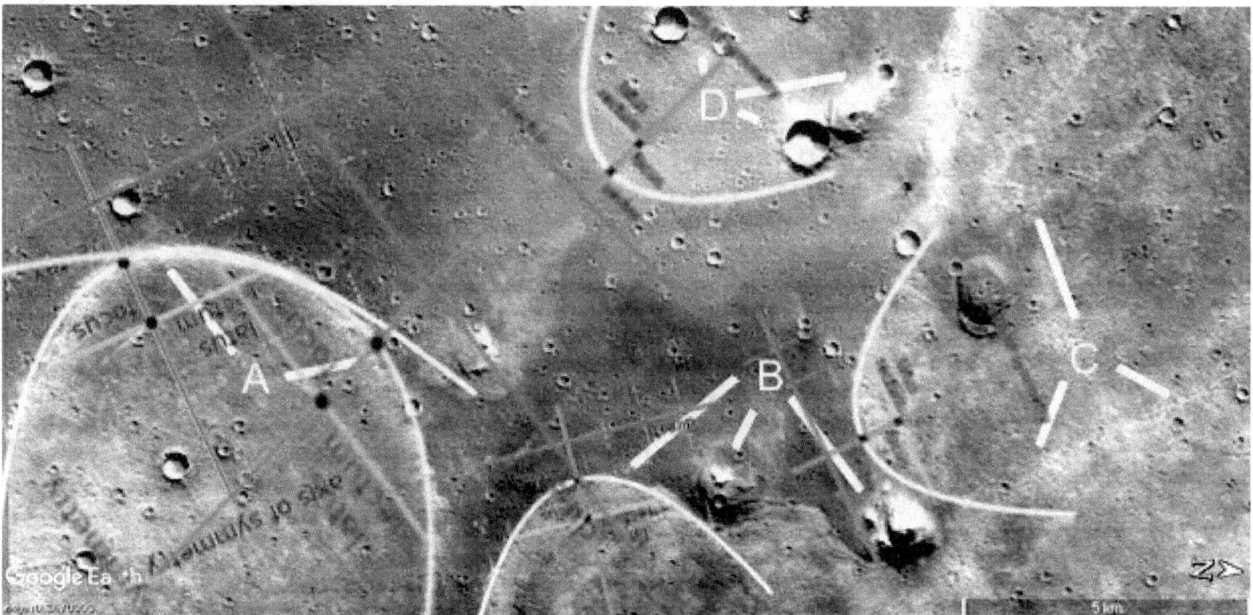

Lakes

The hypothesis is that some water channels and canals connect to larger artificial lakes. This is also something we do on Earth.

Prd886c

Hypothesis

A shows the double walls of this dam at 0 o'clock, also a small cavity in the wall at 8 o'clock. This connects to a star shaped wall from 7 o'clock to 3 o'clock. B shows this dam wall is intact at 10 o'clock, there is a wavy wall like some tubes at 7 o'clock. At 8 o'clock one of the walls is much shorter. C shows this double dam wall continuing at 5 and 9 o'clock, the wall at 12 o'clock has broken up into segments on its end. D shows another walled segment of the dam, below 10 o'clock the wall is more eroded. At 4 o'clock there is a small entrance between the walls.

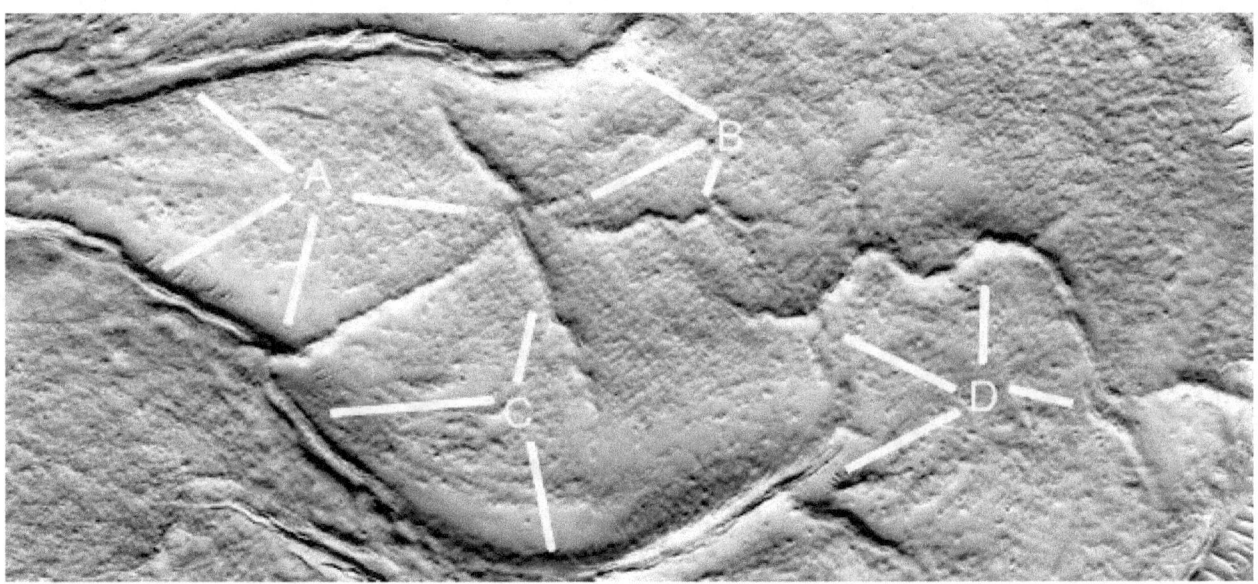

Prd886c2

Hypothesis

A parabola is shown. The axis of symmetry goes approximately through the centre of the star.The focus is also in line with the dam wall between E and F, the latis rectum or line through the focus would then approximately be an extension of this wall. A line is drawn from E to F to illustrate this.

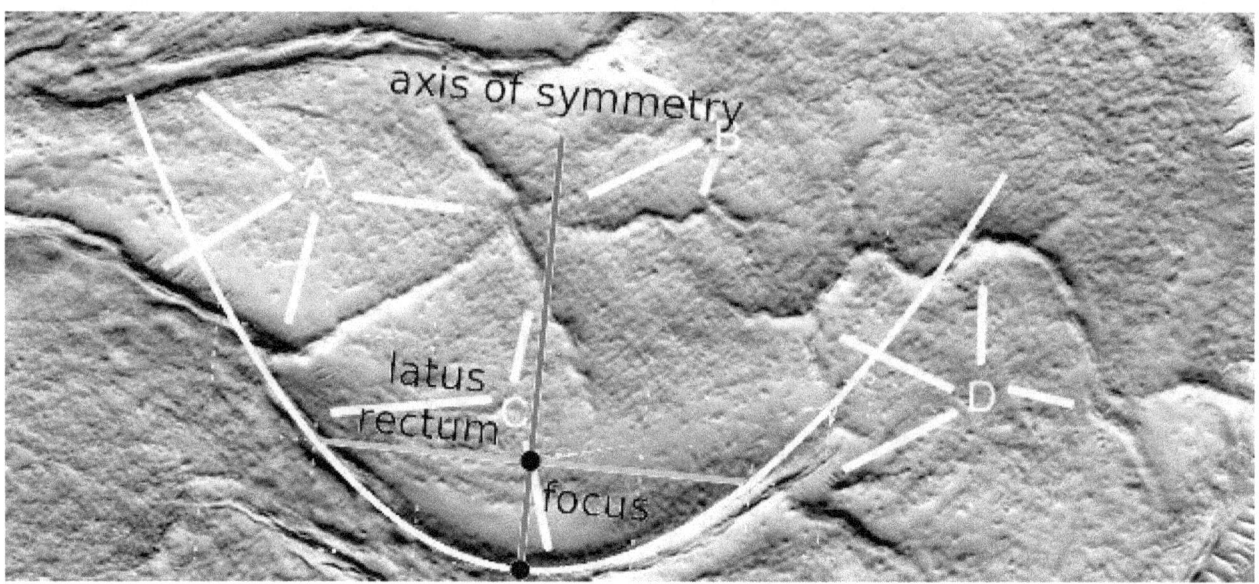

Prd911b2

Hypothesis

Eight parabolas are shown. This is a good example of how natural looking areas in a crater can be looked at more carefully. With a closeup there cold be even six more parabolas here.

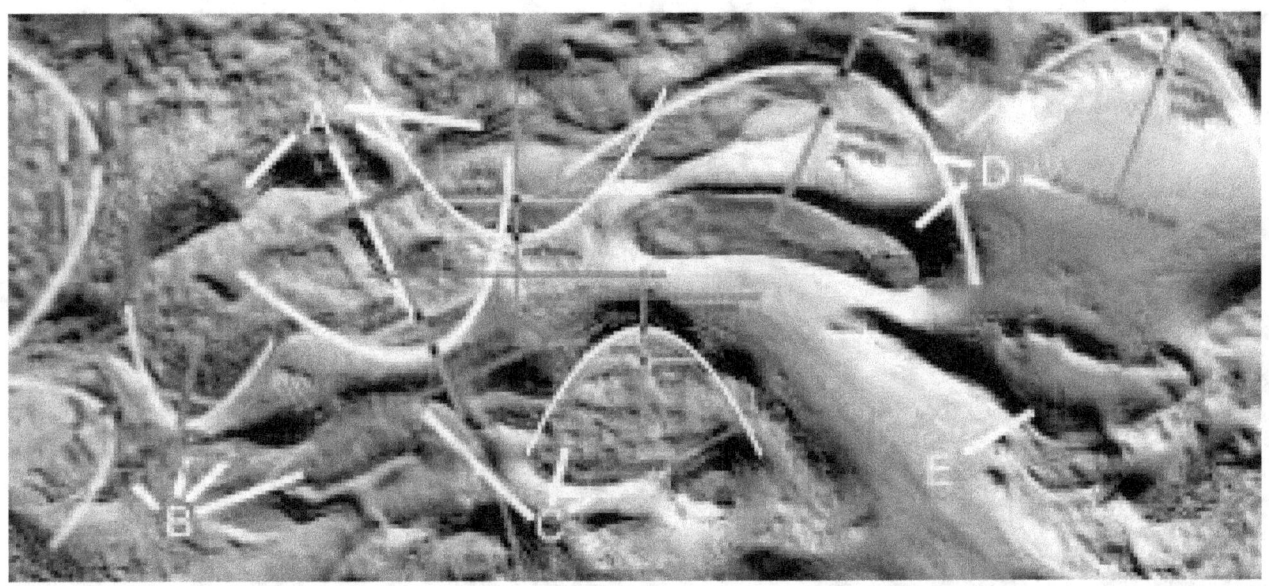

Prhh1018

Hypothesis

Many tubes come out of this formation, A at 8 and 9 o'clock shows a tube intersection. At 3 o'clock is another tube from the pit wall. B shows two more tubes, below the one at 4 o'clock are two small enclosures, also another two between there and C at 8 o'clock. These may all be dams including the large pits. C at 7 o'clock shows many faint tubes coming out of the pit wall. D at 9 o'clock shows the pit wall is doubled with a groove between them. At 5, 6, and 7 o'clock the pit wall is very even and rounded, at 3 o'clock is another tube coming out of the pit wall. E at 12 o'clock shows one of the pale formations inside the pit, these may have been hollow hills and have a similar albedo to parts of the pit walls. At 2 and 9 o'clock the pit wall gets thicker, this part has a roof like a tube but to the right and left it becomes a groove again. It's likely then most of these pit walls are hollow.

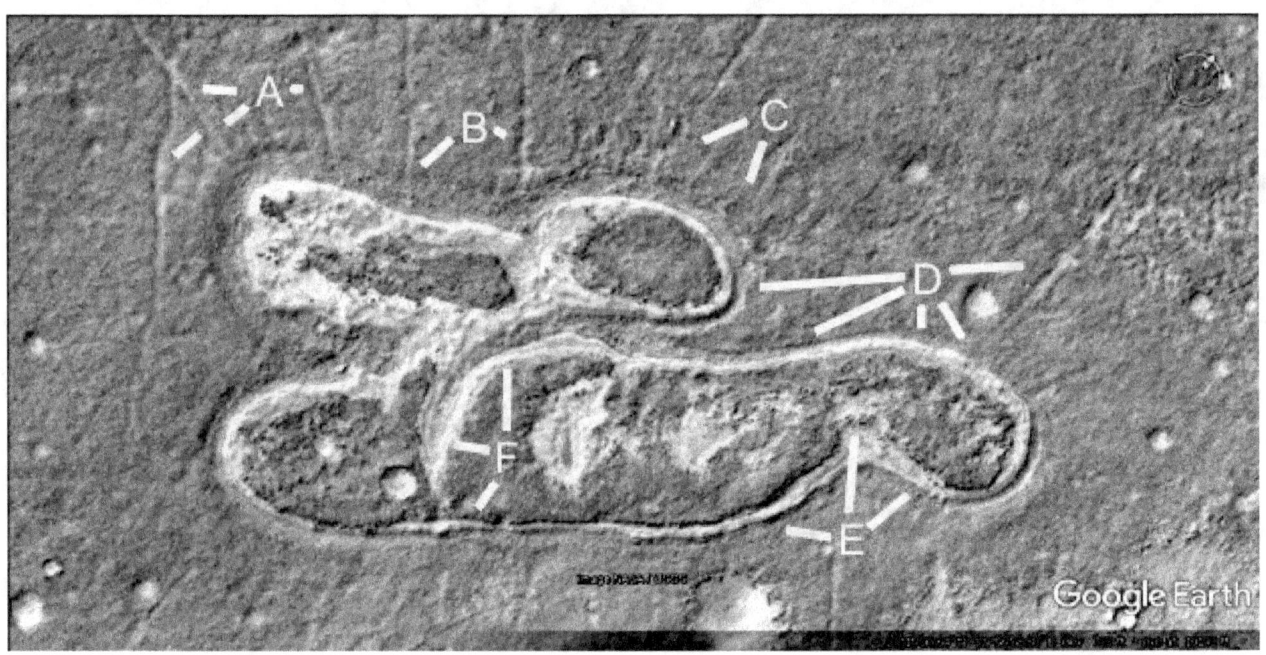

Prhh1018a

Hypothesis

The lines show how straight the tubes are. Also six parabolas are shown to fit onto the edges of the pit dams.

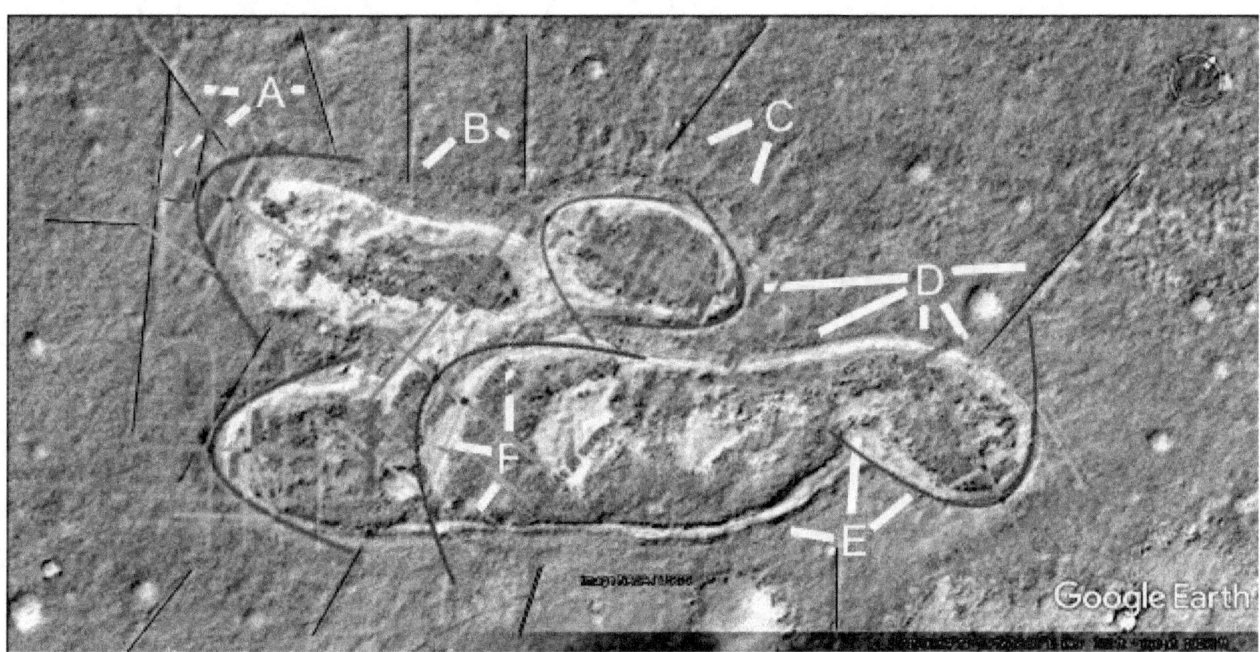

Geometry

The hypothesis is that two hyperbolas were constructed, the one shown here is close to the old Martian equator.

Prt1055

Hypothesis

This shows a nearly perfect hyperbola forming a tangent to the large crater, and to a smaller crater on the left.

Prt1055a

Hypothesis

This shows a hyperbola overlaid onto the formation, it shows it is nearly a perfect hyperbola. It deviates a small amount to the left at A as if affected by the gravity of passing near a planet or moon. B at the top of the image shows two other walls, C shows a road like shape connecting to the crater. B in the crater shows concentric circles which might indicate orbits around the sun, or the surface of a planet with the outer circle being the atmosphere. D is a line or chord drawn as a tangent to the smaller crater, it is at right angles to the vertical transverse axis, the dark line which nearly bisects the large crater. With the inaccuracies inherent from the age of this formation, also in fitting the hyperbola, this may have been intended to go through the center of the crater.

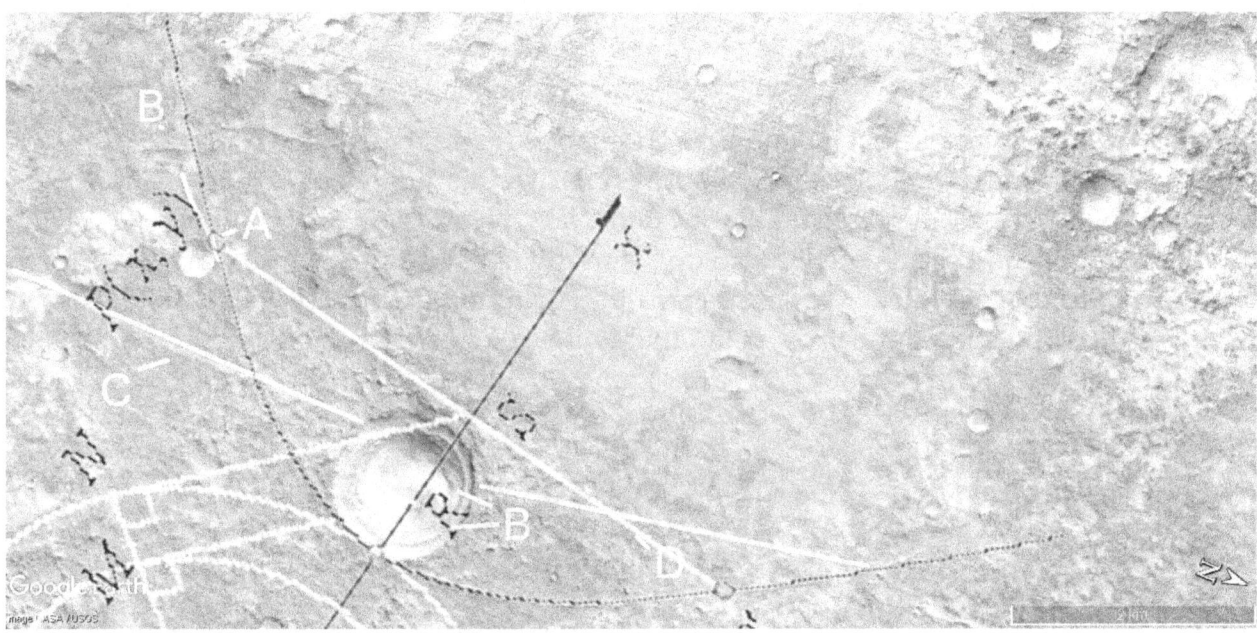

Conclusions

This introduction is intended to show an outline of the global hypothesis, explained in more details throughout the Martian Hypotheses books. There are hypothetical dams in it to collect water, also canals, water channels and lakes. There are two kinds of hypothetical cities, one based on more conventional rooms and walls. The other appears to be based on interconnected tubes. Hypothetical buildings are shown with collapsed areas like rooms. These are often connected with roads and tubes to each other and to farms, canals, dams, craters, and the oceans. With this overview the additional images in these books shows how these hypotheses repeat in many areas and extend into a more detailed global hypothesis. If these are natural then they are highly unusual, the parabolic formations do not appear to occur naturally.

Images, main section

Ht5

Hypothesis

This long ridge may be a tube, it connects into many craters here along A and B. Some of the hills are also unusual, they may be hollow hills. Mathematicians might consider how a parabola would be formed efficiently on such a large scale, engineers on how a long hollow structure could be created. Sociologists might consider how these craters would have been used, to get water for possible farming, drinking, raising fish, etc. Biologists can regard this as a clue to the ecosystem, that water may have come from a water table, from rain, etc.

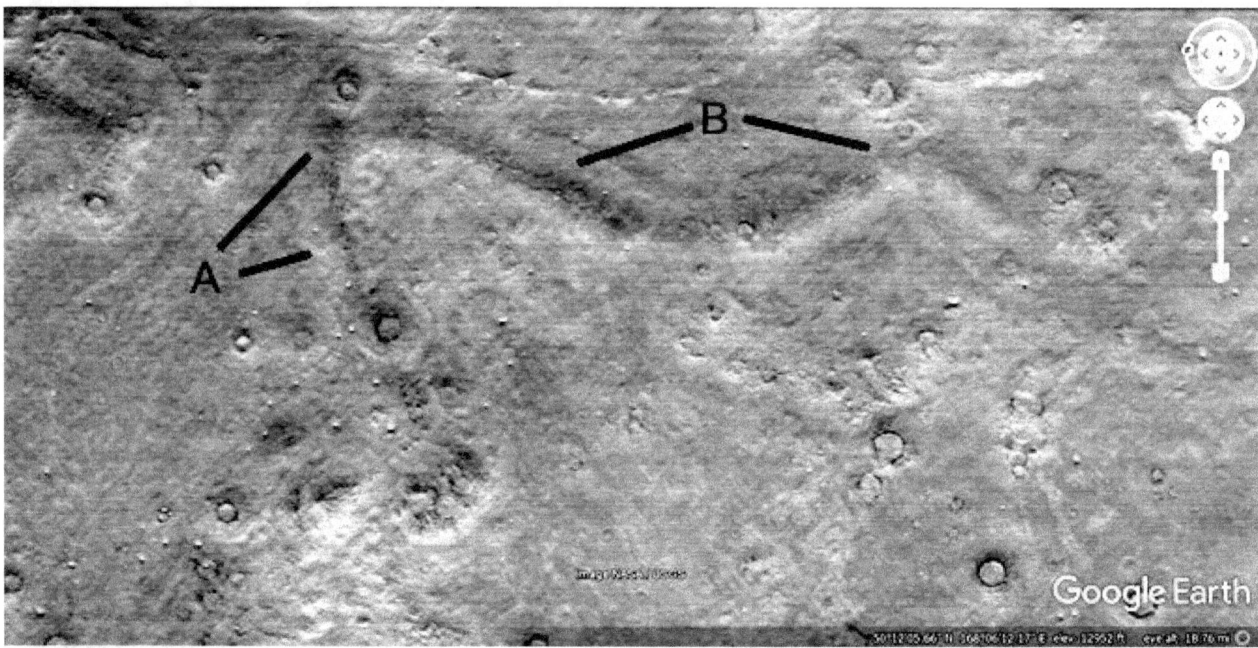

Ht5a

Hypothesis

This shows how part of the formation is close to a perfect parabola, the right curve at B is also close to a parabola. Mathematicians might investigate the other parts of this formation, whether there is a way to create them on a large scale like this.

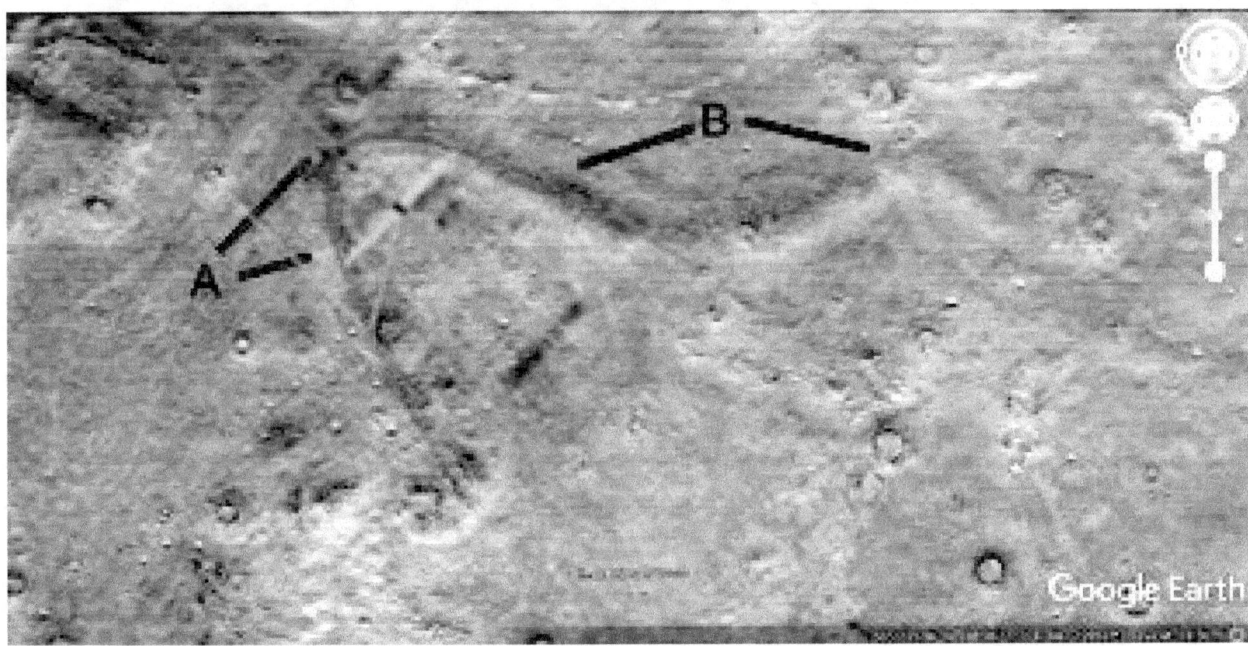

Hhh12

Hypothesis

This hill shows clearer signs of patching, A indicates a darker external wall and a center that has either settled or is designed this way. In between there is a groove that is even in depth and width rather than randomly changing. This may have occurred by design or from the center settling. B shows a lip running around the edge of this interior hollow, it is also even in size and width. C shows a cleared area around the hill as if the material was used from here, or it was maintained to be cleaner. For example it may have been used for crops. Horticulturalists might consider whether some cleared areas could have been used for crops. D shows a ridge like part of an interior support.

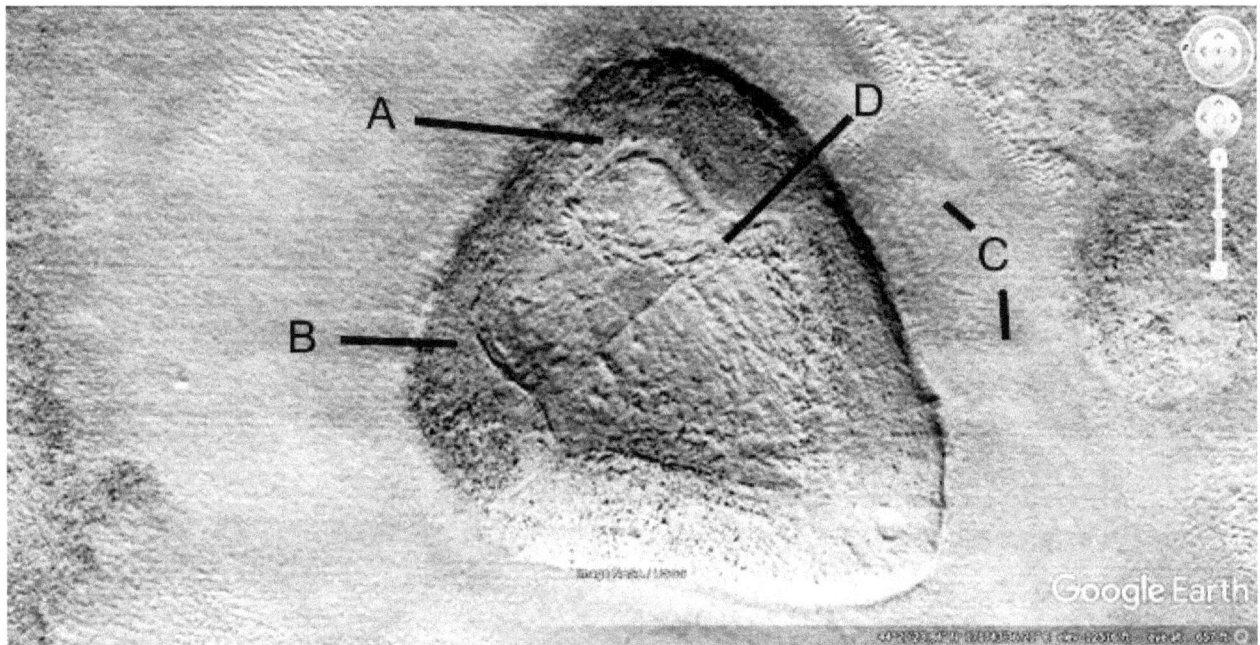

Hhh14

Hypothesis

This hill also appears to have been altered. A shows a wall on the upper side which is fairly regular. B shows a pale section which make have lost an outer skin, accounting for the change in shade. It may also be a patch. C may be an area that is settling, there appears to be tubes on the surface which may be a lip around a patch. The area is quite chaotic as it if had been patched many times. D shows another patch like shape like a parallelogram. It has a pale groove going off it to the left all the way to the edge of the hill. E shows a dark area at 9 o'clock, at 8 and 10 o'clock there is a step between the higher outer area and this section. The step at 10 o'clock appears to continue to the left to the line from B. This lines between the shades give the impression of patches or a deliberate design. This also has a cleared area around it, another possibility is dust storms might threaten to bury some hills and so the ground around them was cleared regularly.

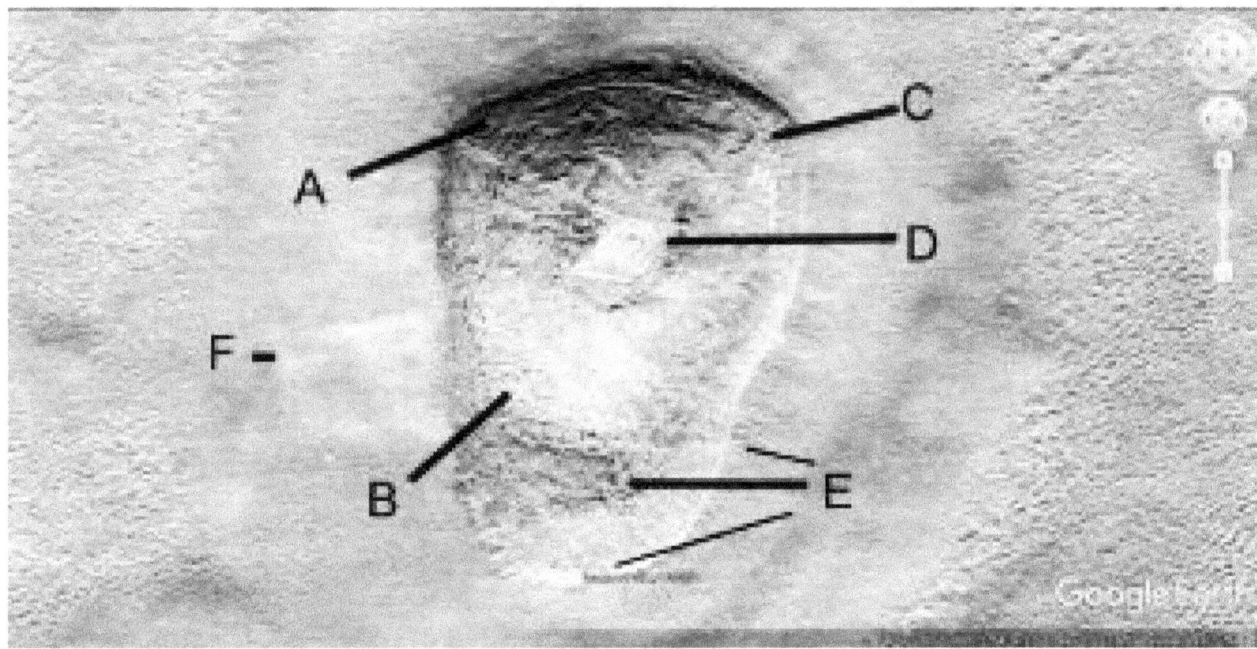

Hhh18

Hypothesis

A shows a line narrow pale area, it may be a patch or a collapsed cavity. B is a similar stripe. C shows an irregular dark shape above the pale material, it is however only around the hills not further out. So a sandstorm could not do this without covering the area above A,B, and C. D shows 4 more possible patches, E is all dark on one side and F shows another long cavity. G shows a dark hill above the same kind of pale material, this would not have eroded from the hills because they are not big enough.

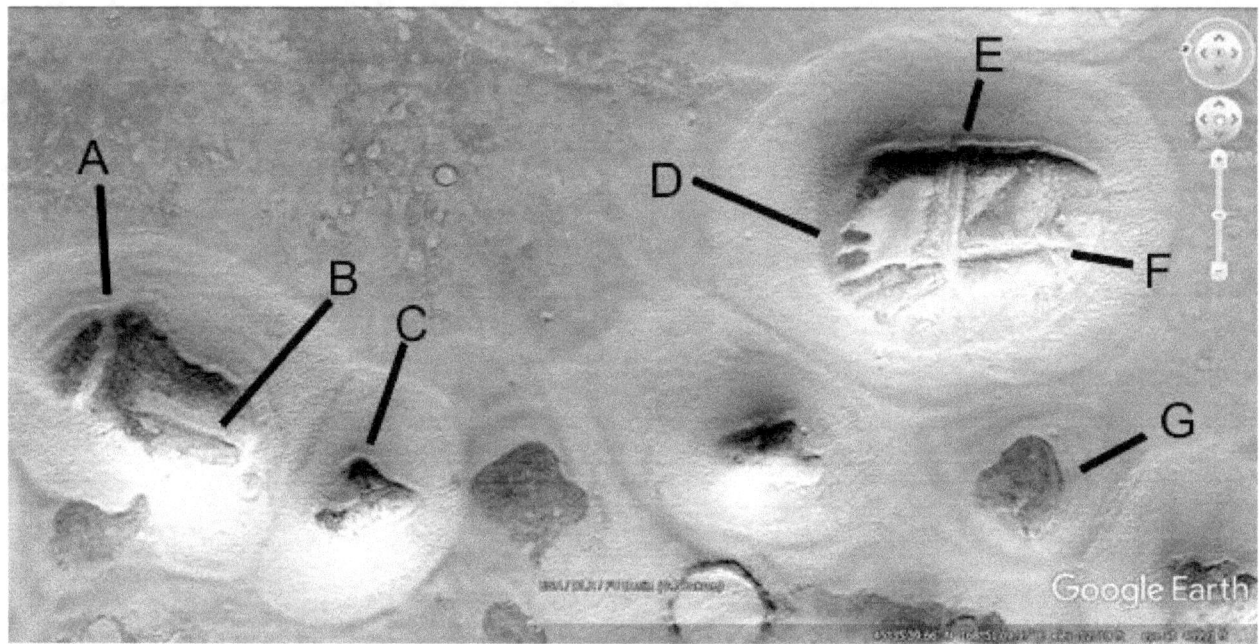

Hhh19

Hypothesis

This shows possible roads leading between hollow hills, it is logical that there would be movement between them. Sociologists and perhaps anthropologists may examine these routes to understand their use. It might for example be trade on the longer roads, movement to and from crops, etc. Some hills may be more important as communal meeting places, have government functions, house animals, be markets, etc. A shows something has worn the surface down to paler material, it looks like many lanes of a freeway. It may then indicate a lot of traffic, forming separate roads to avoid congestion. B shows a larger road shape going into smaller ones, it may also be the space between the roads has eroded making it look larger. E shows a hollow hill that has collapsed, there is a pale track from the roads at A and C to it. Roads appear in many routes to the right, a larger one at D. C shows a more separate road, it may be crops were planted in this direction.

Already the components of a possible civilization are starting to emerge. There are tubes connecting craters for water, hollow hills for dwellings, and roads between them. These represent a priori predictions, with more images the concepts of a civilization should be repeated to be real.

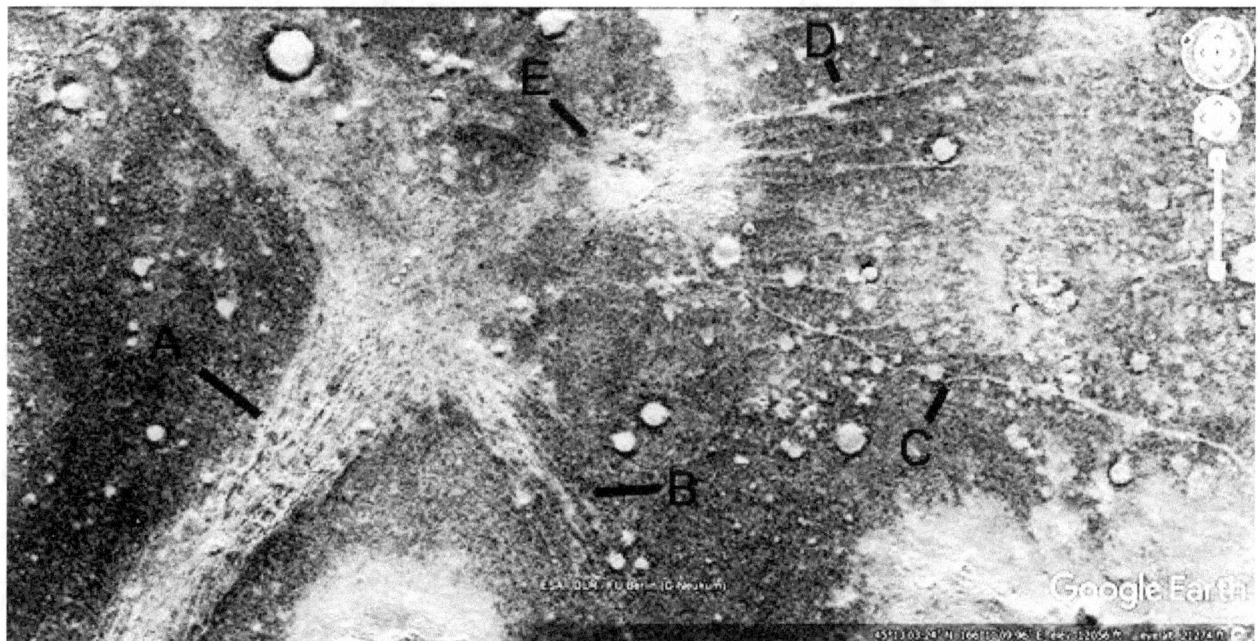

Hp26d

Hypothesis

This hill shows signs of decay on its roof on the second section of A at 5 o'clock. B shows a possible tube going up the side of the hill and inside it. Some appear to end at the edge of hills at ground level, others seem to enter further up. A at 4 and 7 o'clock appear to show another tube.

Hp50

Hypothesis

A shows a hollow hill with cracks or bands of darker material along its roof, perhaps in the process of collapse into a pit like C. It as a crater on the roof which may have caused the cracks. B shows possible patches on its roof and perhaps has collapsed on the left side. There is a lighter area from A to around B, perhaps caused by traffic. D is another pit, to its right is probably another collapsed hollow hill. E may be a very large hollow hill with some interior supports showing, only the lower part appears to have some elevation. F appears to have an interior support pointed at by the F line.

Hp51

Hypothesis

Many pits, probably former hollow hills are shown. A has a large interior support still standing but the rest seems to have fallen into pits. C and F have some structure remaining, H appears to be largely standing with a patch or cavity on its roof. It may be some shapes lasted longer, like a more conventional rounded dome here. When the shape becomes too elongated it may have points of weakness where it will crack first, bringing down the rest. L is more rounded and has survived, as well as parts of O. Engineers could consider how to model this terrain, then construct dome shapes on these pits. This may give insights into which would have been more stable. There are few interior supports here which may have caused the problems, however these may well have collapsed after the atmosphere had frozen.

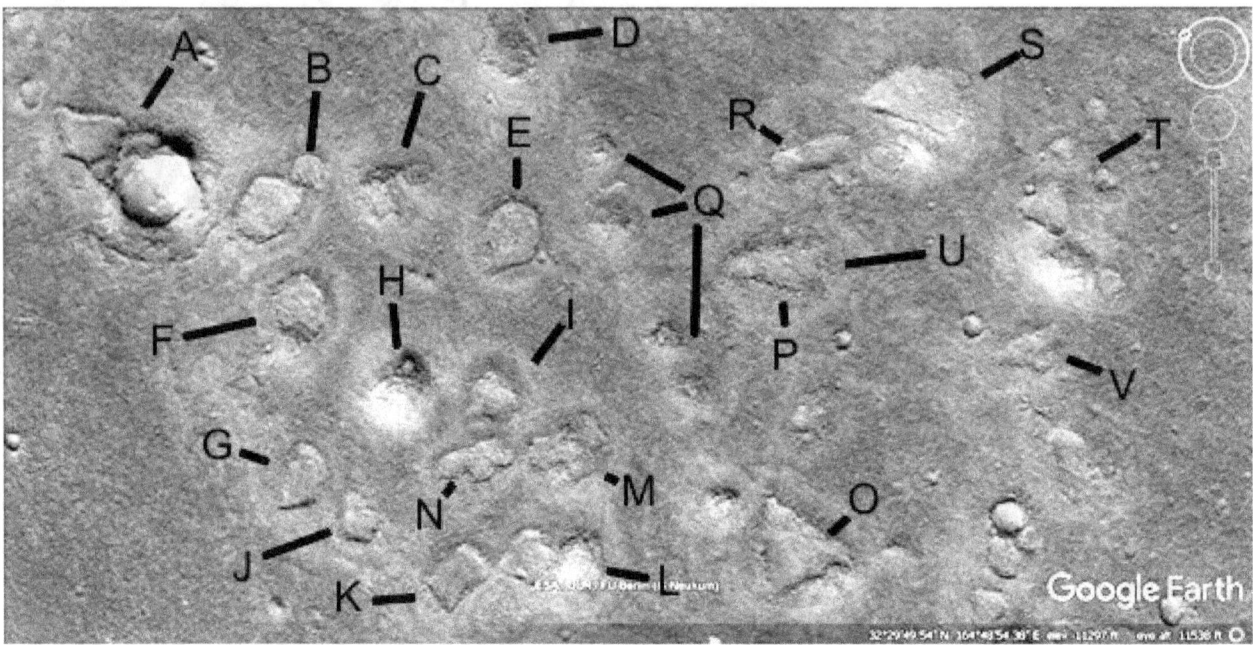

Hp52

Hypothesis

So many hollow hills have collapsed here, it is on the eastern edge of the Hecates area nearer the ancient ocean. A tsunami from a meteor impact may have caused this. B and C show interior supports, A may have faint signs of one. E seems to have 4 interior supports but still collapsed, with no debris it may have been washed away. H and I may have had an interior support, J up to 6. Sociologists might consider why some hollow hills were much bigger than others, whether this was for the creatures to gather in or because some were wealthier. Often there are very large hills with settled sections on the roofs, then there are smaller hills around them.

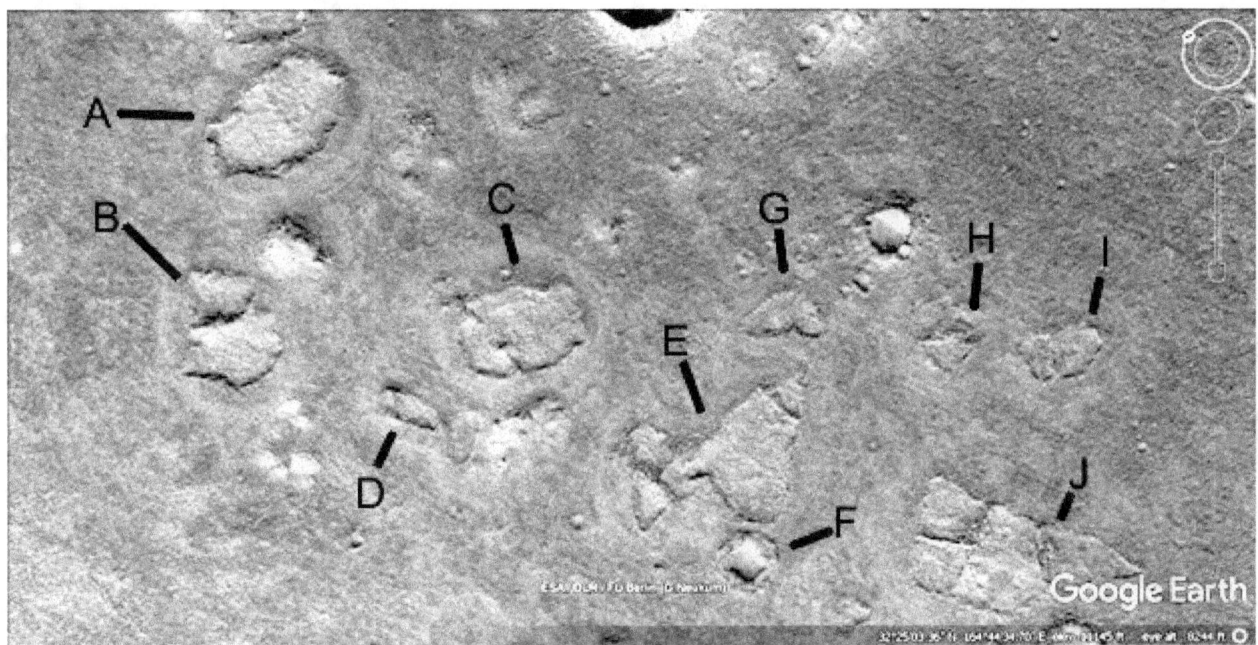

Hp53

Hypothesis

Many more pits here, to the left of the A line there appears to be an entrance. B and C are somewhat heart shaped which is a recurring motif with these pits. H shows an intact hollow hill that seems to reach out and join to a crater, called a hilled crater here. This may have been to use the water in the hill, perhaps there were tunnels directly into the side of the crater. It is also problematic for geologists, a crater is supposed to impact randomly. This hill could not have known to be exactly in this position, as if it is catching a ball in a mitt. Also the crater is flattened on the hill side like it has been rebuilt there.

Hp57

Hypothesis

A has a sharper point than most pits, the narrow pale area may have been an entrance. C is also built up more on the lower side. Some of the pale mounds in this area might be small habitats.

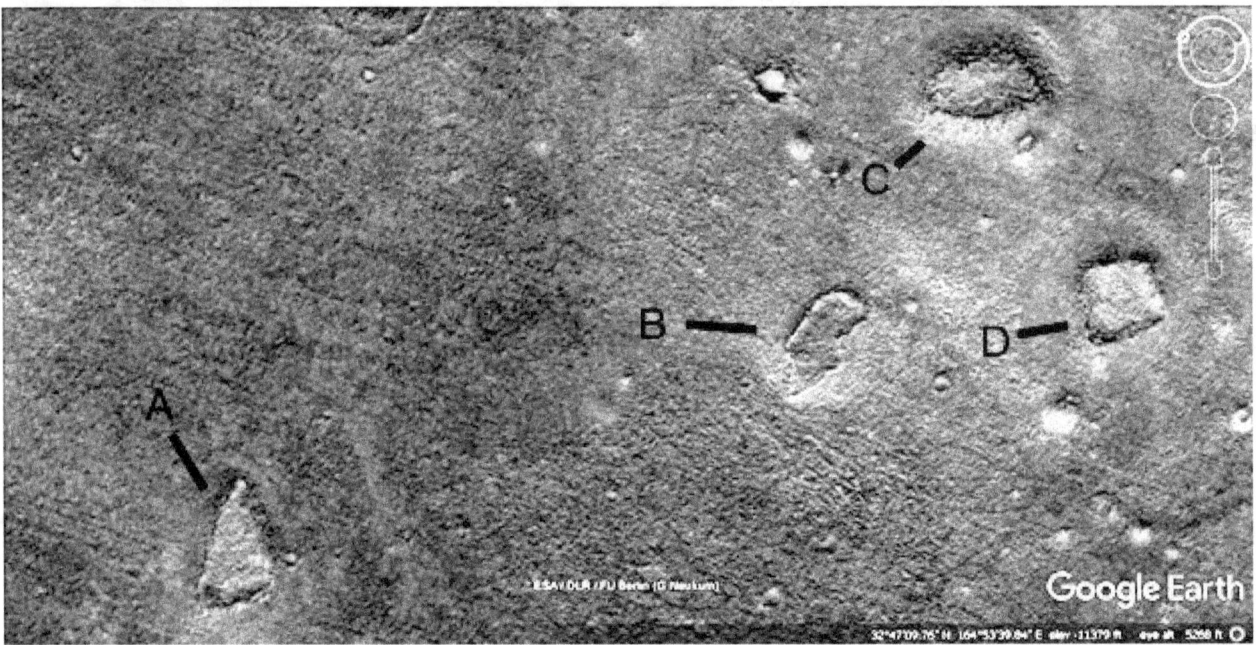

Hp65

Hypothesis

A has a straight edge on its left side, the right side of it seems to be full of material perhaps from the collapsed roof. C has much thicker walls so this is more of a cavity than a pit. D shows a walled hill on the lower side and a pit above it. F shows a pit with a tube going into the walled hill, also called a tubed hill here, it has a similar flat roof to D. It may have been part of E as one large hill.

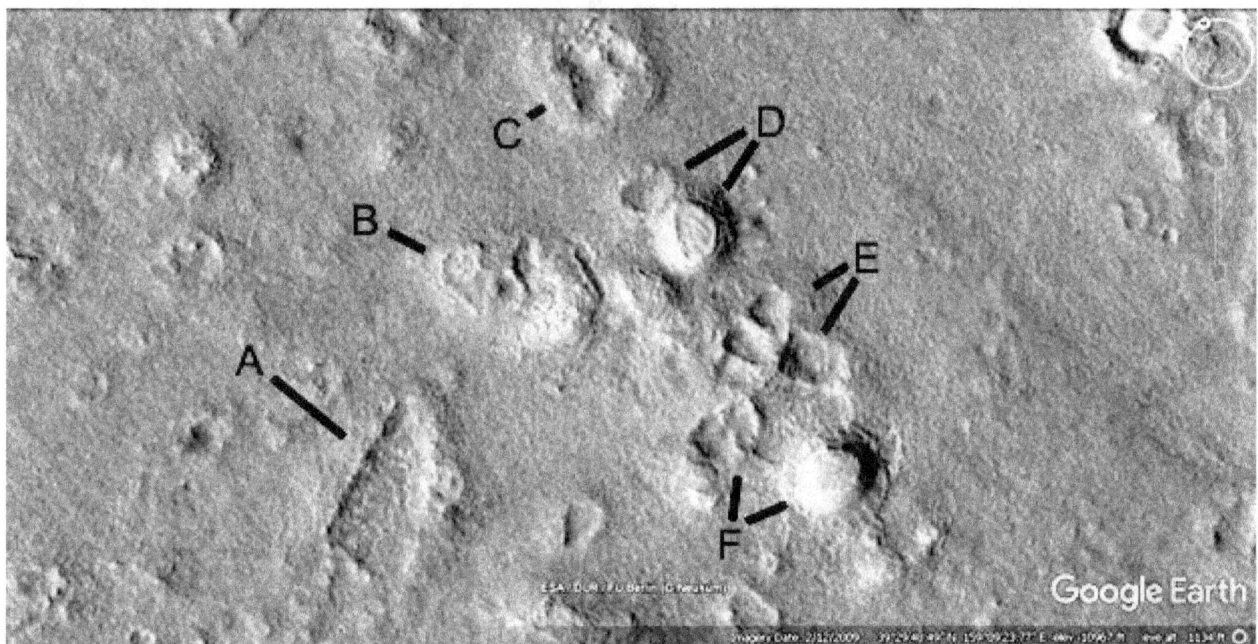

Hp68

Hypothesis

A appears to show the interior supports sticking through the settled roof, as if ready to collapse. D looks like a very artificial pit with straight walls, though it may have had a dome for a roof, it also seems to be connected to C. There also seems to be a connection to the crater E. The mottled surface is also seen in many areas, it appears in some cases like many corridors are going through it. This may have been a construction technique to make much wider habitats by using many more interior supports.

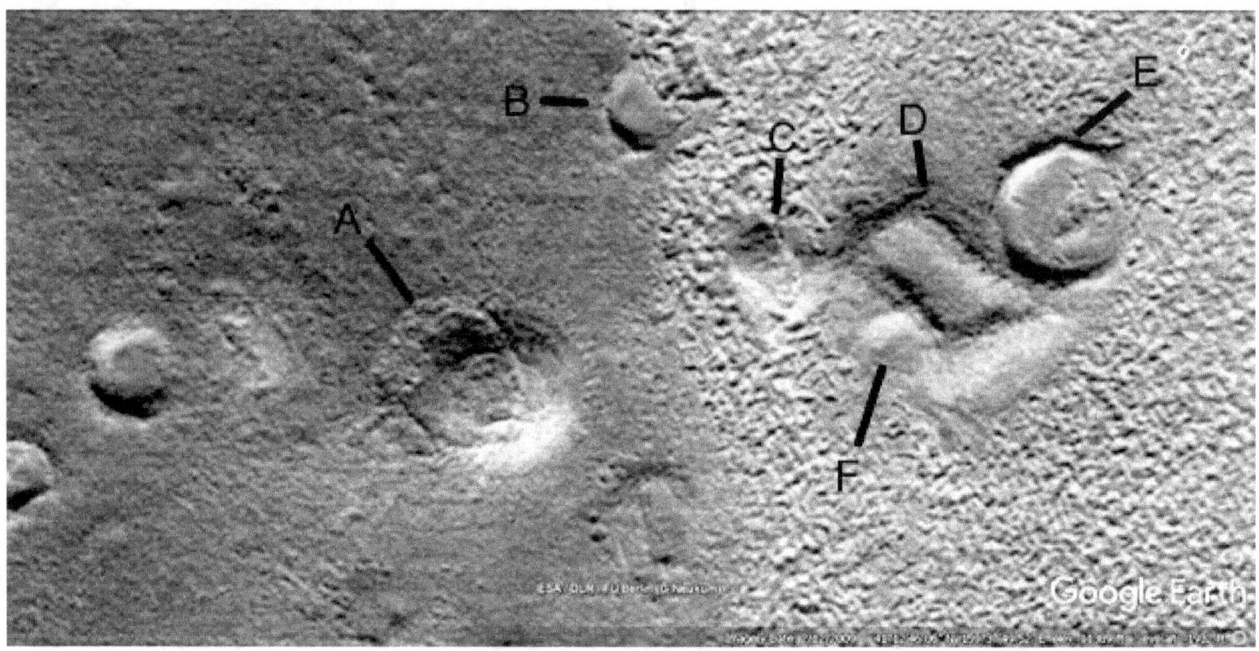

Hhh87a

Hypothesis

This shape would appear to be natural, but it closely follows a parabola as in many other areas.

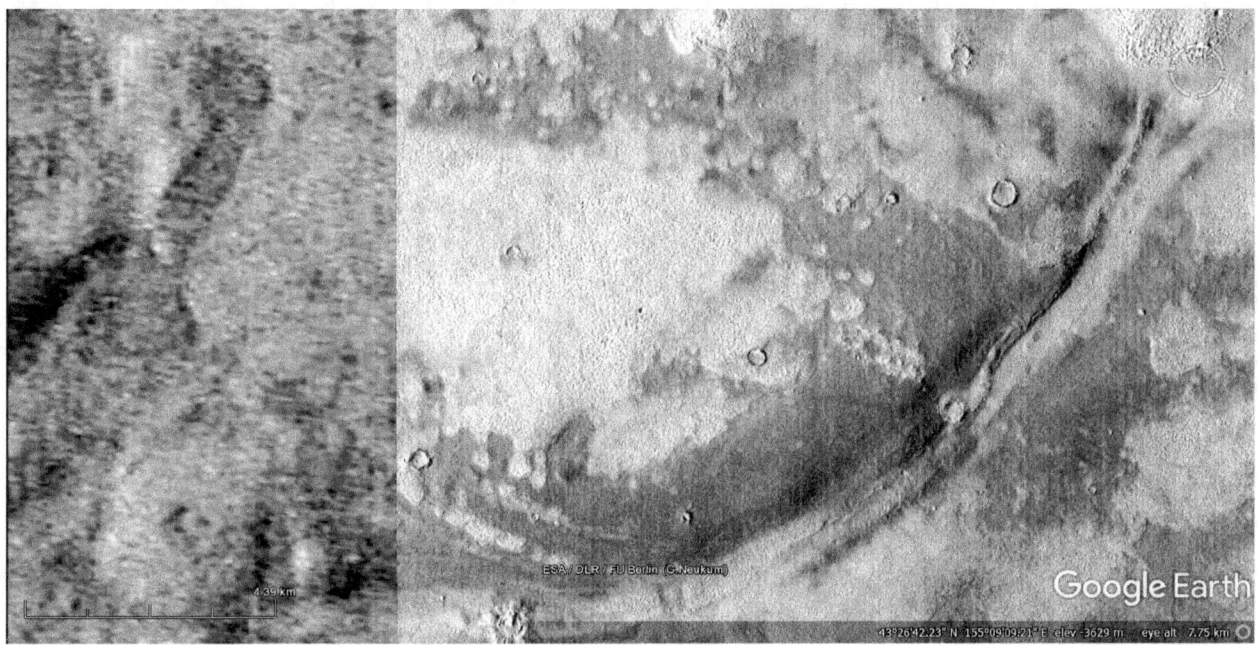

Hhh87a2

Hypothesis

Here is a parabola superimposed on the ravine around this hill.

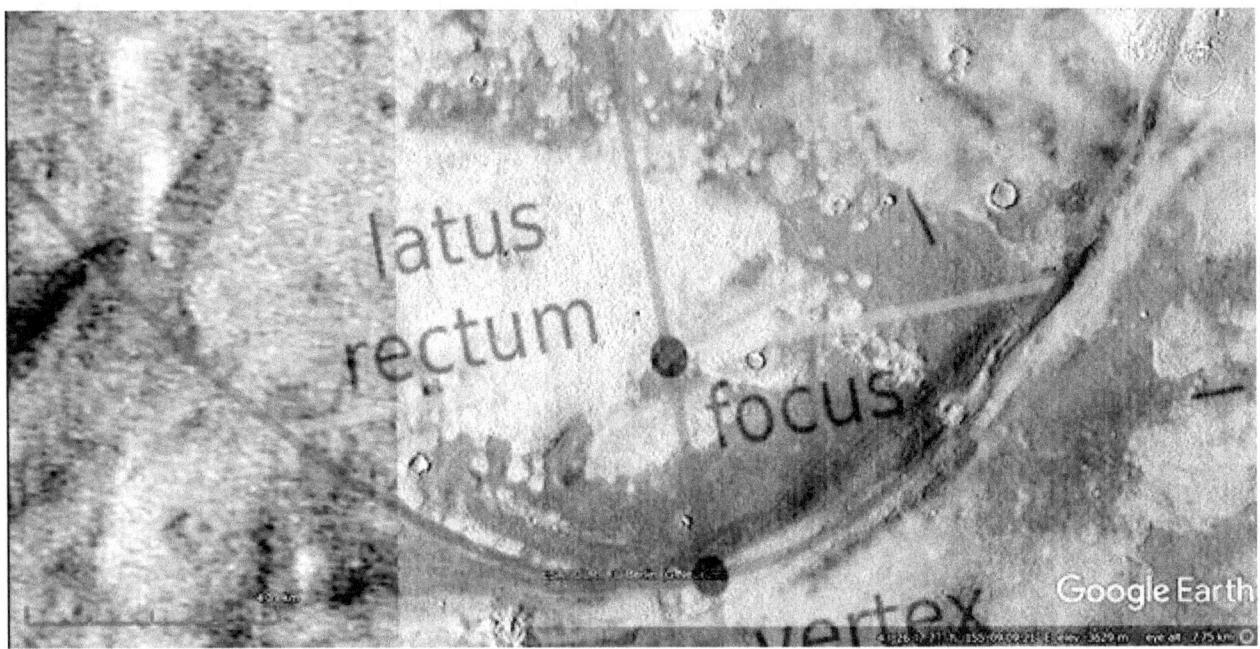

Ht89

Hypothesis

A shows a small hill with cracks and perhaps a patch on its roof. B shows a tube going from the crater on the left to the collapsed hill in the middle. Then it continues on to C with a pale spot at 10 o'clock like a collapsed hollow hill. At 12 o'clock and 2 o'clock C shows a tube like shape that expands to two other collapsed hills at D. The whole area under B and C may be a large low hollow hill, there is a mottled appearance perhaps from collapses.

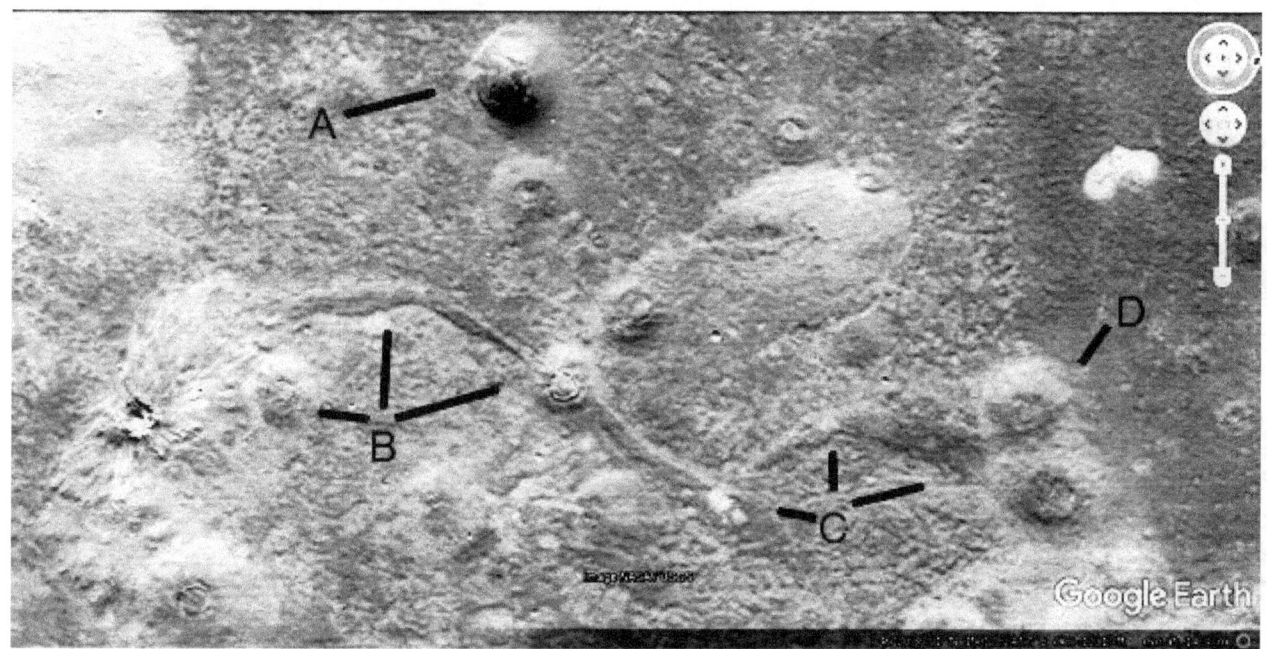

Ht89a

Hypothesis

A parabola is shown.

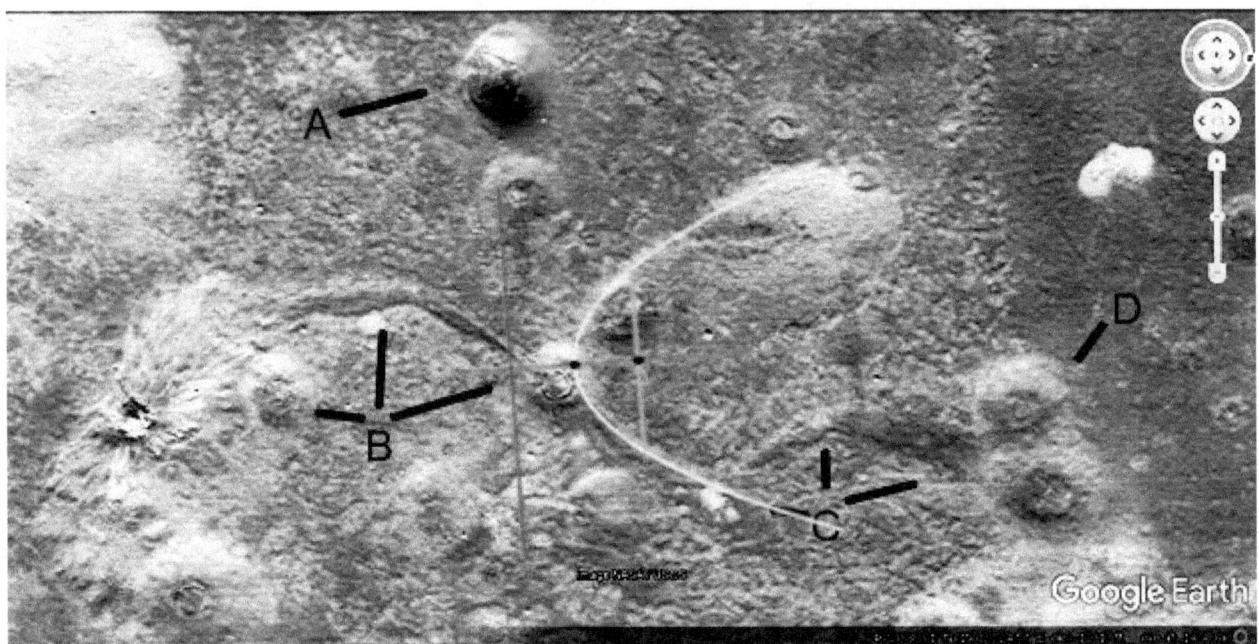

Hhh97

Hypothesis

A and B show a thick wall that extends upwards to the next hill, partially like a tube. A also points at a settled cavity on the roof surface. There is a mottled skin here exposing a smooth core as in Hhh95, perhaps the same construction technique. There may have been a smooth skin like cement over this that was peeled off. C shows a walled hill with the concave roof perhaps settled. D shows a smaller hollow hill perhaps patched.

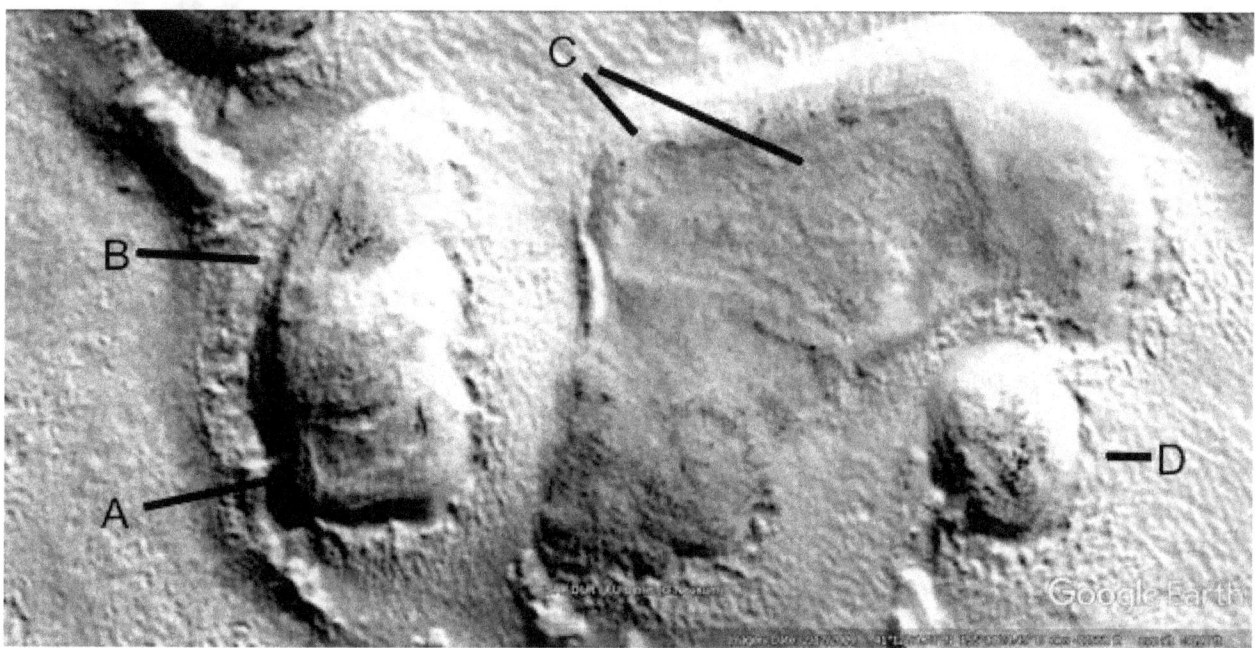

Hhh98

Hypothesis

A is a walled hill with a concave roof, this also has bands of darker material. The crater there does not appear to have collapsed it. B shows this is connected to C by a tubed hill, in this case tubes come out of both sides to connect the outermost hills. C has possible patches on its roof and skin on its left which may have originally covered the hill. D shows a larger tube going down to a crater, this merges with many smaller tubed ravines. They can be seen by their lighter left side like with the hills showing they are not ravines. E shows a possible tube on the side of the hilled crater F at 11 o'clock. At 9 o'clock the hill may have a cavity or a large crater on its roof. G shows a peeled skin on the hill. Hi, I, and J show a tubed hill connected to a tubed pit. H is very angular and artificial looking.

Hhh99

Hypothesis

A shows a mottled appearance perhaps a collapsed section. B shows a hollow hill fading down perhaps to a pad rather than a pit. C and D show a walled hill.

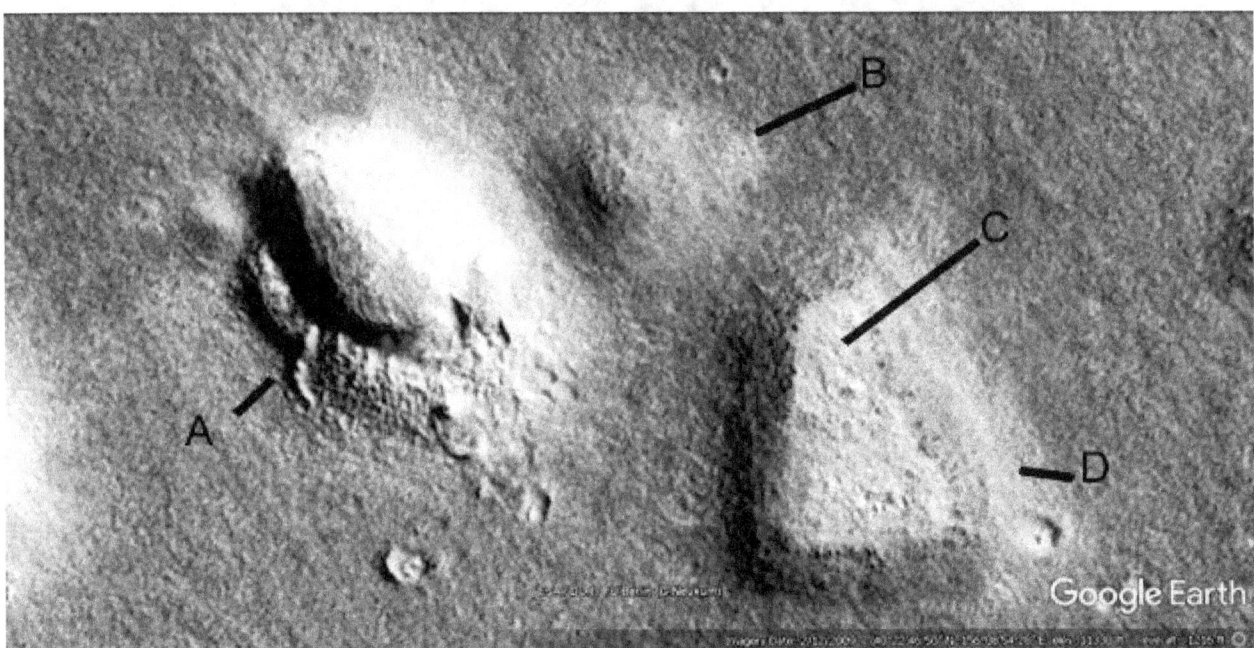

Hhht104

Hypothesis

A is a tubed hill connected by a tube to B, also to a decayed hollow hill at C. It also connects to a decayed hilled crater at D The tube at F connects to another crater, E also appears to be a tubed crater with the tube coming out of it down the image.

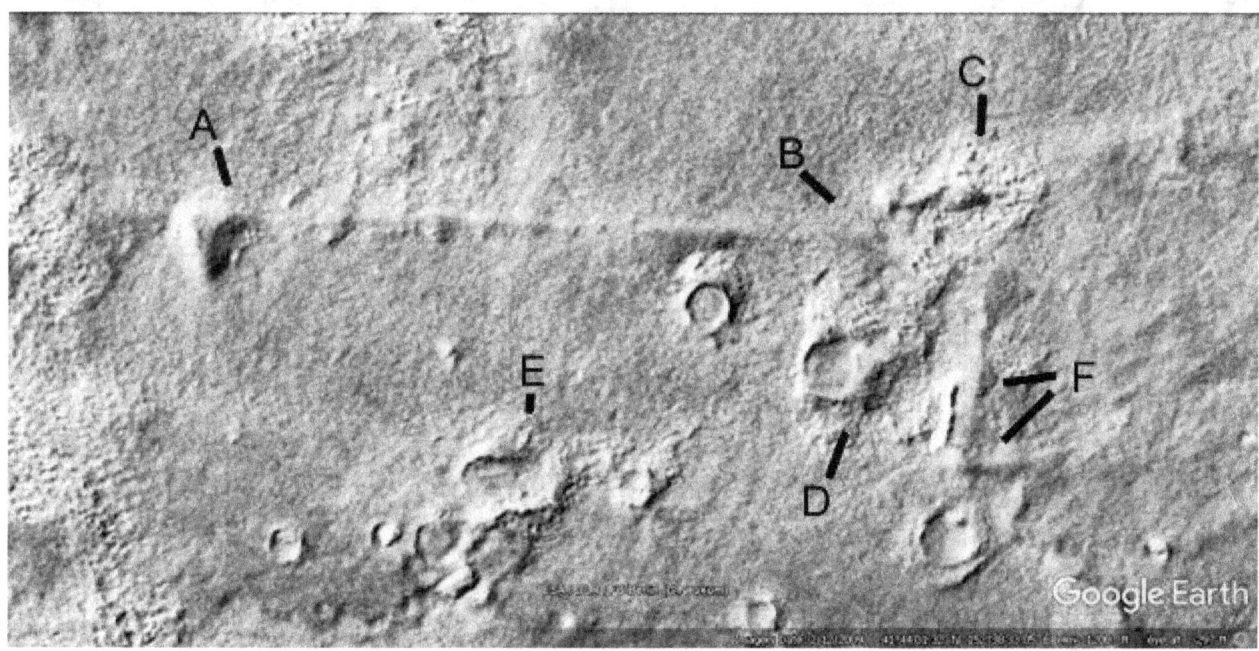

Hhht106

Hypothesis

A shows how this tube like hill connects to the hill at D. This also connects to E and to a road at F going through the dark hill I to the road at K. I has 2 lines, the first shows a road or tube coming out of J and the second line further on is a darker hill road with the road K coming out of it. D has a collapsed section on the left, perhaps some patches on its roof. E appears to have settled on the peak of its roof with some cracks, similar to a pingo. H looks like a hollow hill with the skin peeled off except for its left side. The connection to D is smooth like cement though very ancient, other materials appear to have collapsed or disappeared. J appears to have missing ejecta perhaps to build the other hills. B has a skin on the upper side on both hills, the left hill has a cavity probably as the hill collapses. C is probably a collapsed hill, there may be another one between D and E. G shows a road or tube between a small hill and a crater. L may be a large hollow hill.

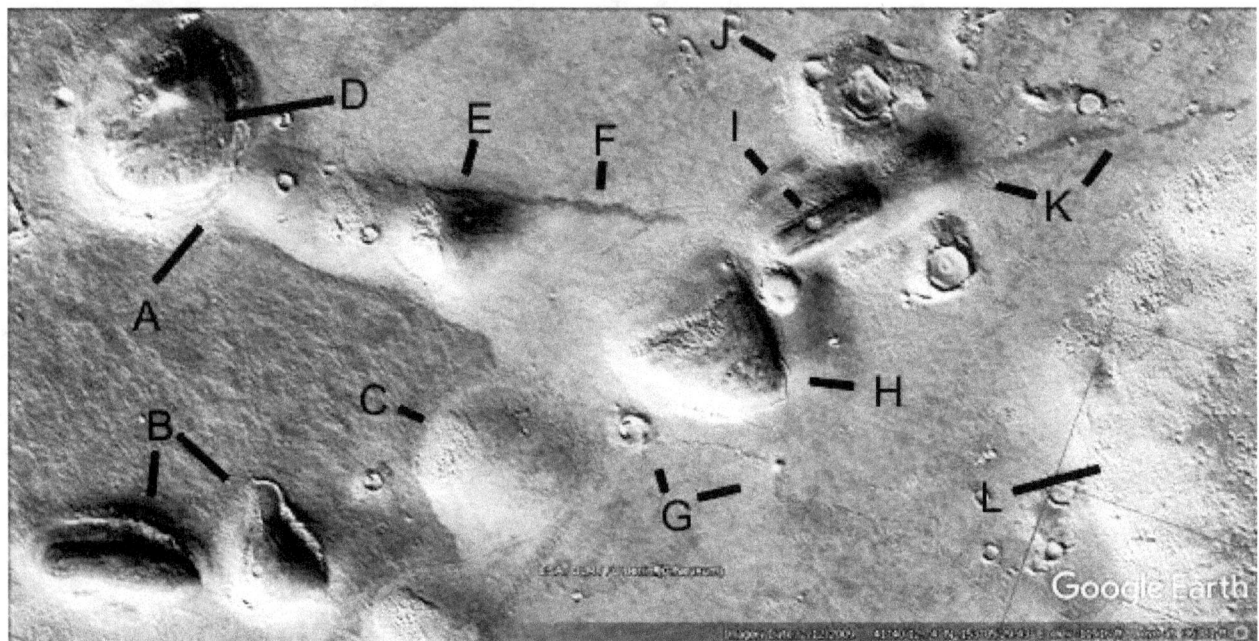

Hhht128

Hypothesis

A appears to be a tube, B is an unusual hill with a straight flat section of roof. C appears to have cavity shapes on its lower side and perhaps a dark patch on its upper side. D has many cavity shapes as well, the crater may have collapsed part of the hill. Under where the line comes from D there may be another crater that caused a collapse, there is a circular arc to the cavity here. The right side of the crater appears to be missing so perhaps it was repaired.

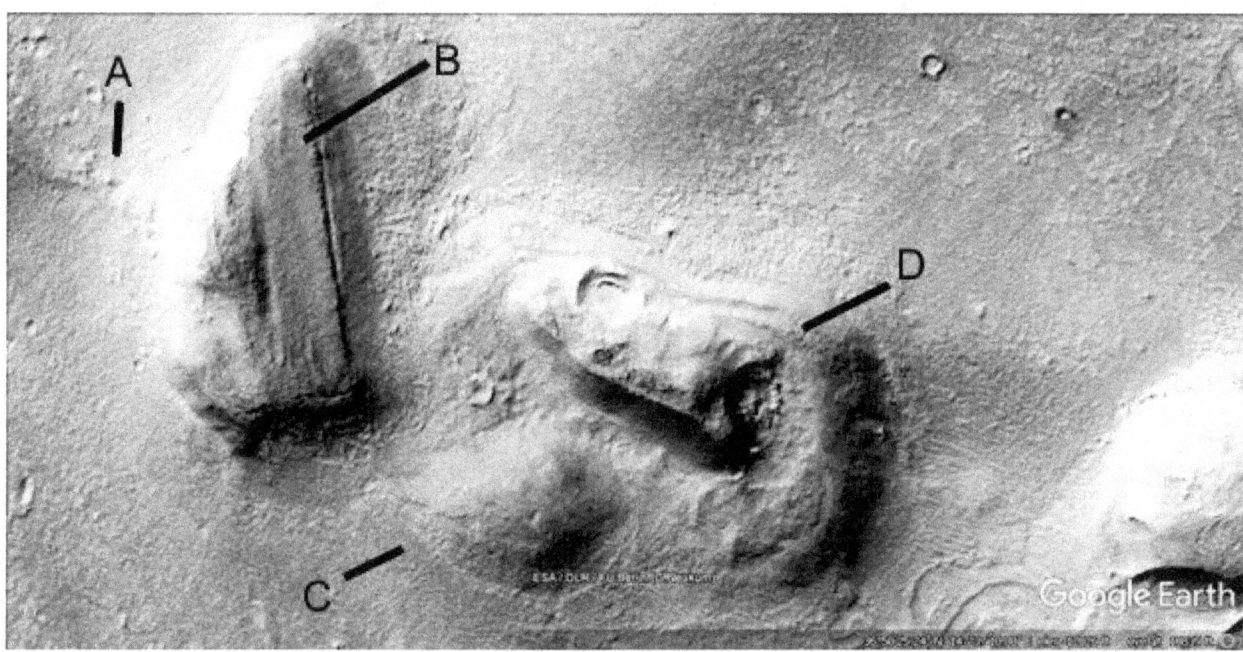

Hhht129

Hypothesis

A shows a small hill with a tube or road going down to the left, and perhaps a pad where a hill was. It goes down to the right through another small hill, then a large hill then to B at 2 o'clock. B at 4 o'clock shows a wide tube going to a collapsed hill to its left. C s a long tube perhaps connected to two craters on its way to a collapsed section at D. E shows many tubes like a mesh around small hills with a cavity on their roofs.

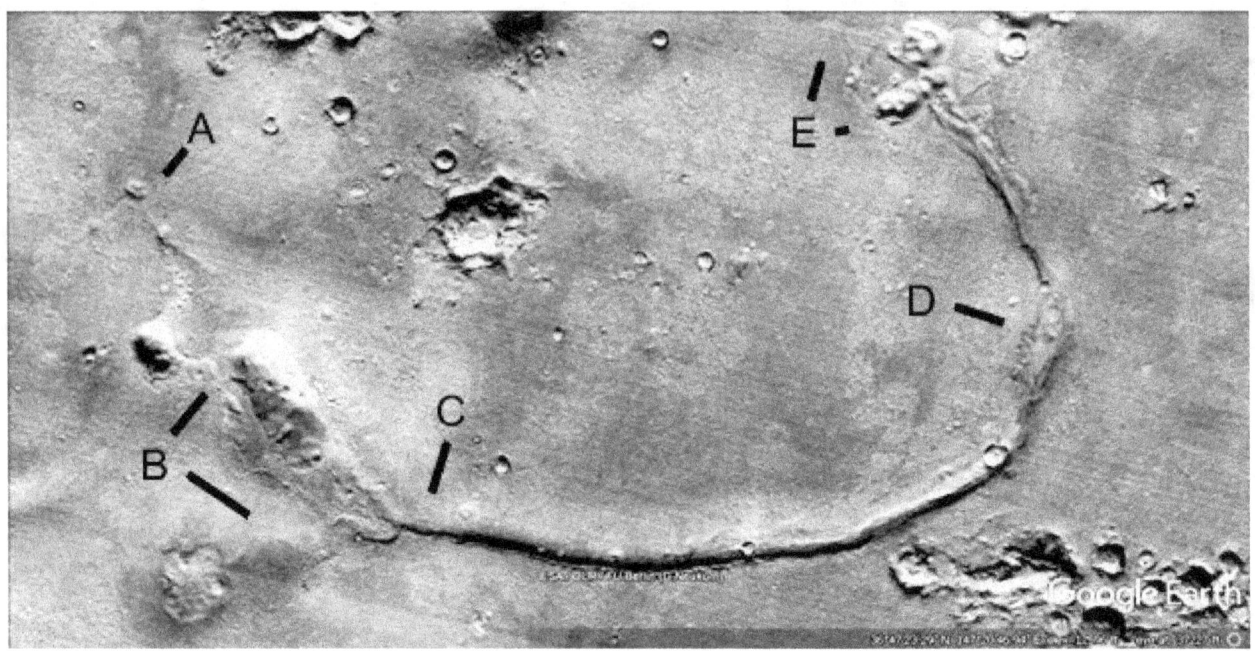

Hhht129a

Hypothesis

A parabola is shown.

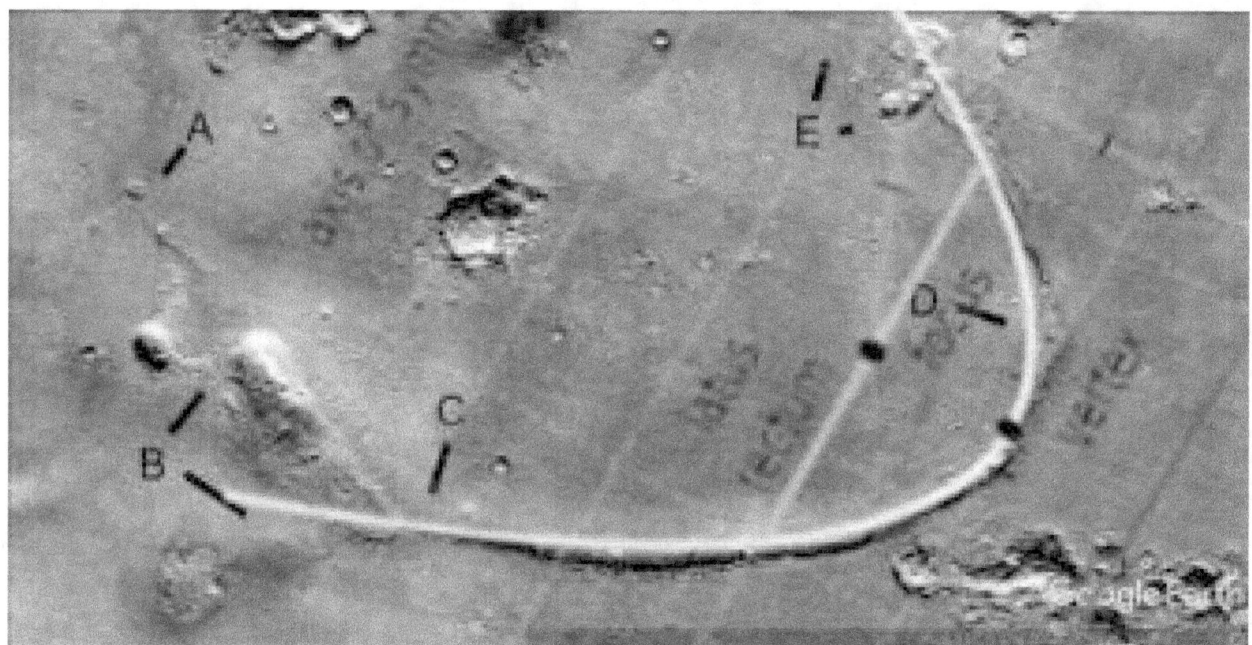

Hhht130

A shows a partially collapsed hill, a tube comes out of it at B where it crosses a groove at D. This may have been a small river that cut the tube, it appears unlikely because there are no signs of pooling on one side before it broke through. It may also be a collapsed tube or road shown by C. The tube goes to the right through E where it appears to have collapsed, this may also be an interior support of a former hill. F shows a series of small hills on a tube leading to a partially collapsed hill on the right with a dark patch on its roof.

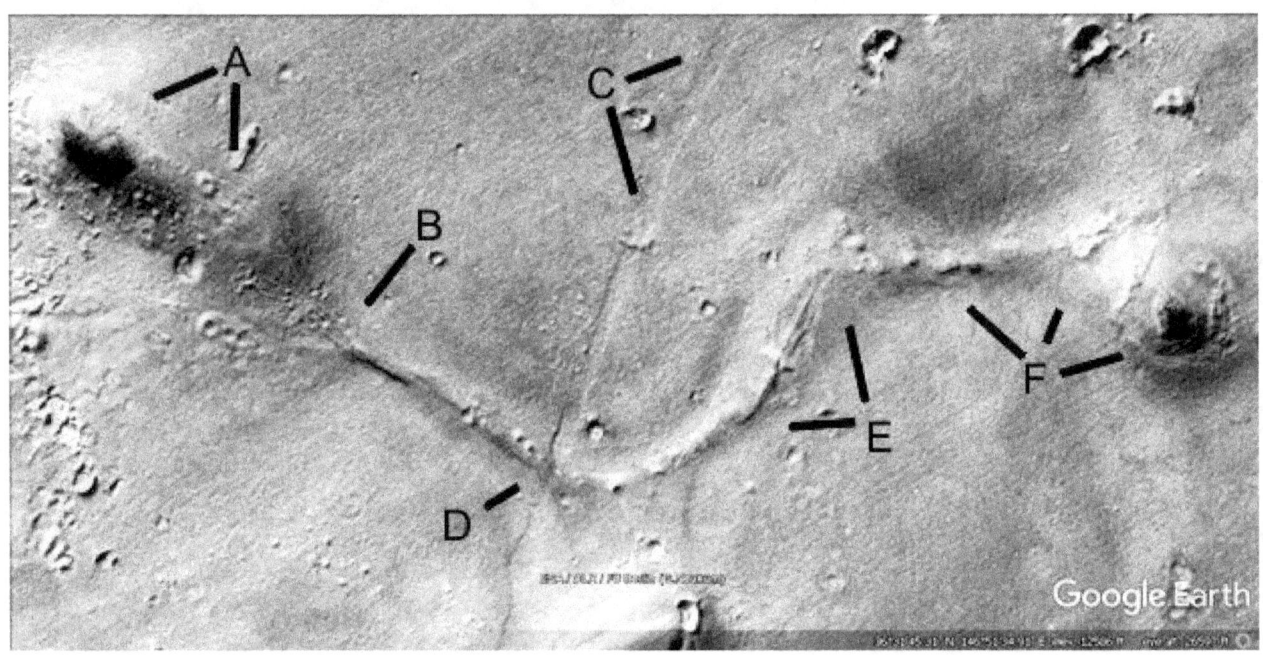

Hhht130a

Hypothesis

A parabola is shown.

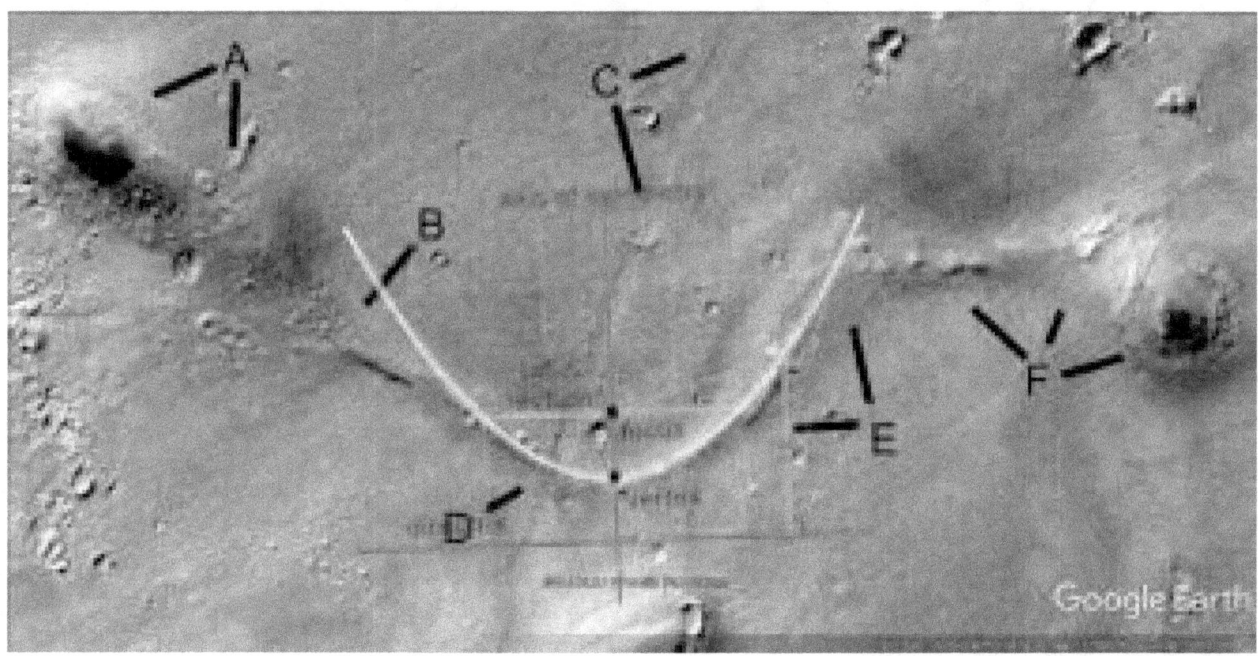

Hhht132

Hypothesis

A appears to be a groove like a collapsed tube, it goes to a hill at E. B shows two hills with craters on their roofs. These have not collapsed the hills or they were repaired to act as dams. C at 8 o'clock shows a crater connected to one of the hills, C at 4 and 7 o'clock shows two partially collapsed hills. D shows another collapsed hill. F shows two hills connected be an extension of this groove, showing it cannot be a river. G shows a partially collapsed hill, at 9 and 10 o'clock there is a dark area connecting the two hills. H at 2 o'clock shows another partially collapsed hill.

Hhht134

Hypothesis

A shows a walled hill with a concave roof, there are the interior supports showing through at B perhaps as the roof settled. C shows a section of the wall that is steep, with a lip on the lower side down to the roof. Such a roof might itself have been a dam to collect water, even to grow crops on it.

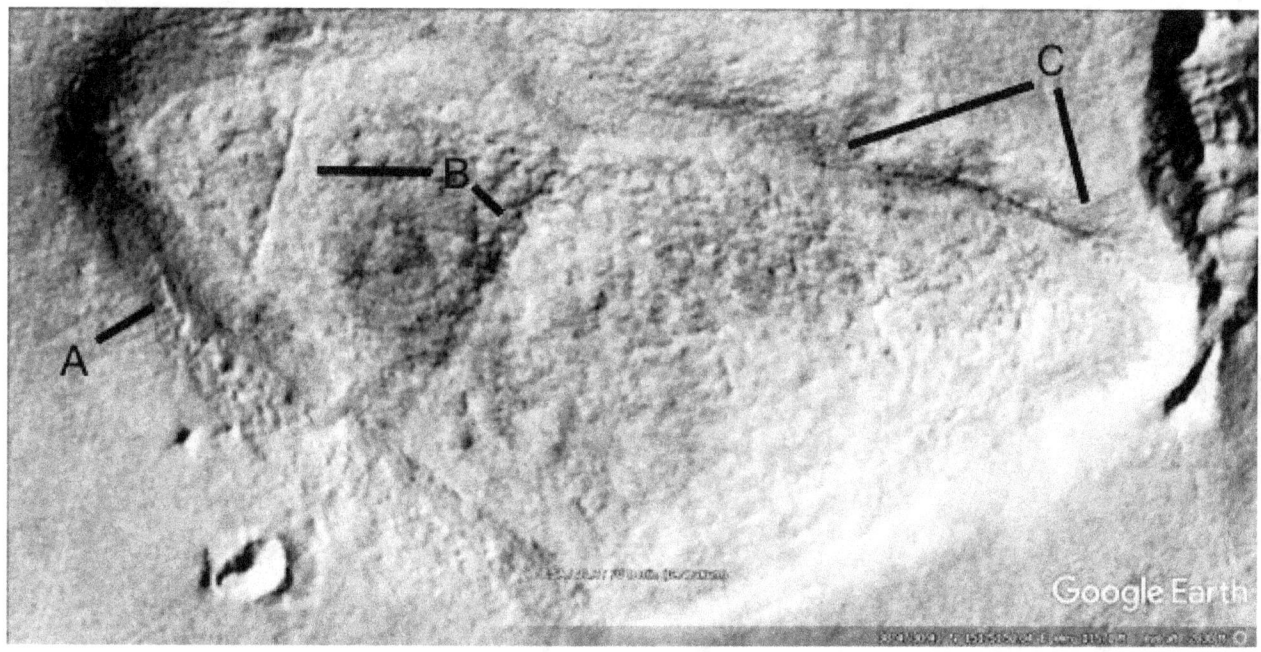

Hhht139

Hypothesis

A shows a road going into the walled hill, the pale concave roof looks very unnatural at C. The walls as shown at B are very steep, the crater on the left again appears molded into the hill with more roads coming out of it to the left. D connects to the walled hill with roads, it also has some settled areas on its roof.

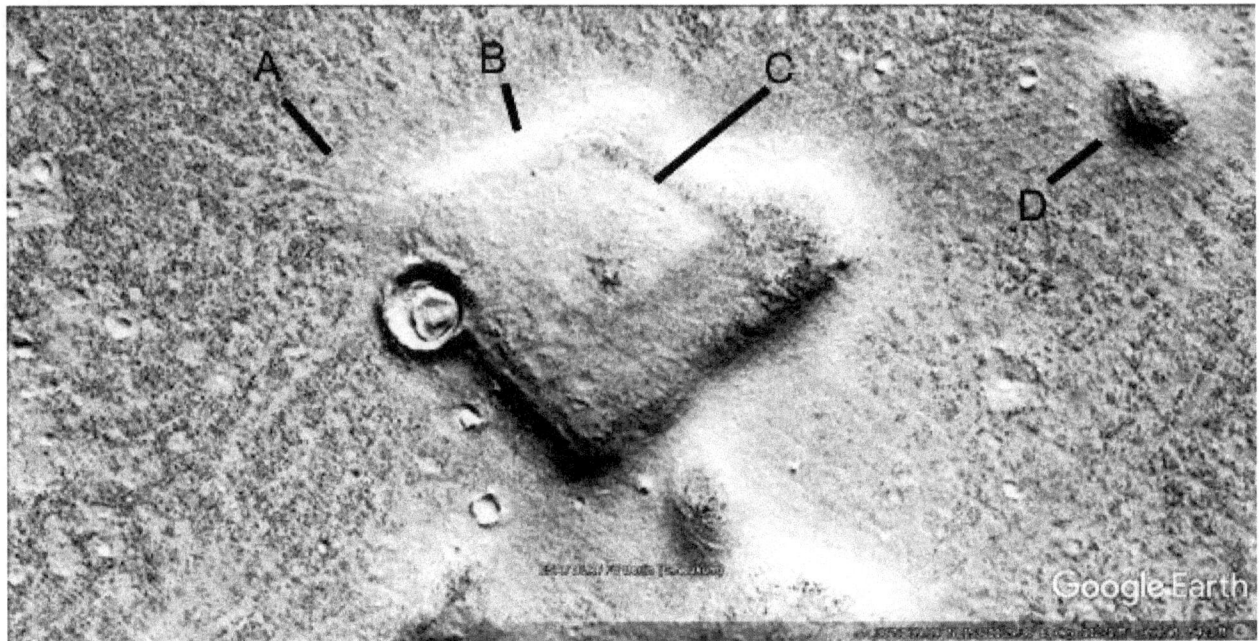

Cymd151

Hypothesis

A and B appear to be degraded dams, many of these are found in Cymmeria though this is far away from the ocean. They would then have been fed from the Artesian basin or precipitation.

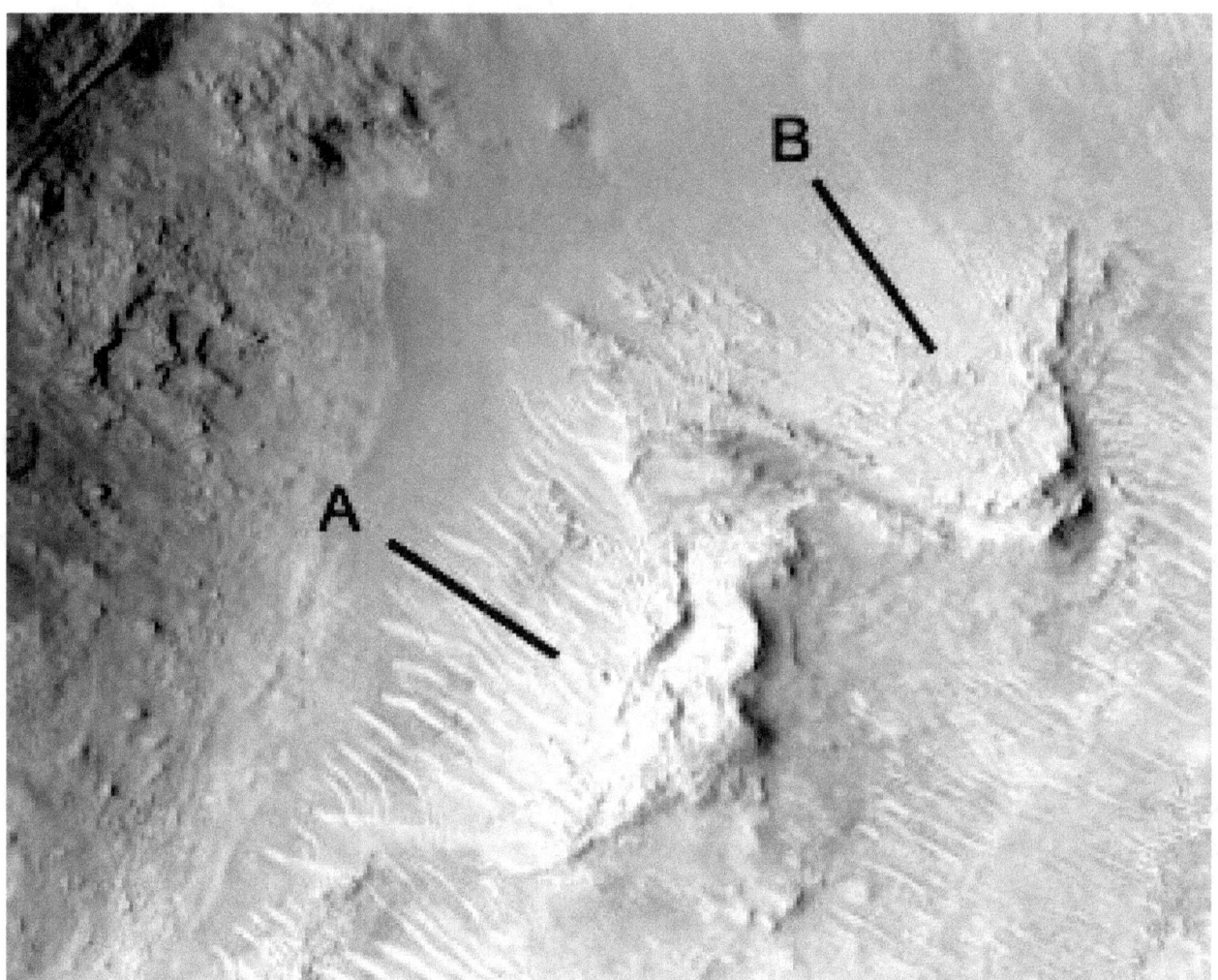

Cymd151a

Hypothesis

This shows a parabola superimposed on the right dam, the fit is very close and many dams will be shown to have this parabolic shape. Geologists might consider how a mud slide, as the only known natural explanation would form near perfect parabolas so often. They would be a strong way to build a dam, parabolic arches[i] are used in some buildings. Parabolic shapes are also used in many Earth dams[ii].

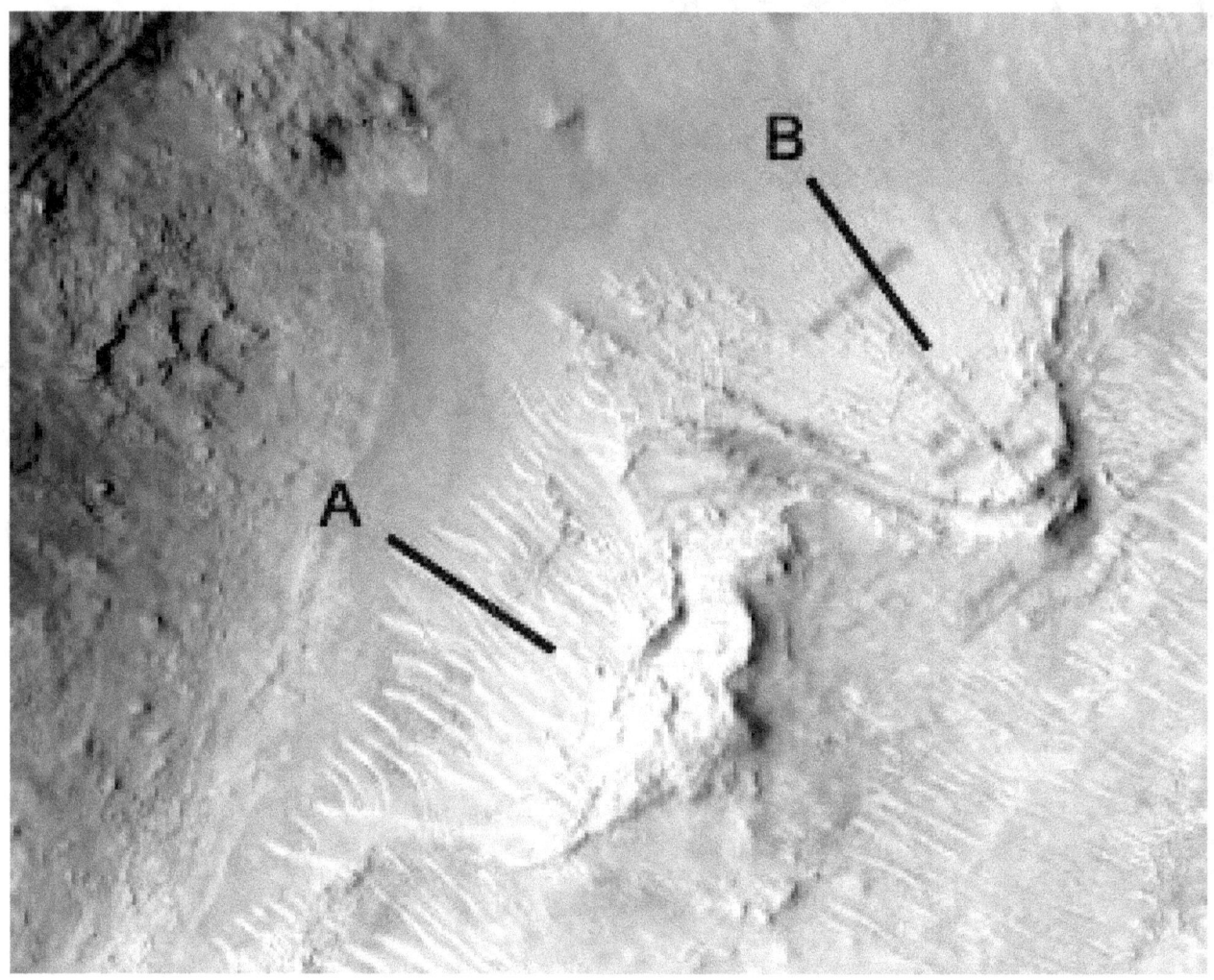

Cymd155a

Hypothesis

A shows a channel for water, it may have flowed through the two curved openings at 11 and 12 o'clock. At 3 o'clock there is a small dumbbell shaped formation. At 5 o'clock there would be the bottom of this channel, which may have directed water to a lower dam. A shows a small dam shape, the point at 6 o'clock at the bottom of the dam and the bottom point of the formation at A at 3 o'clock are on the latis rectum of a parabola fitting the dam shape under it. The point at 6 o'clock at E then is the parabolic focus of this. B shows a smooth bottom edge of the dam wall probably also a parabola, like the inner and outer parabola used in Earth dams. C shows signs of the dam cracking with cavity shapes appearing. D is also a dam like shape.

Cymd155a2

Hypothesis

This shows a parabola superimposed on the image, the horizontal line in it is the latis rectum which goes from the A formation point to the focus at the E formation point. This would be a logical way to create a parabola. In some cases the directrix is also in the formation.

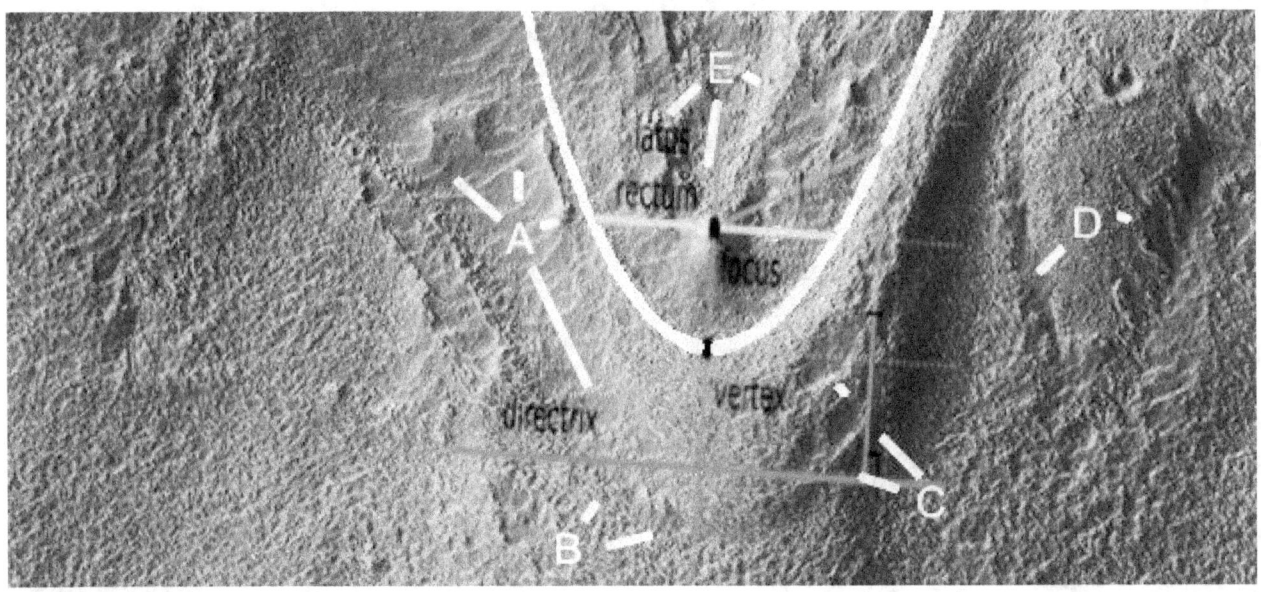

Cymd155a3

Hypothesis

This shows the bottom curve is also a parabola, using the Earth technique of a dam with inner and outer parabolic shapes.

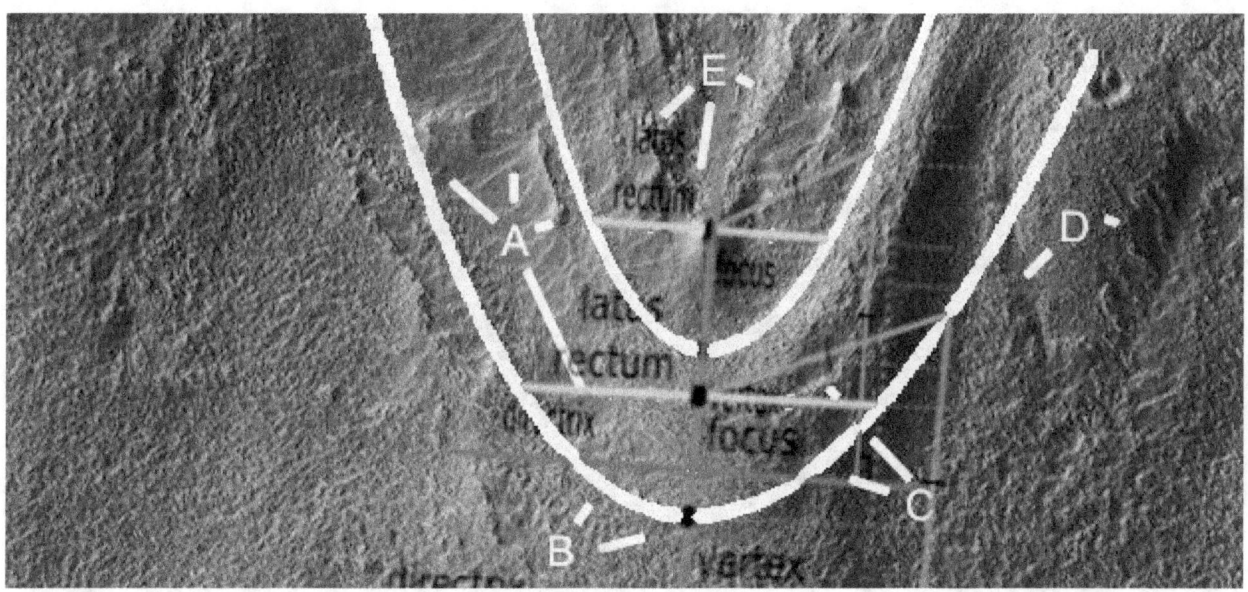

Cymd155b

Hypothesis

A and C show the edges of a parabola, B shows where the surface is breaking up.

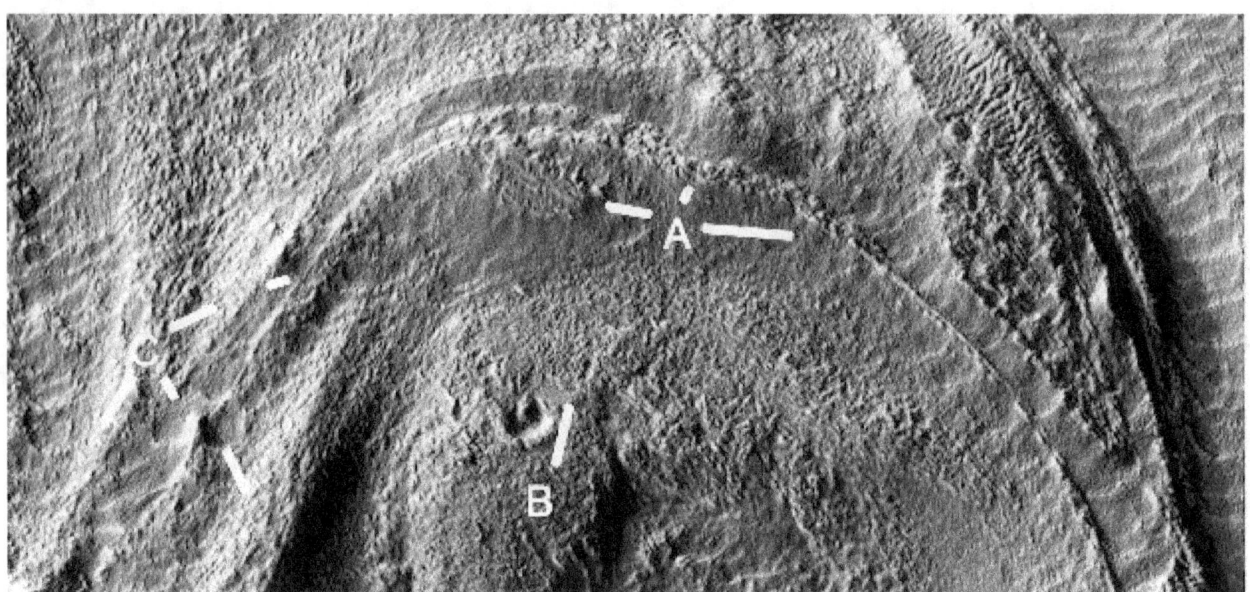

Cymd155b2

Hypothesis

This shows a parabola superimposed on the formation, it appears to be a parabolic arch for strength. The point C at 5 o'clock is the end of the latis rectum line. B shows a show like a small parabola where the focus appears, the apex of the parabola is close to a mark at A at 10 o'clock.

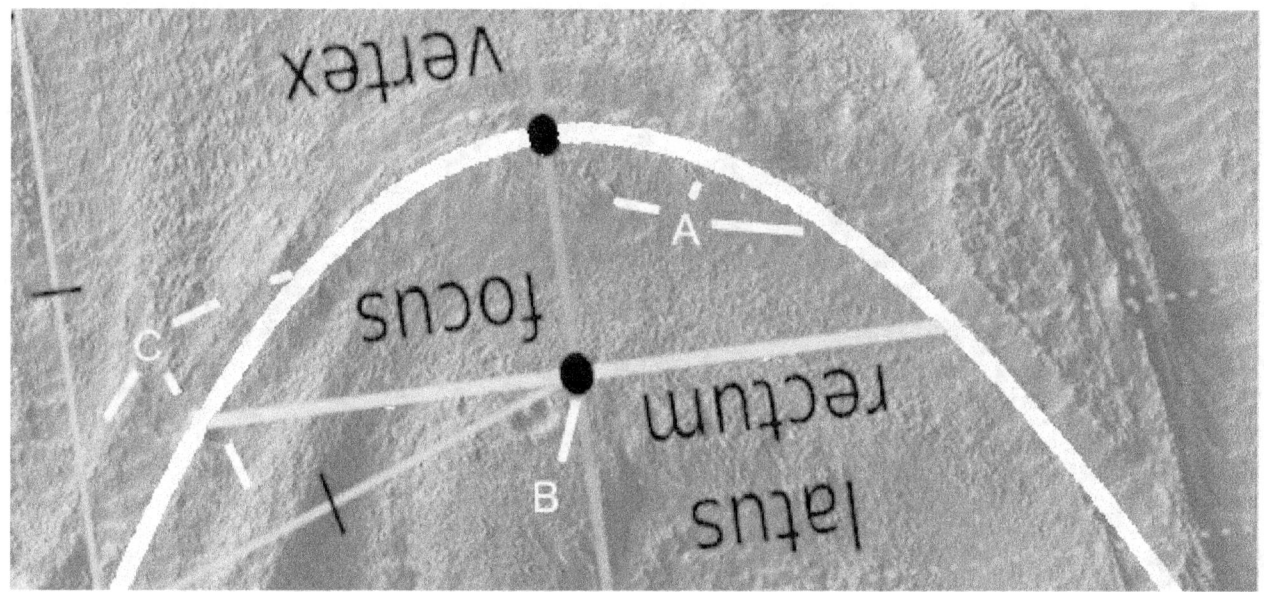

Cymd155b6

Hypothesis

This outer shape is also a parabola as shown.

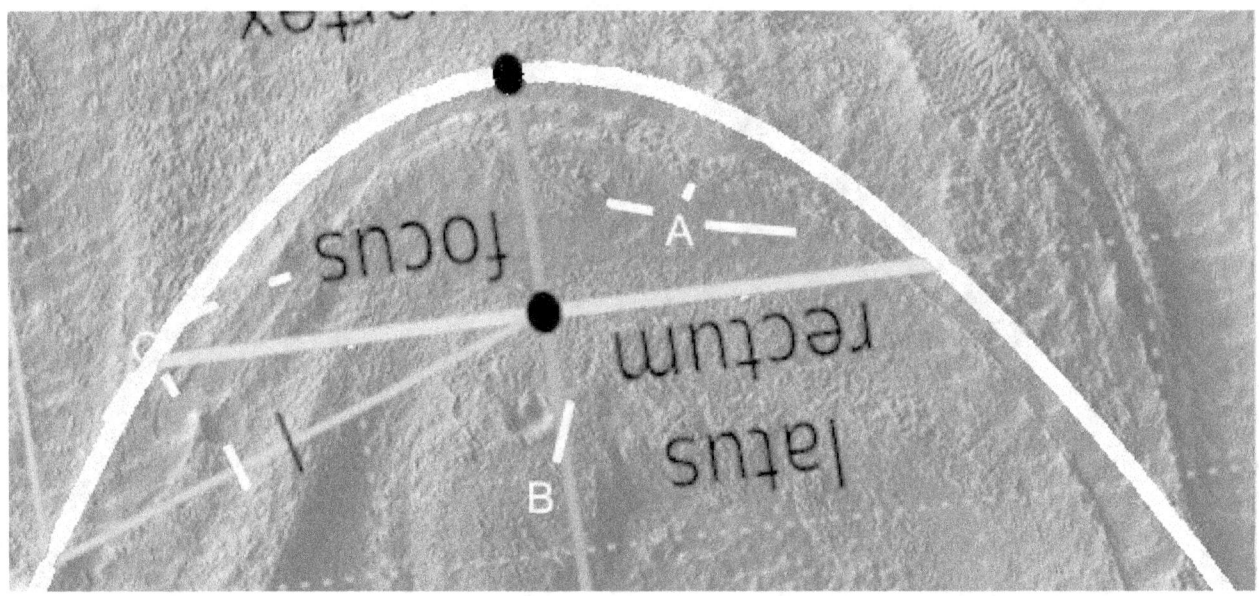

Cymd156

Hypothesis

These are two more parabolic dams. A and B show a straight edge under them.

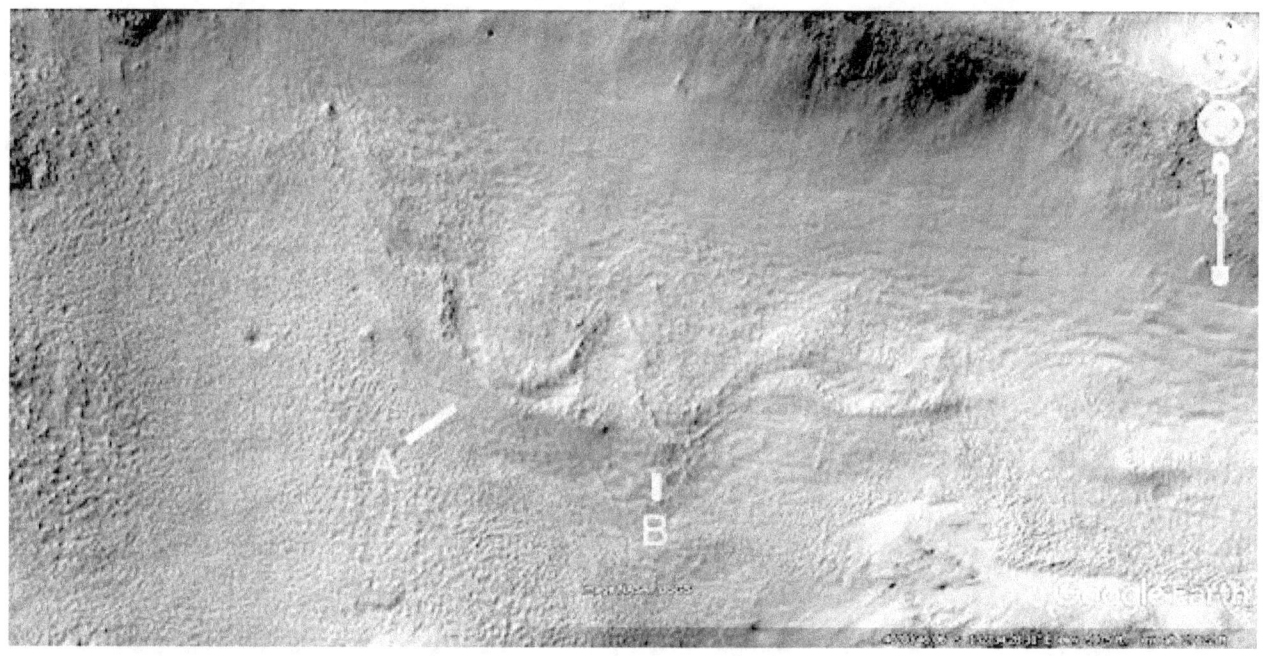

Cymd156a

Hypothesis

This shows a parabolas superimposed on each dam, the fit is very close. Each of these examples is very different from the others, it might be expected here the flat lower section would give a flat mud slump. But above this is two parabolas, a logical way to make strong dams.

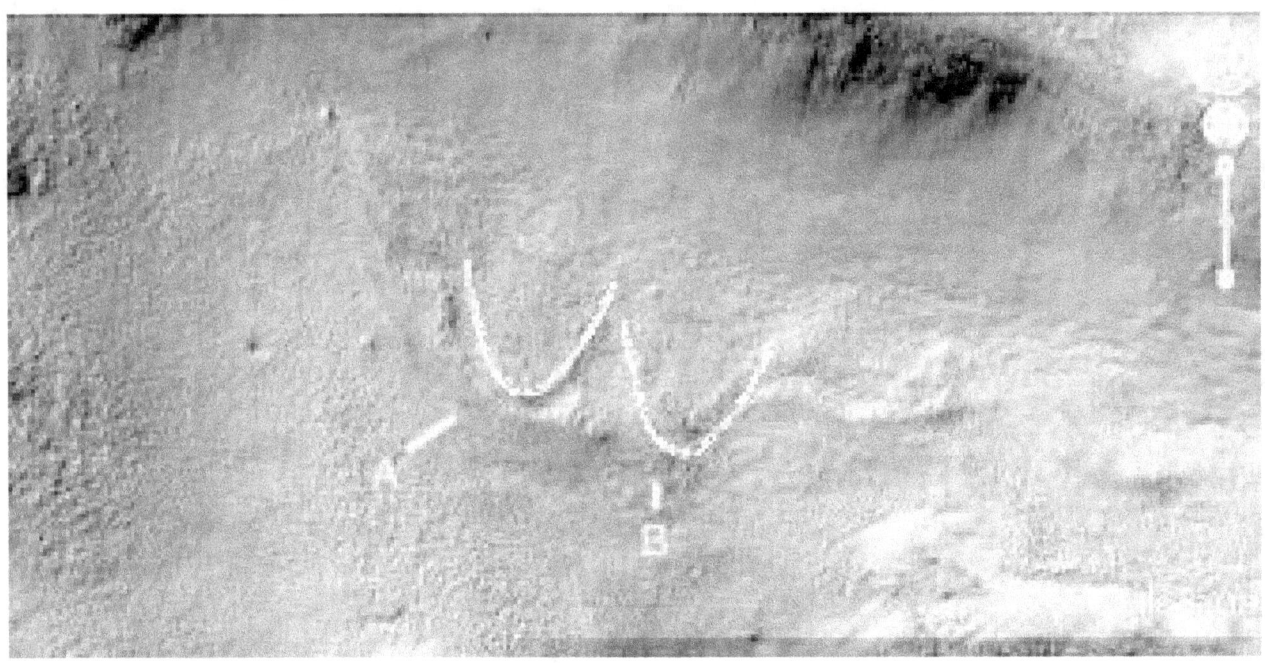

Cymd157

Hypothesis

A has a walled dam rather than a parabolic shape. B shows another parabolic dam, this is separated by a parabolic arch to a second parabolic dam at C and then another parabolic arch.

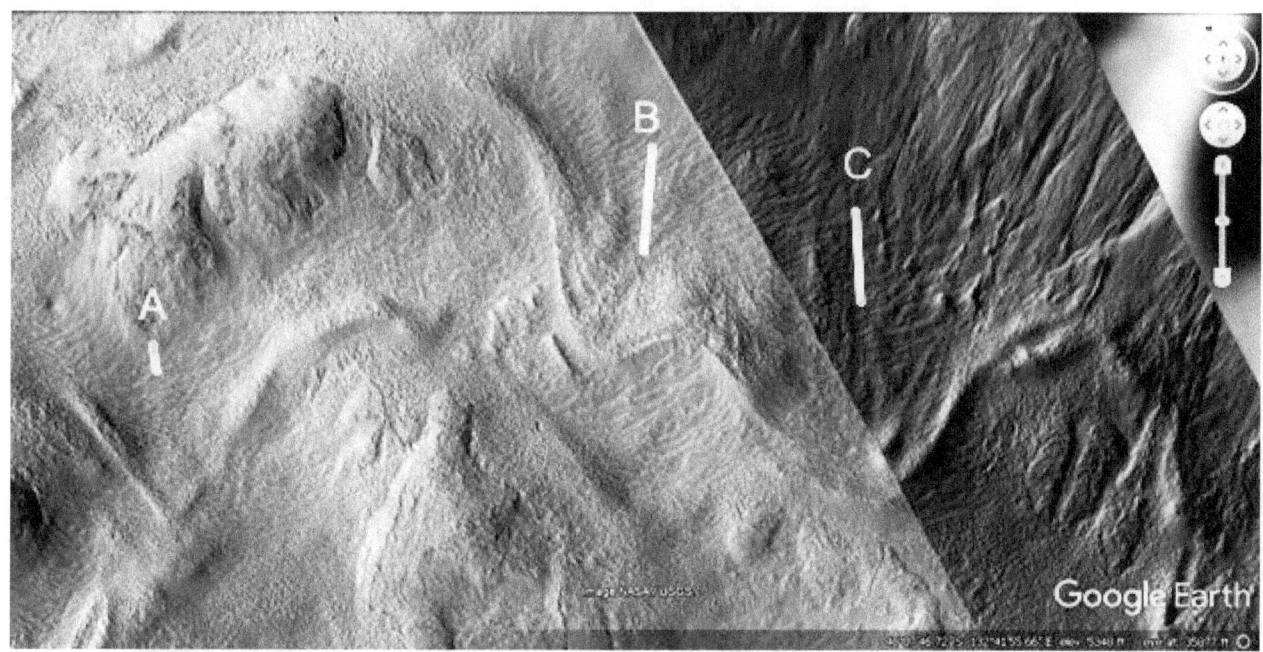

Cymd157a

Hypothesis

This shows a parabolic arch and parabolic dam outline, though the other two are the same shape.

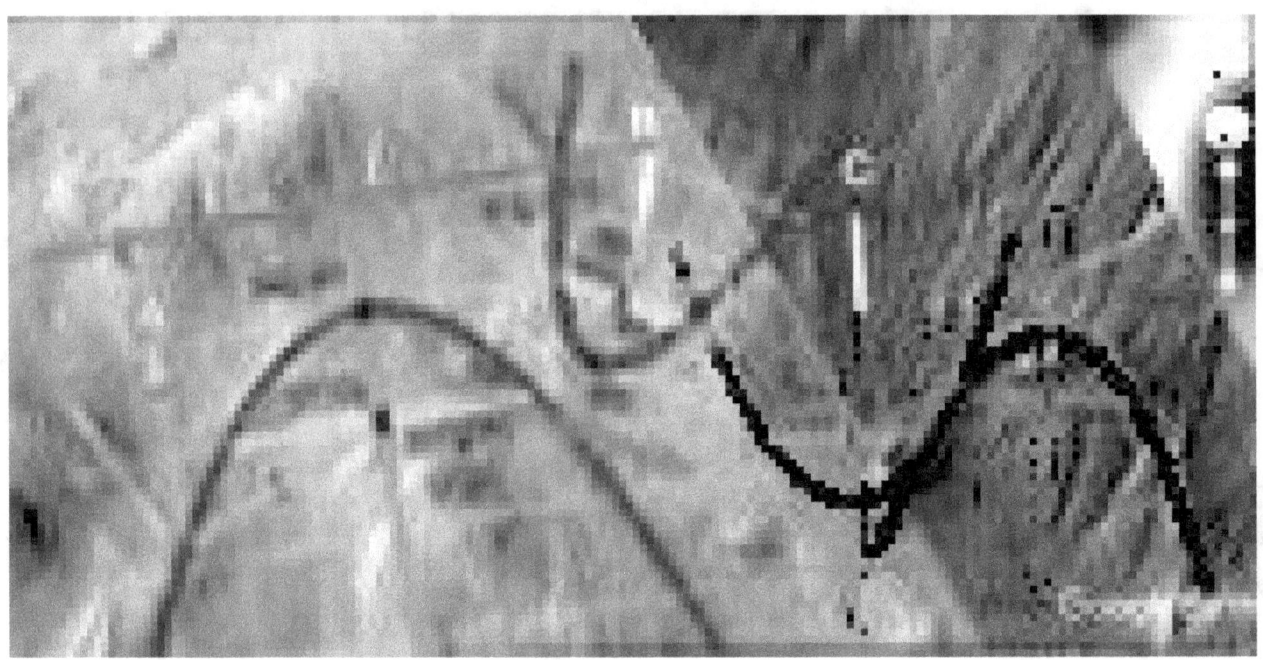

Cymd158

Hypothesis

A is probably an eroded dam, B is another parabolic dam.

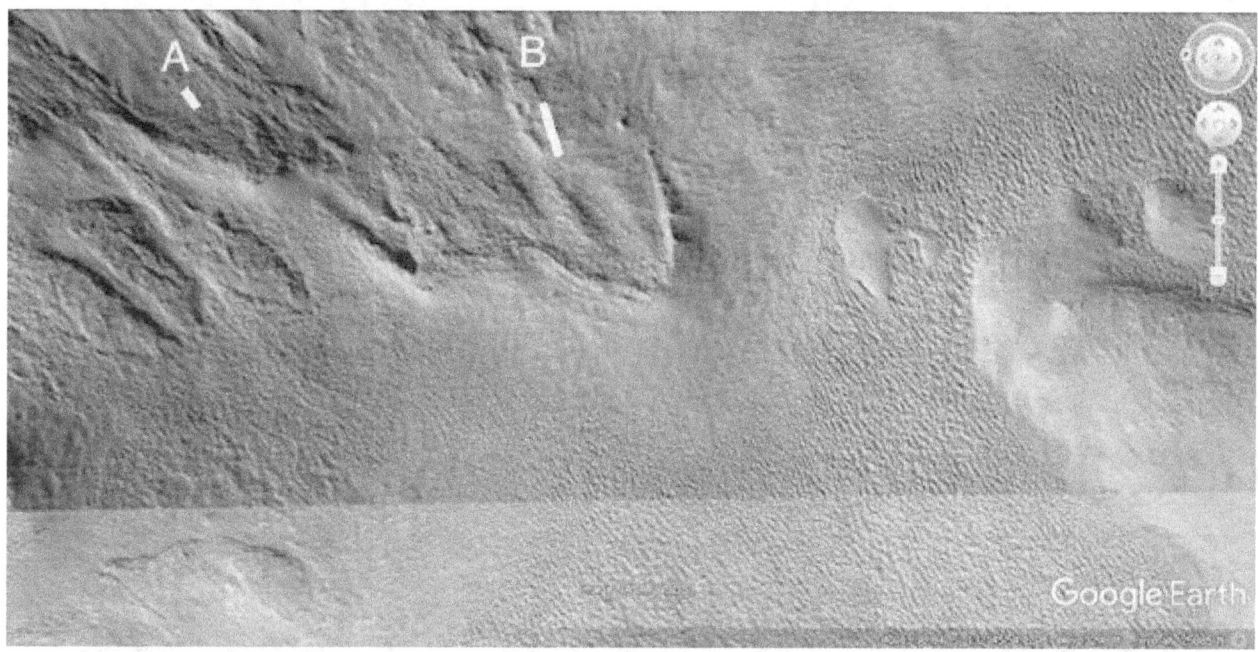

Cymd158a

Hypothesis

This shows a parabola superimposed on the second dam, the fit is quite close.

Cymd161

Hypothesis

These three dams are also close to parabolic, some creep in the materials or other erosion in the crater could account for the difference. There is also no need for the construction technique of the dams to be exactly parabolic, this might change according to the shape of the crater wall there. A may show an inner and outer parabola, B also probably shows an inner and outer parabola. C shows two parabolas closely connected like there is a wall between them. D shows a dam at 10 o'clock and then a channel at 8 then 7 o'clock going into the dam at A. It also shows two dams at 12 and 1 o'clock. E shows a dam wall at 8 o'clock, then where it has collapsed at 5 and 7 o'clock.

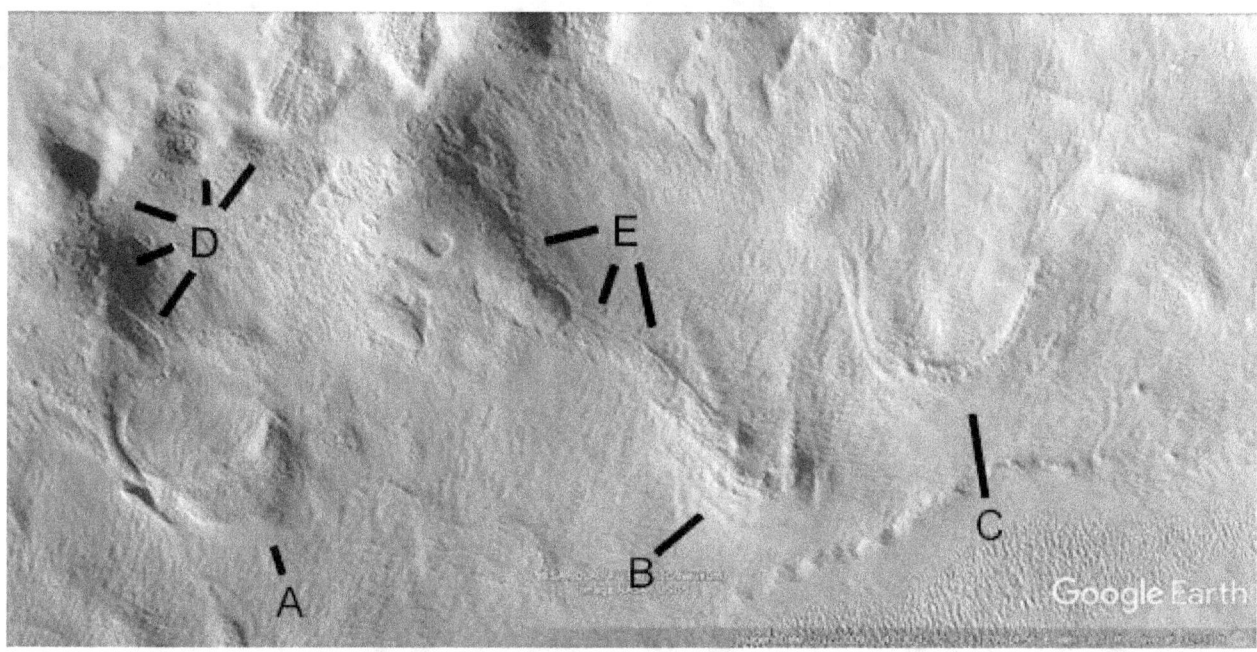

Cymd161a

Hypothesis

This shows parabolas superimposed on two of the dams. It is not an exact fit but when these were constructed there may have been no need to make them exactly parabolic, or the materials experienced creep over time. The parabolic dam C has its latis rectum lined up well with a ridge to its right.

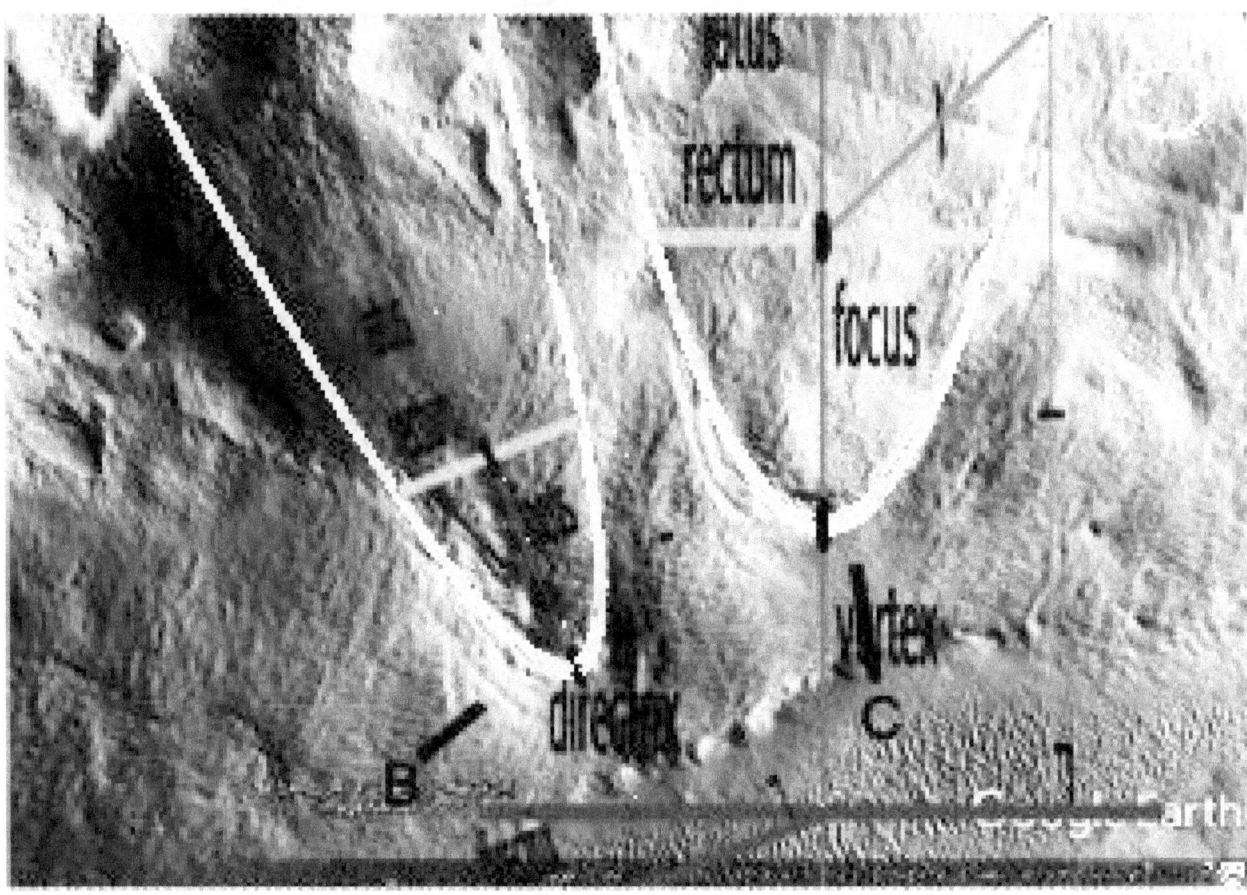

Cymd161b

Hypothesis

This shows the dam at A is close to parabolic, the inner parabola also has its apex close to the focus of the second parabola. B and C are shown here as different fits for a parabola. B is lower down with a layer going around the dam, C is also lower down

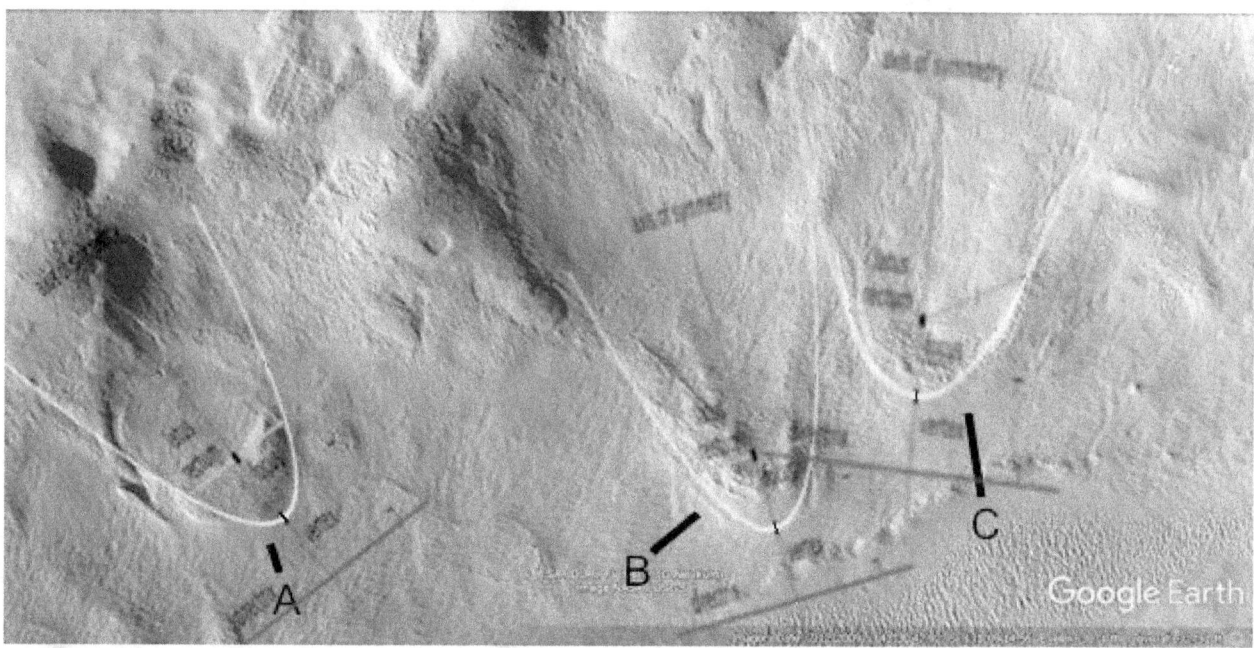

Cymd162

Hypothesis

A shows an eroded dam, probably parabolic. B shows an inner and outer parabola. S shows another probable parabola, with a channel down the middle and perhaps another parabola on the right. D shows other smaller dams highly eroded as does E.

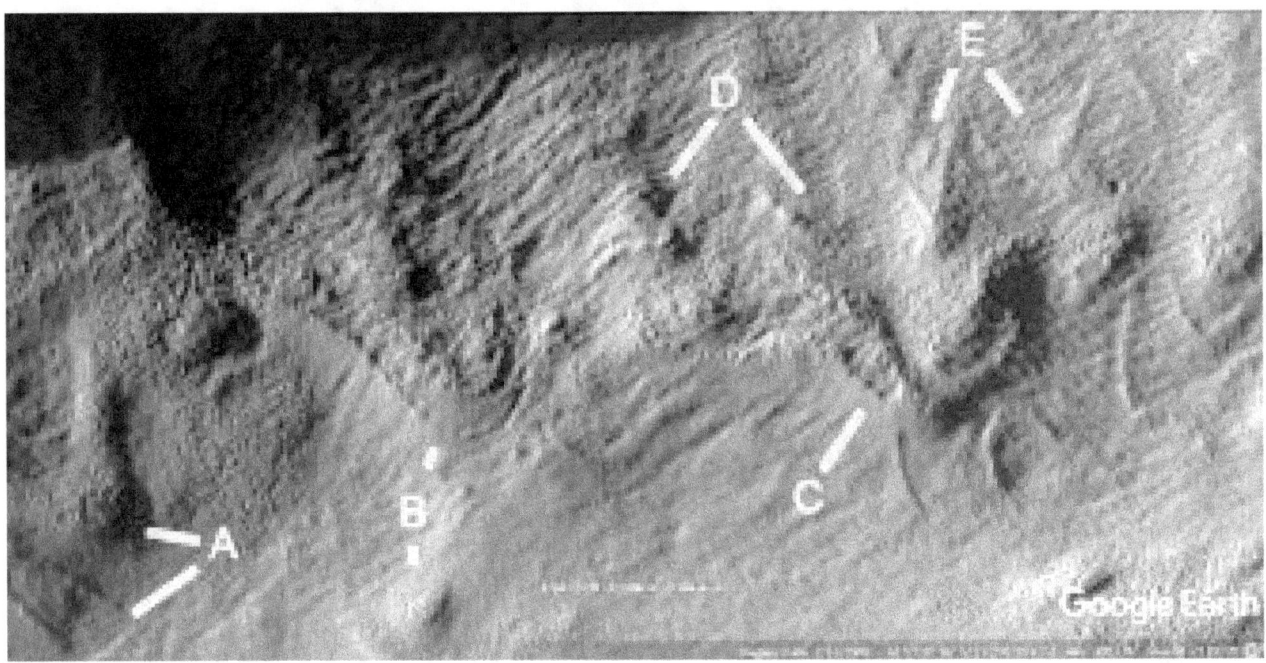

Cymd162a

Hypothesis

This shows the parabolic dam at B.

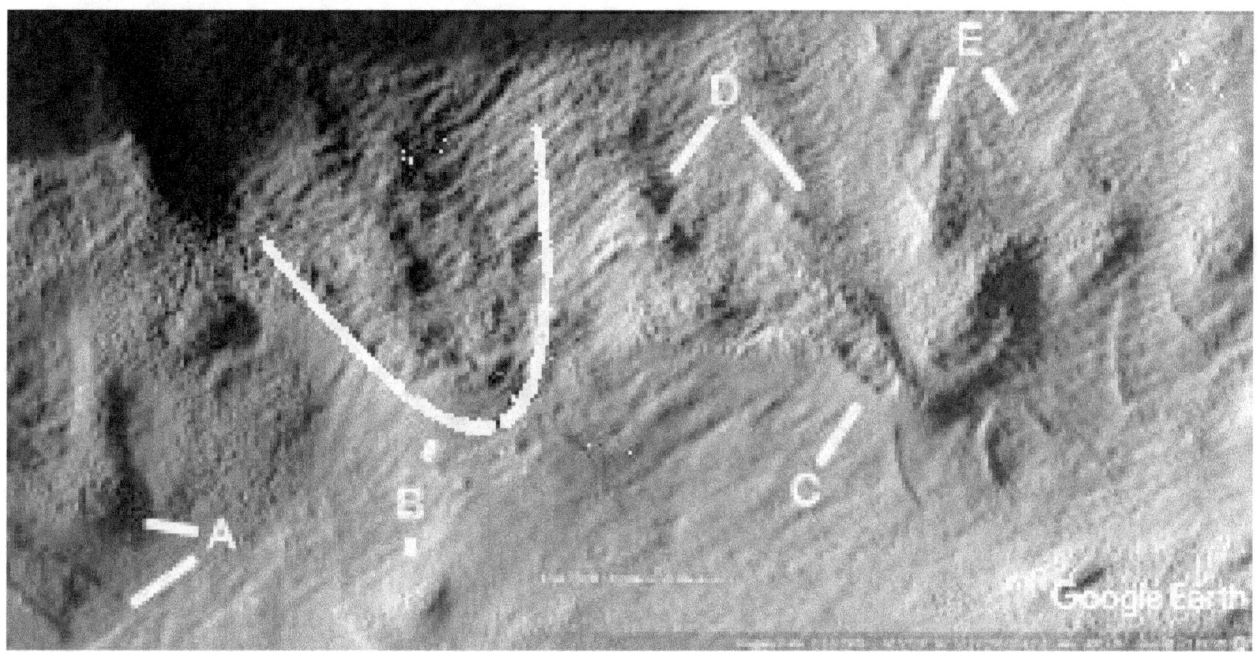

Cymd163b

Hypothesis

A shows an excavated damthat appears to have been carved into the rock, at 5 o'clock it shows a channel probably formed from erosion. It indicates water probably flowed here for a long time after there were creatures to maintain the dam. B shows a dam with a smoother gradient at 6 o'clock, it seems to be starting to form a channel. This may also have been choking up with silt from above. C shows 2 other dams.

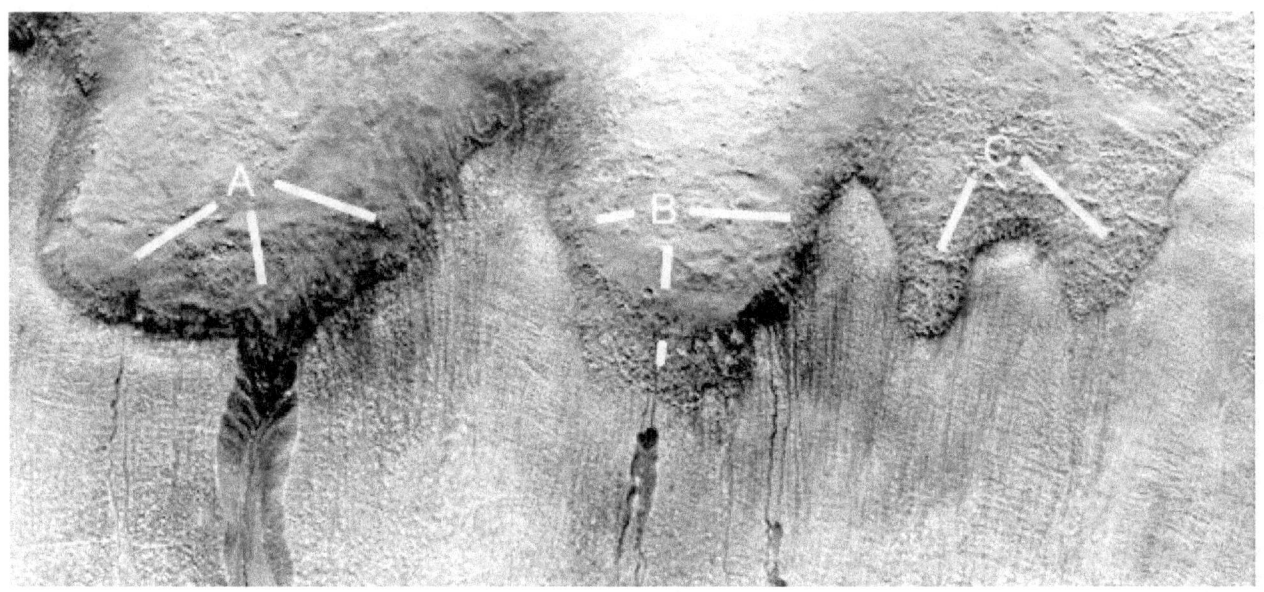

Cymd163b2

Hypothesis

Two parabolas are shown.

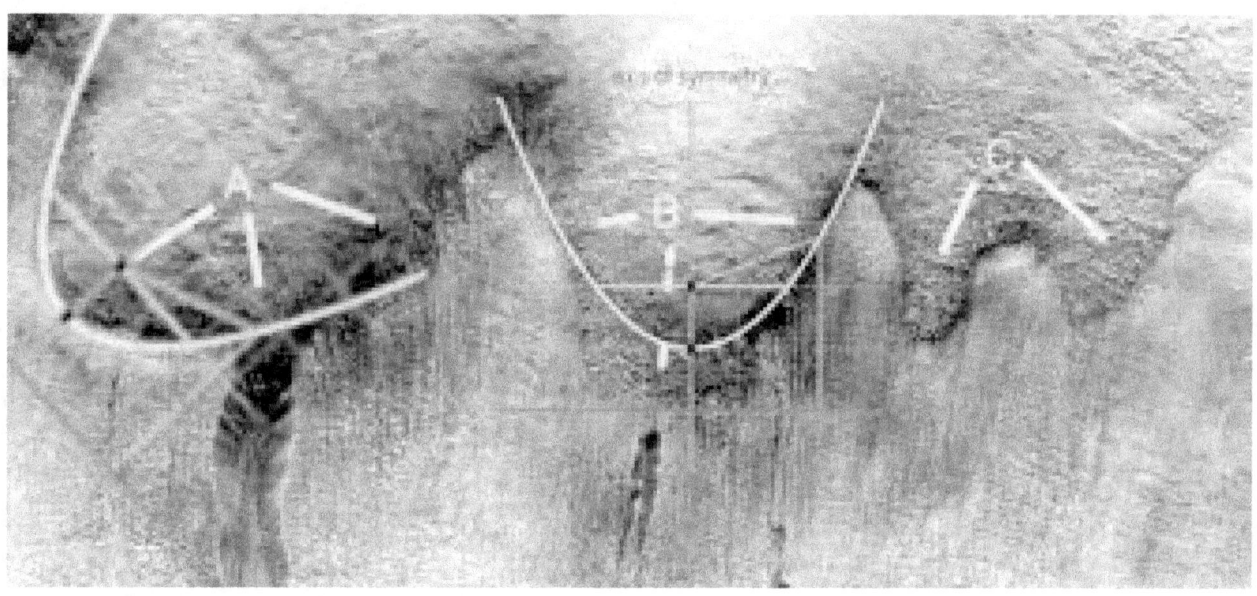

Cym164a

Hypothesis

A is like an excavated dam on the left side, on the right there is a walled dam at B. the cavity under B at 7'clock is hard for geologists to explain, why the ridge above it did not erode and collapse as the cavity formed. Overall B is similar to a parabola.

Cymd164a2

Hypothesis

Two parabolas are shown.

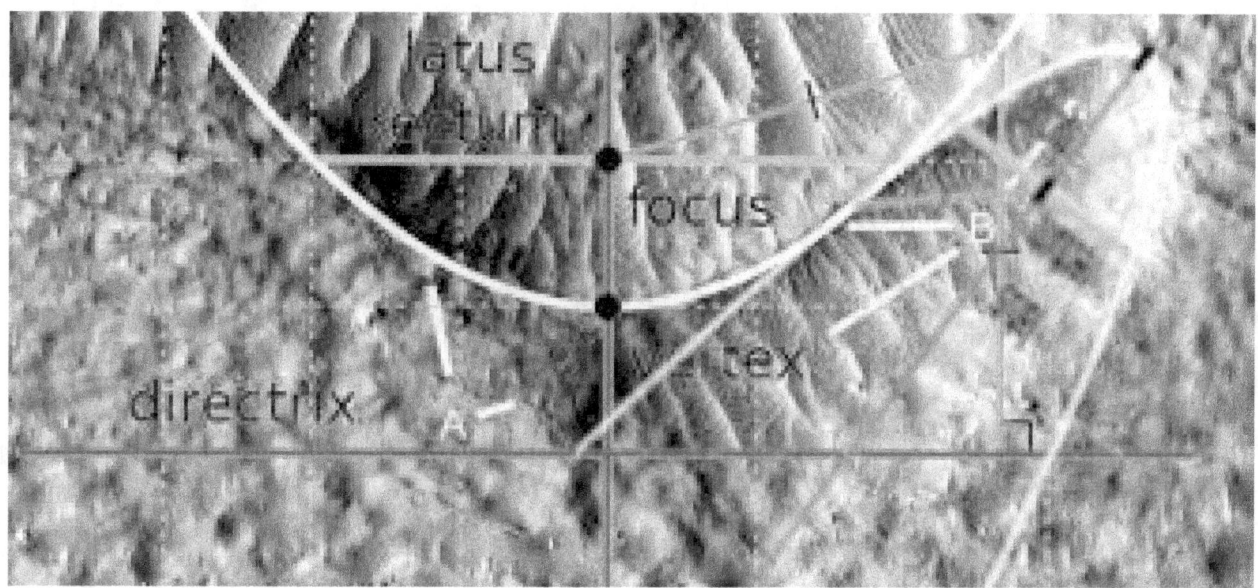

Cymd164b

Hypothesis

A and B are both excavated dams, B at 4 o'clock also has a wall to make part of its shape.

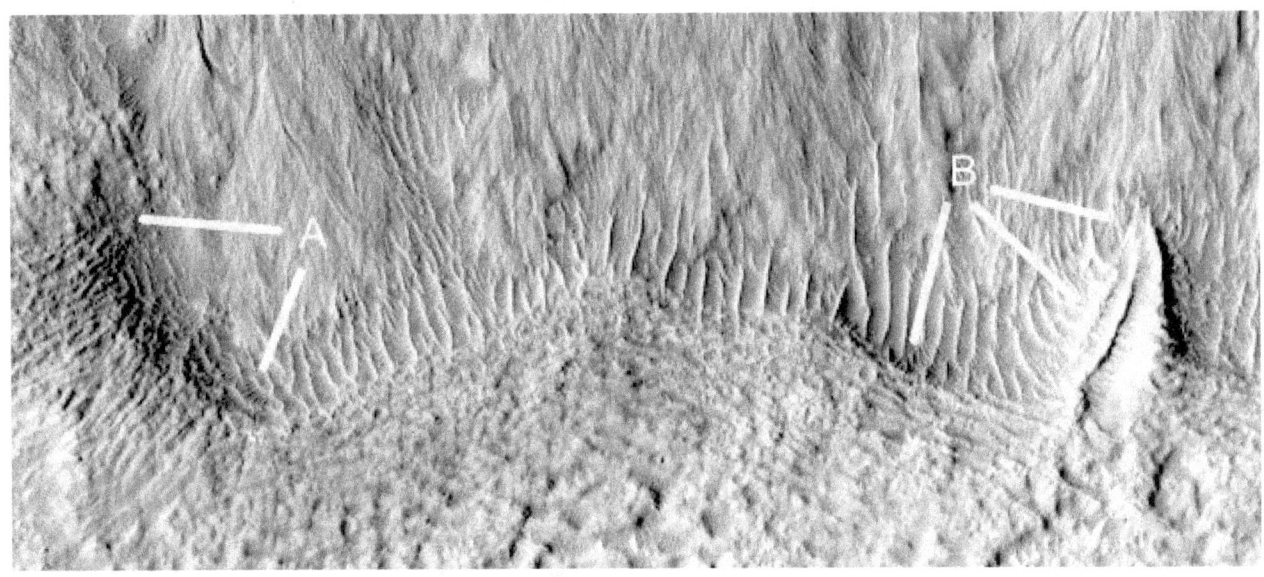

Cymd164b2

Two parabolas are shown.

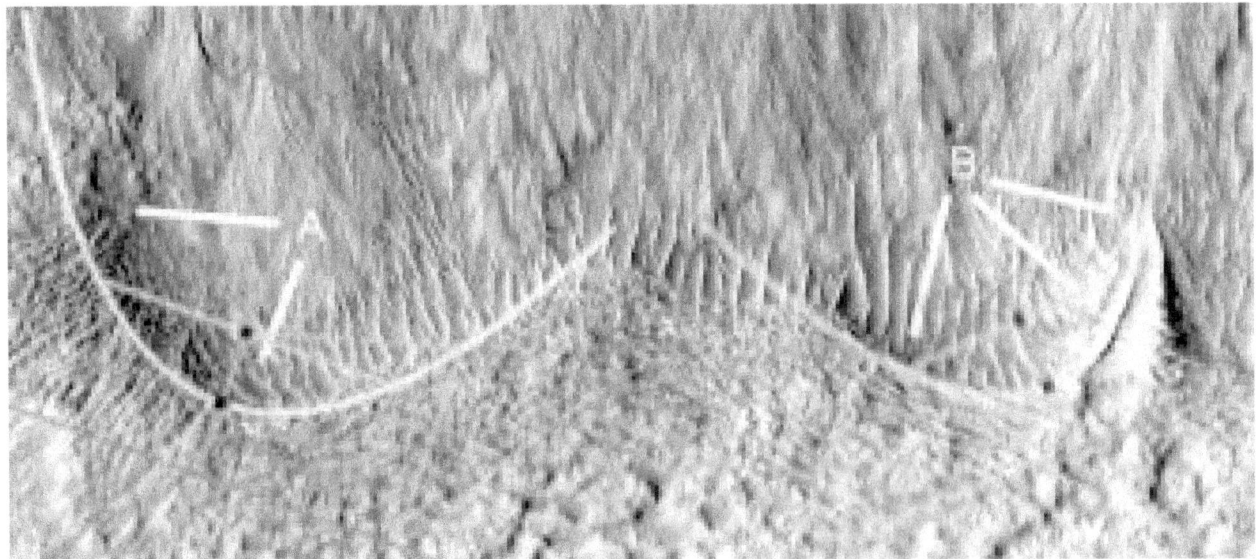

Cymd166a

Hypothesis

A shows two eroded dams, at 1 o'clock it appears to be a parabolic dam. A at 7 o'clock also appears to be parabolic. B at 6 o'clock shows another small dam. B also shows at 1, 3, and 4 o'clock appears to be another long narrow dam with perhaps a smaller dam half way down it. D shows another long narrow dam with multiple walls. C shows the bottom of two dams. F shows a parabolic dam, at 10 o'clock this ridge is approximately the latis rectum of the parabola. F at 11 o'clock and E at 4 o'clock show a long dam wall. E at 7 and 8 o'clock are another long dam wall. These overall appear to be more eroded than other dams, this might also indicate a lack of maintenance. Some areas might have been withdrawn from, for example as the climate changed some areas might have had their water table drop and their dams dried up. Climatologists then can consider whether the signs of water in these dams correlate to the ocean that was in Utopia, Elysium, and Isidis Planitias.

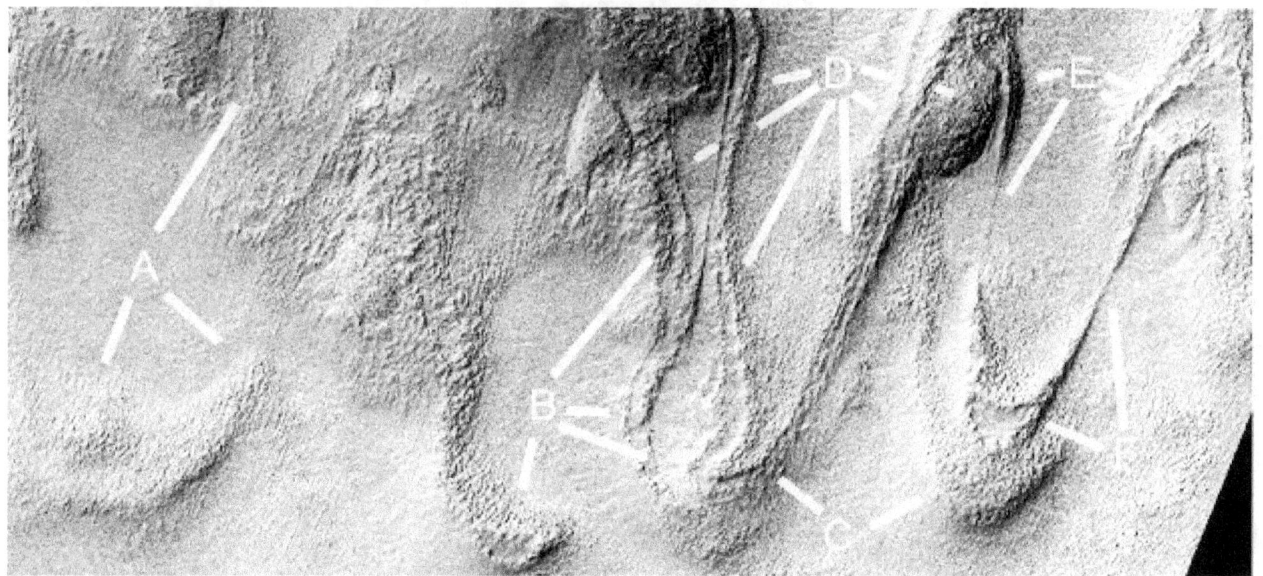

Cymd166b

Hypothesis

This shows the inner and outer parabolas at C and F to be highly accurate. Also the apex of the inner parabola approximately is at the focus of the outer parabola.

Cymd167b

Hypothesis

A shows how the inner and outer parabolic shape is decaying, as if the top of the wall has broken off exposing the insides. Engineers might consider how this shows a construction technique. B also shows a cavity there is usually a peak on this part of the dams. C shows two decayed areas at 4 and 8 o'clock. D and E also show decayed areas, in each case the rounded top of the wall seems to have broken off like capstones.

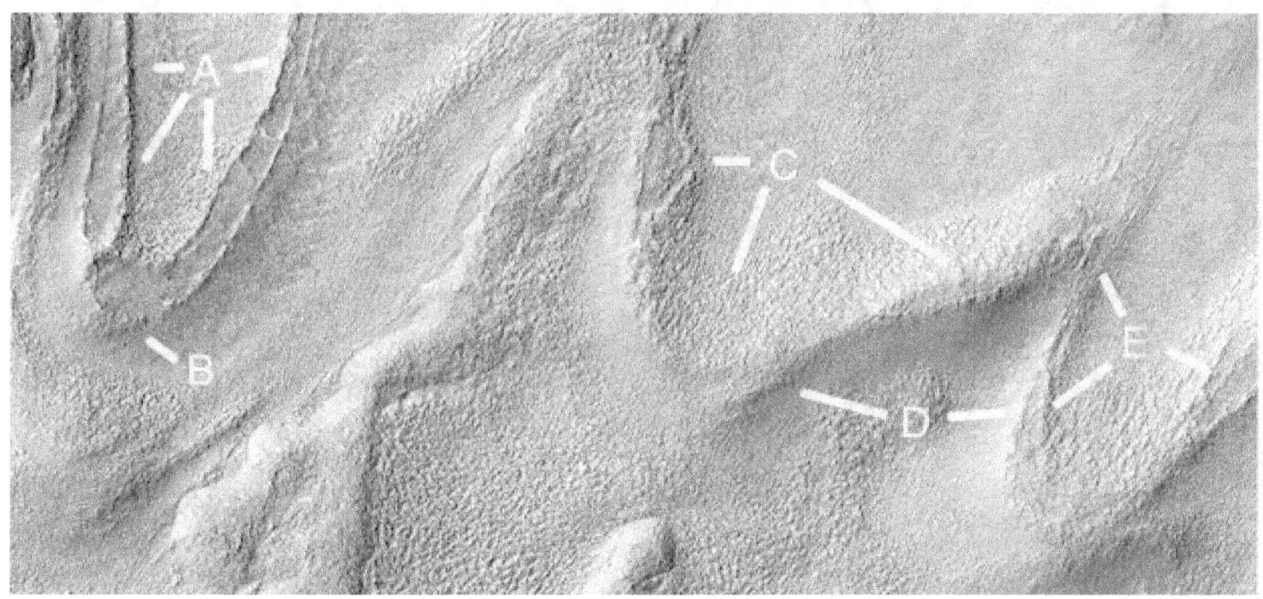

Cymd167b2

Hypothesis

Four parabolas are shown.

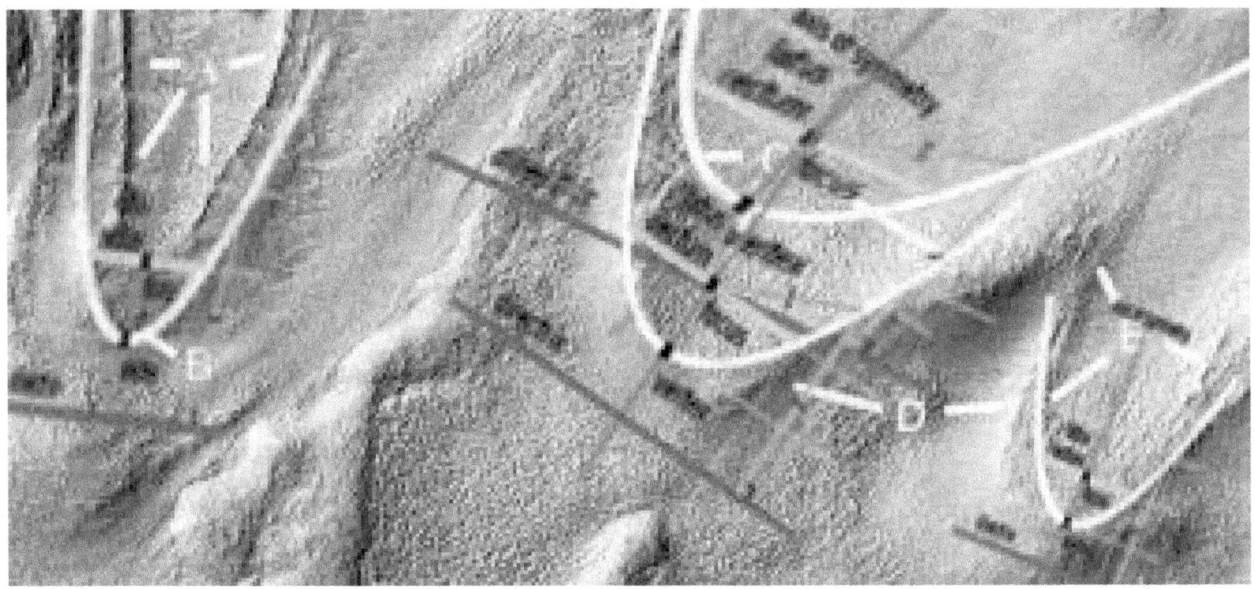

Cymd167b2

Hypothesis

A shows how the inner and outer walls are also collapsed here as does B. C shows another inner and outer parabolic dam. D shows an inner and outer parabolic arch. E and F show an inner and outer parabolic dam with some cracks in the walls.

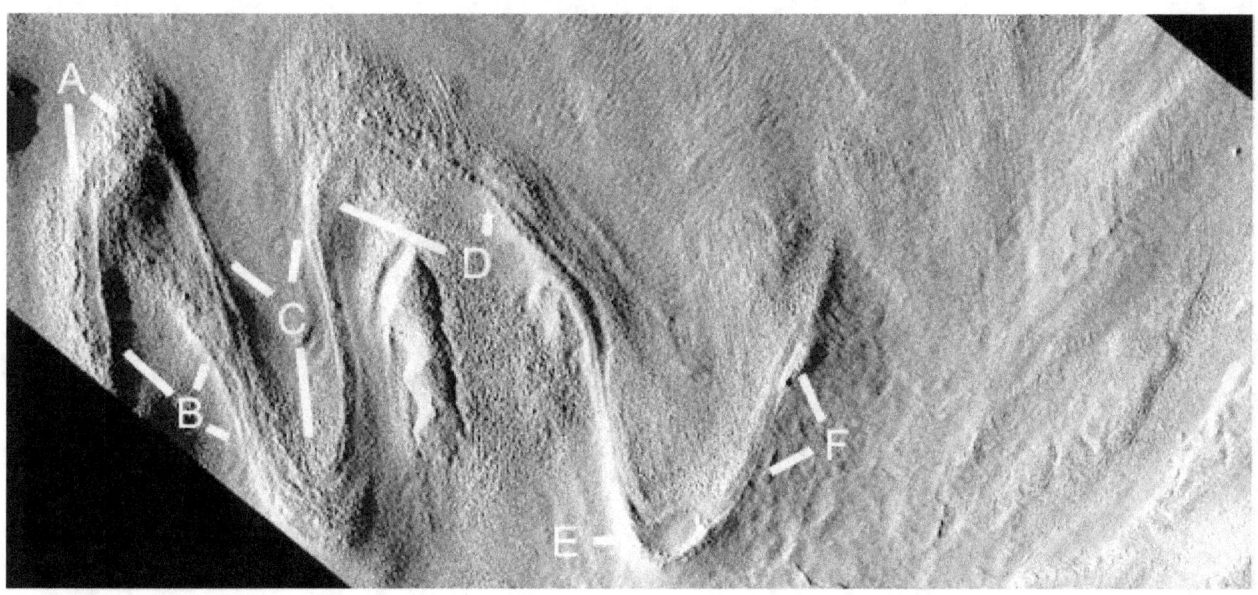

Cymd167b2

Hypothesis

These show three parabolas superimposed on the two dams and one arch, the fit is again very good considering materials may have been creeping downwards with a cold flow. For possibly hundreds of millions of years to have passed it would indicate a kind of concrete to resist this. Engineers might consider this rate of cold flow to work out relative ages of dams, and also what materials may have been used.

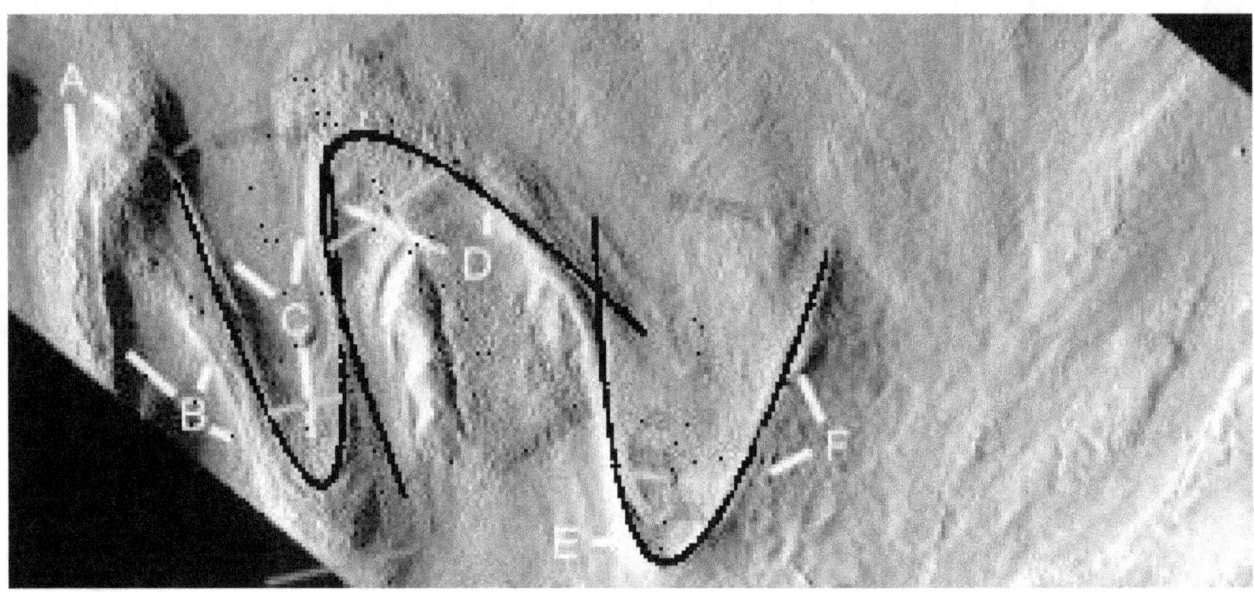

Cymd167d2

Hypothesis

A shows the outer wall of the previous parabolic dam, B shows at 7 o'clock part of the inner wall. At 5 o'clock it shows a small cavity, this might have accumulated water or a piece of the wall fell off here. At 4 o'clock it shows a groove inside it. C and D show an inner and outer wall with a small groove between them.

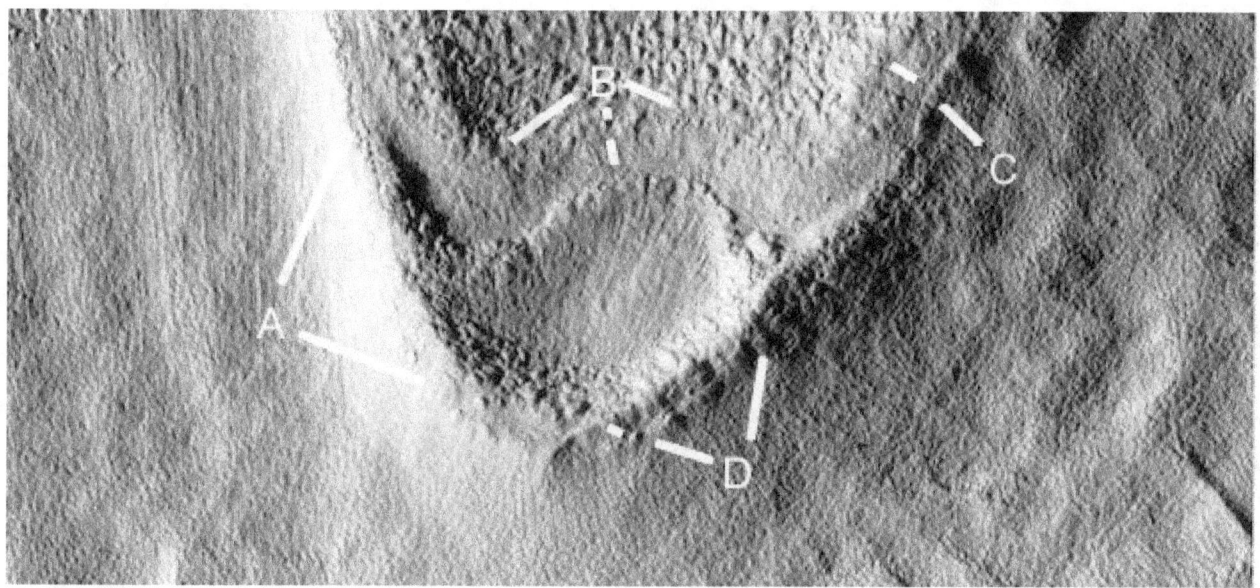

Cymd168

Hypothesis

A shows a walled dam, as if this was cemented onto a smooth crater wall. B is more parabolic shaped and appears to have an inner and outer parabola with a cavity in between. C shows another inner and outer parabola that appears to have filled up with silt, D may be a tube connected to it. C then also might indicate a lack of maintenance.

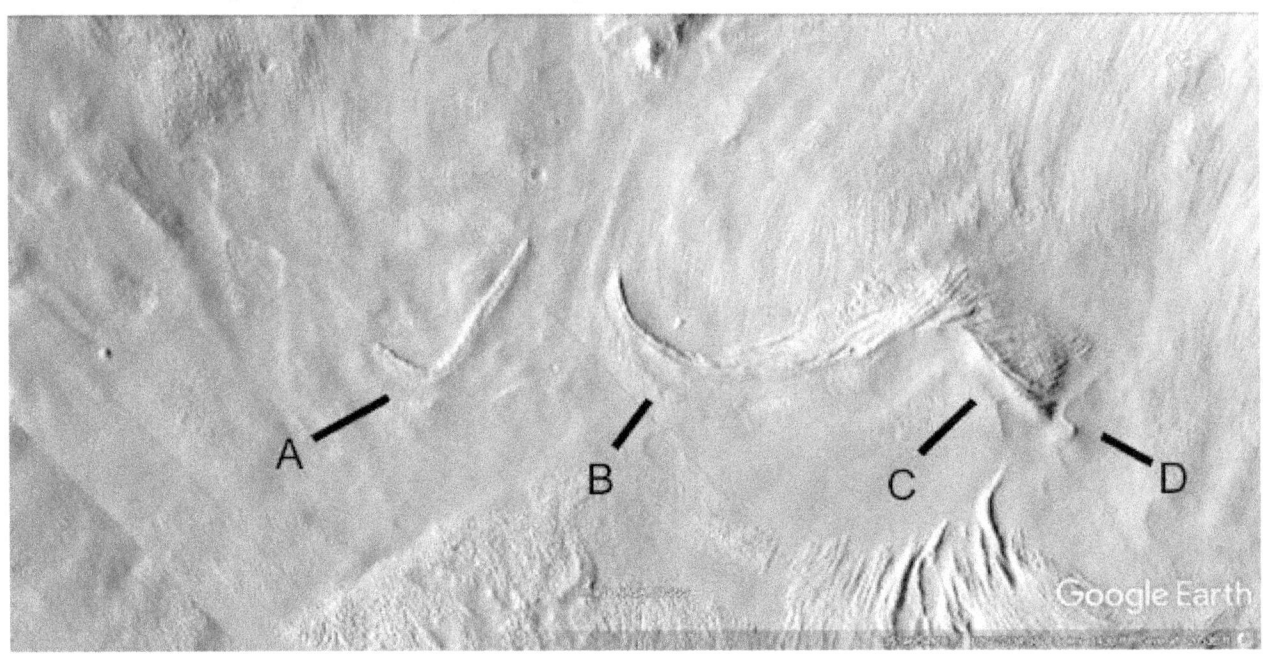

Cymd168a

Hypothesis

A parabola is shown.

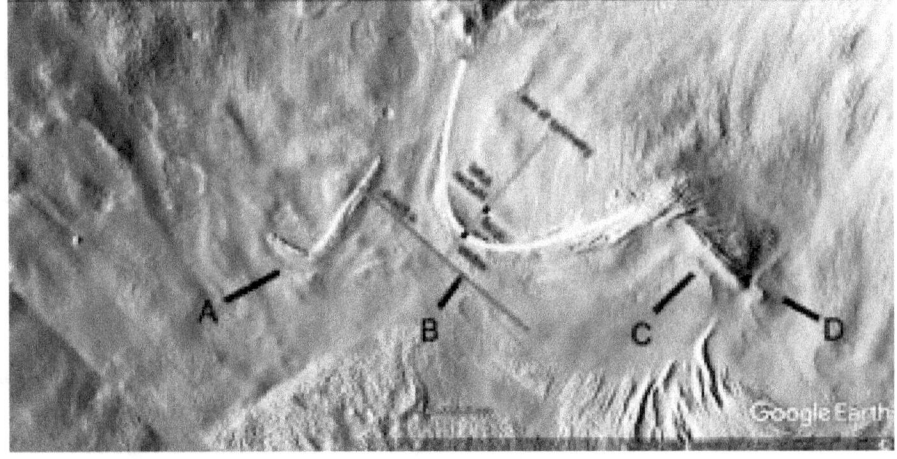

Cymd171

Hypothesis

All these dams are approximately parabolic, B shows an inner and outer parabola. The forces of mud slumping down a hill should be random, not form the same geometric shape over and over. E appears to be more choked with silt though this could from dust storm that blew into the crater. The other dams are clean and perhaps well maintained from the time they were used, E may also have had a mud slide from water coming out of the dam wall.

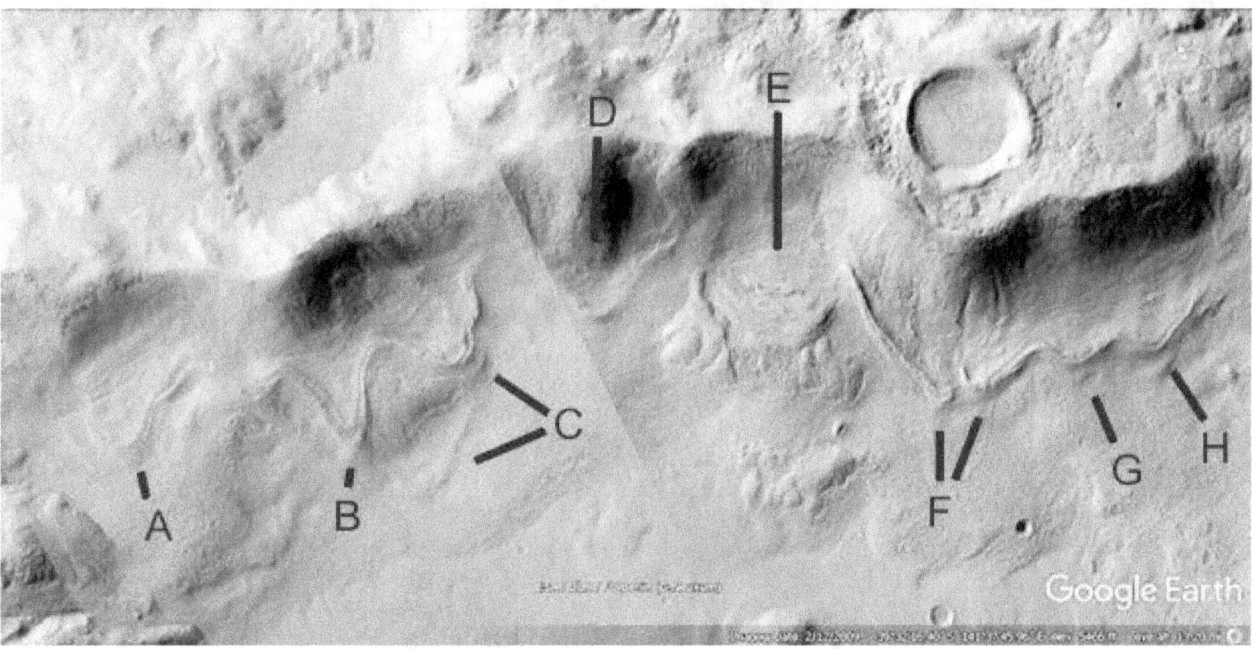

Cymd171a

Hypothesis

This is A and C in Cymd171, A shows an inner and outer parabola as used on Earth to make dams. E shows the inner parabola which becomes the outer edge of the inner and outer parabolic arch at B. C shows another inner and outer parabolic dam, the wall at D appears to be collapsed at 10 o'clock compared to 12 o'clock.

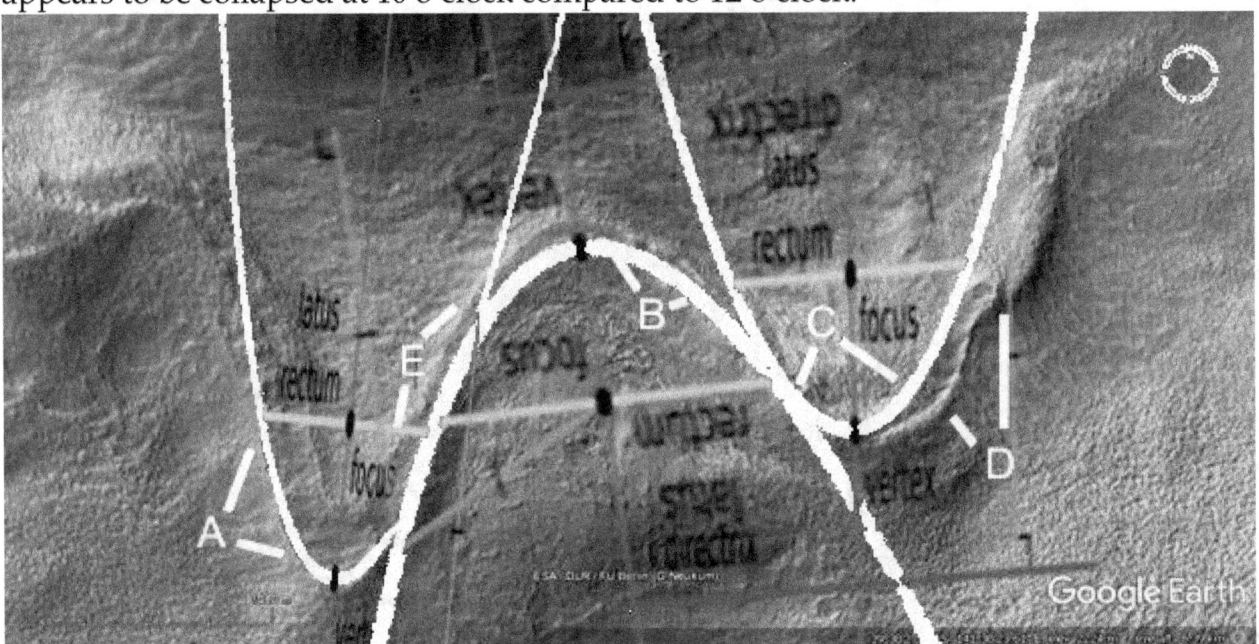

Cymd173c

Hypothesis

A shows steps as if creatures may have used these to move down the crater. B shows another smooth parabolic dam.

Cymd173c2

Hypothesis

This shows a parabola superimposed on the dam.

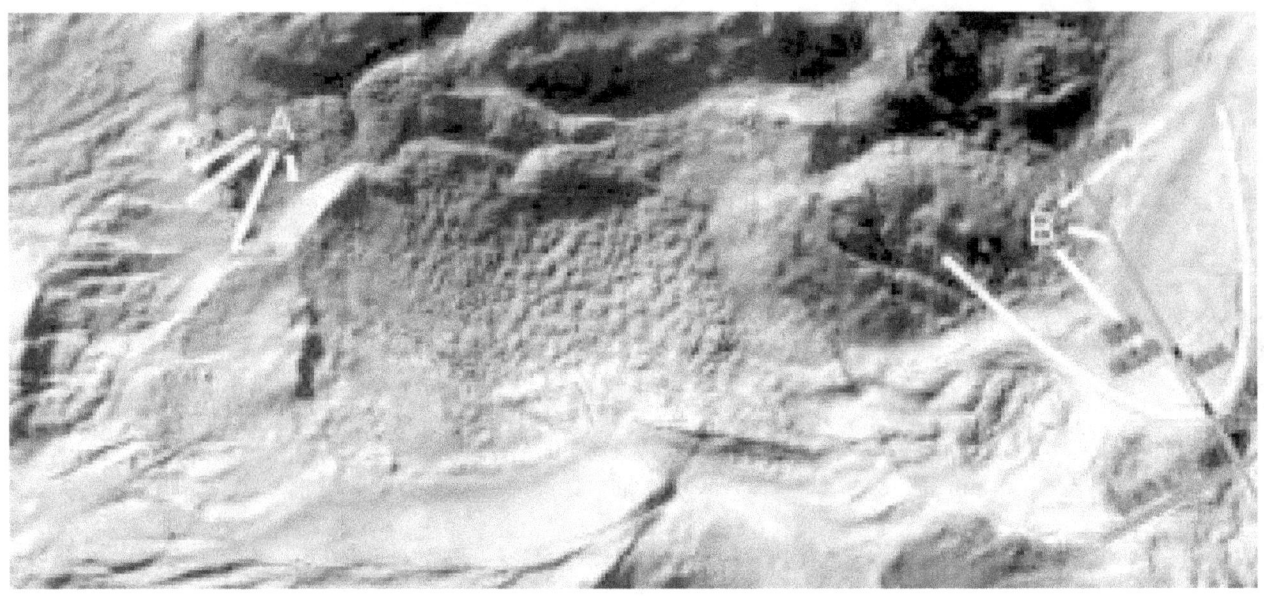

Cymd173d

Hypothesis

A shows signs of cracks developing as if from the dam material creeping downwards. B and C show the upper edge of the smooth dam, how it contrasts with the rough material above it. E shows a groove between the inner and outer walls of the dam. D shows the smooth interior surface which looks like new at this resolution, it would be a kind of cement. This also has a groove between the inner and outer walls perhaps as a sign of erosion.

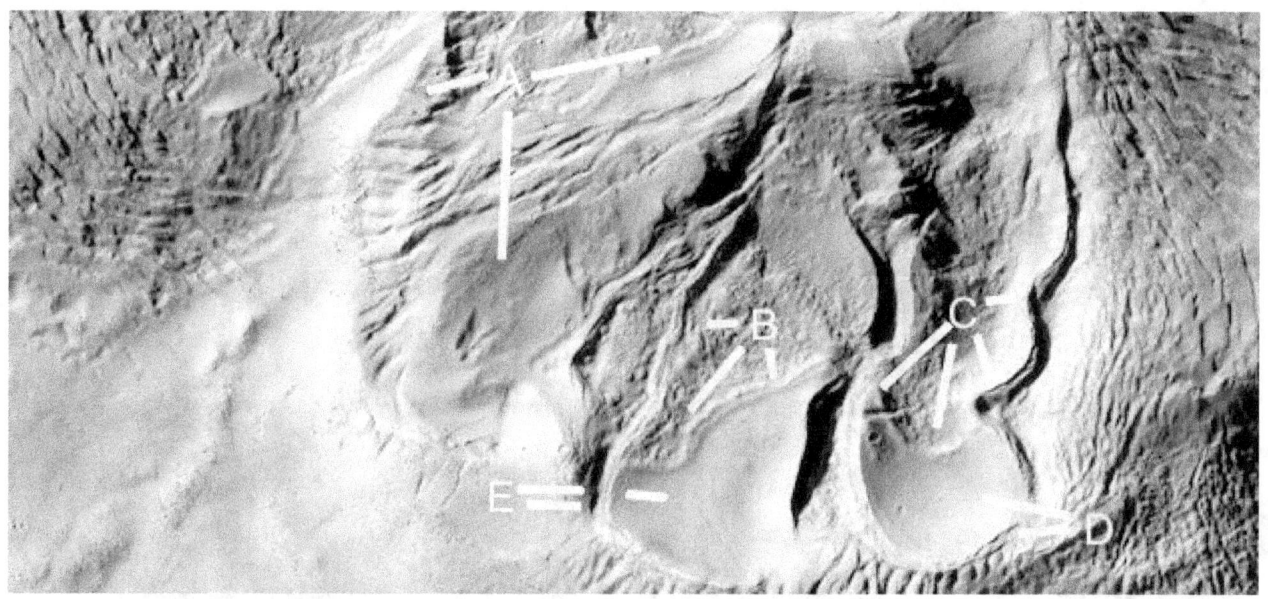

Cymd173d2

Hypothesis

A parabola is shown.

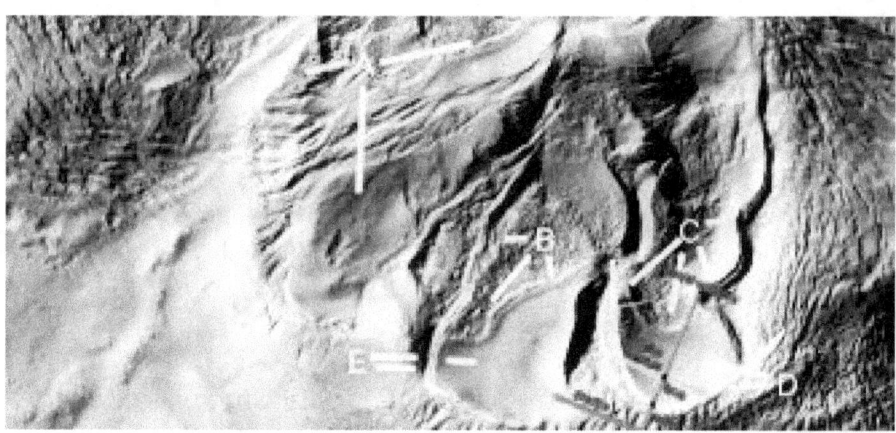

Cymd174

Hypothesis

A, B, and C show more excavated dams.

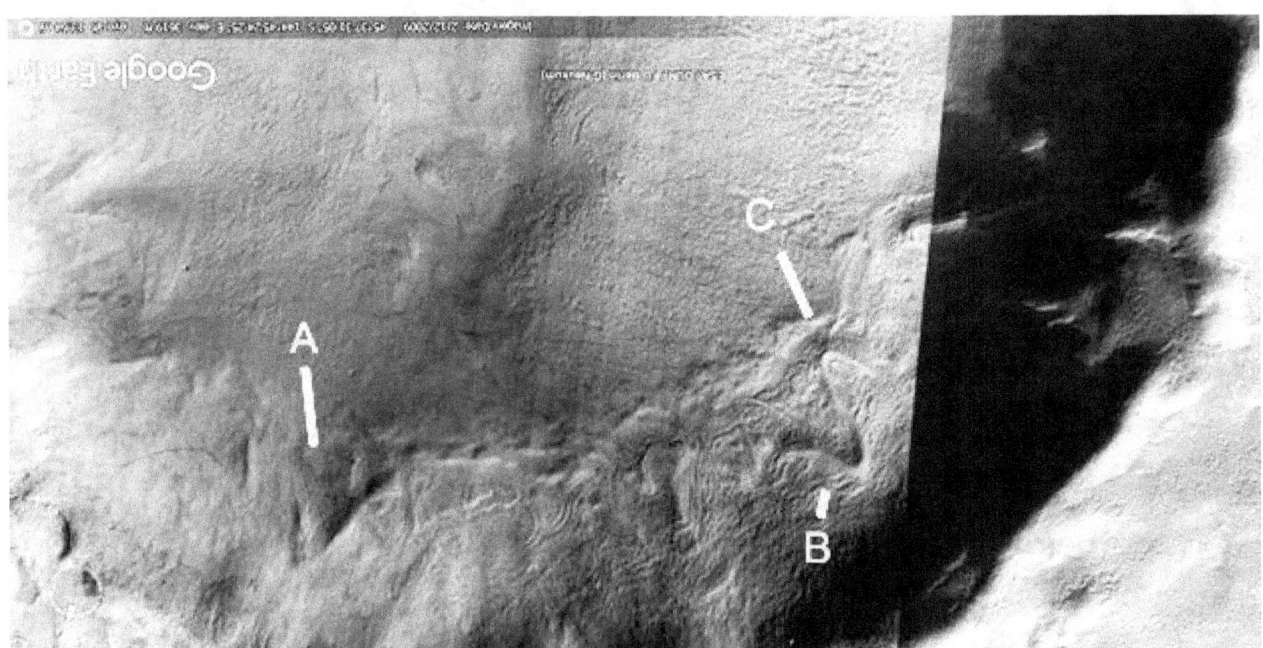

Cymd174a

Hypothesis

This shows how the two parabolas A and B fit together like a parabolic sine wave. The latis rectum of each is parallel to the other. C, D, and E are unusual shapes apparently carved into the crater wall.

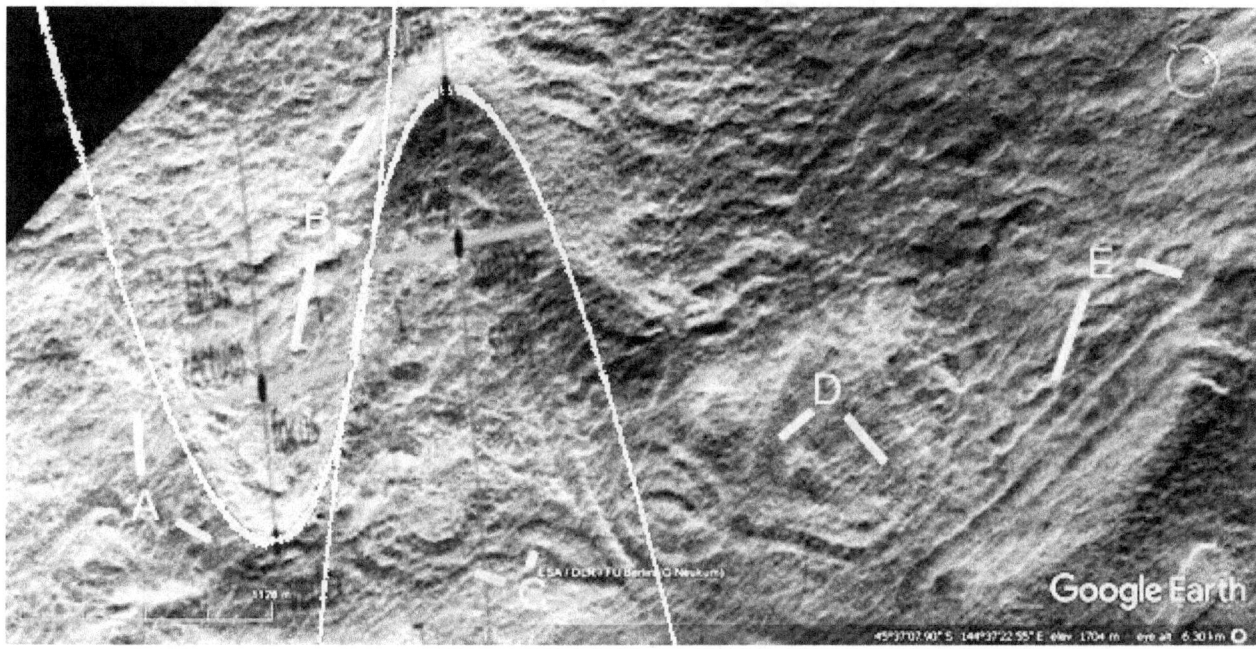

Cymd175

Hypothesis

This shows a walled dam with an inner and outer wall, A and D show the ends of it. C shows how this cavity ends with a straight edge. B shows the apex of the dam is also a sharp angle.

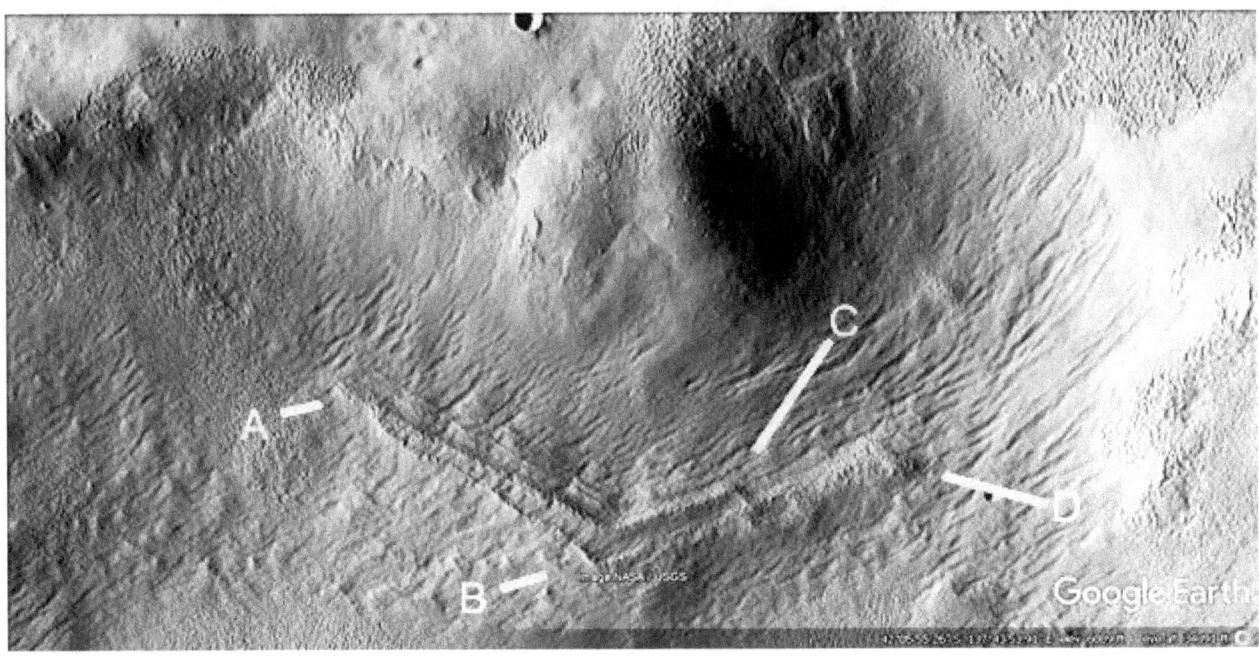

Cymd175a

Hypothesis

This shows how straight the walls are, even though the rest of the crater is very rounded.

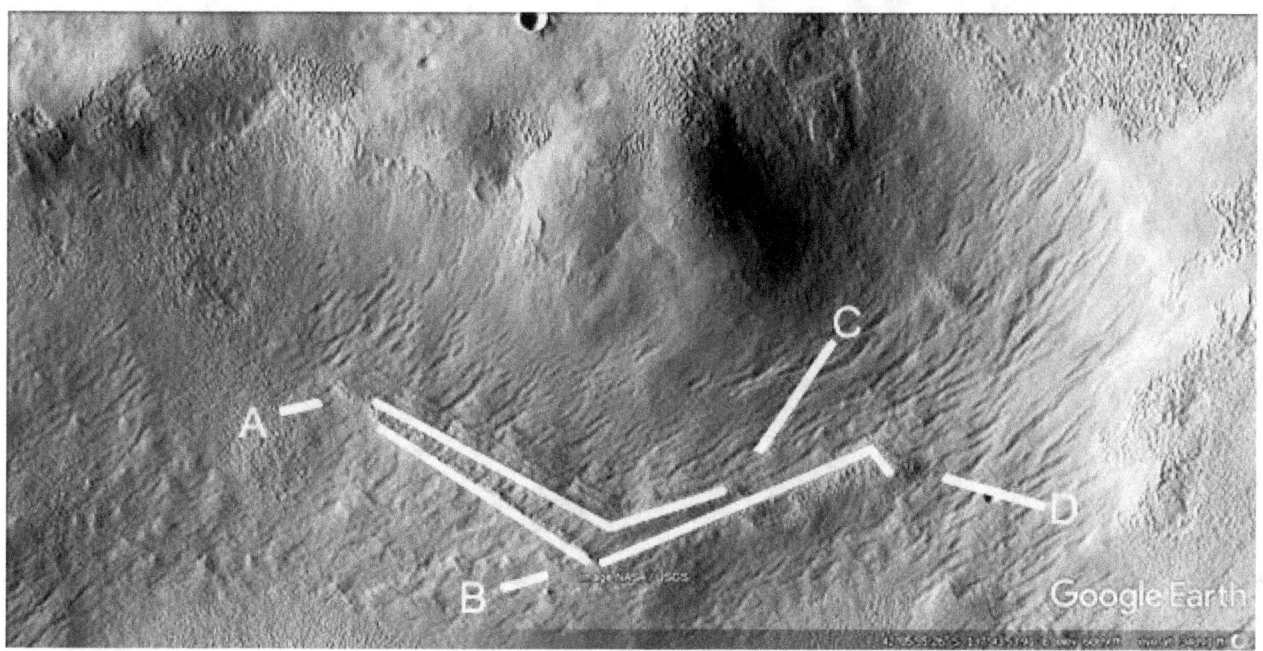

Cymd176b

Hypothesis

A and B show the two catchments of this dam, A at 3 o'clock and at 8 o'clock appears to be tiles or bricks. While it could be polygonal cracking it is very regular in shape. This would be a logical construction technique, to use either smooth cement or bricks to form the dam shapes. C shows another dam, D at 10 and 1 o'clock shows how smooth the upper part of this dam wall is. D at 6 o'clock shows a shallow dam. E shows a channel. G shows how this layer is breaking off.

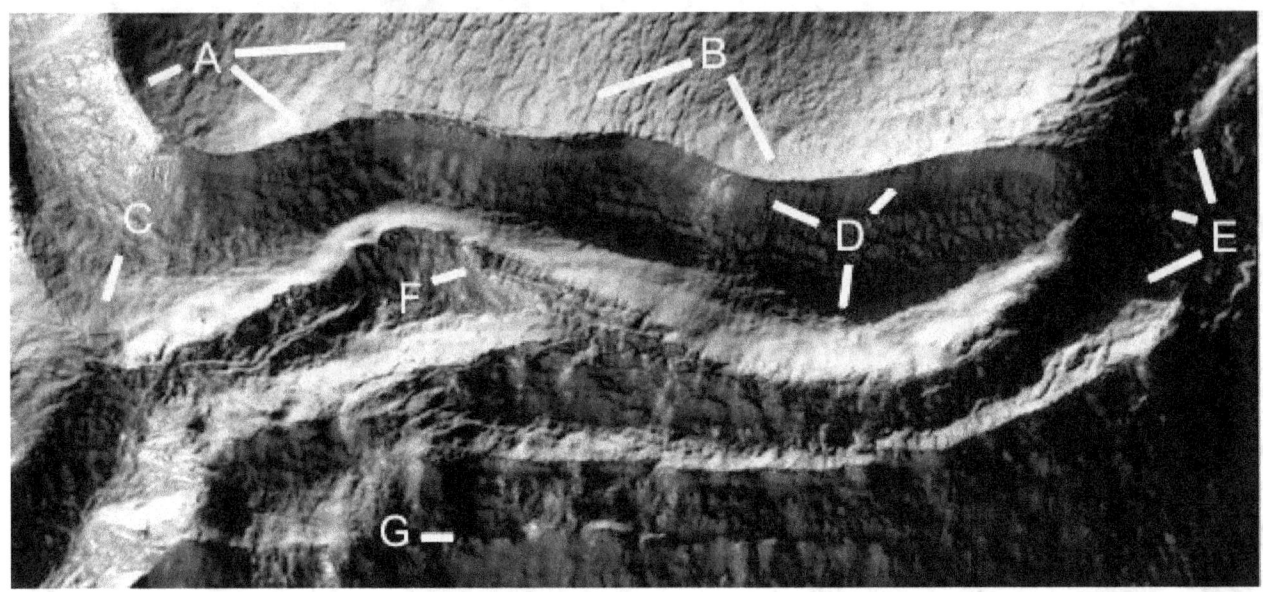

Cymd176b2

Hypothesis

Three parabolas are shown.

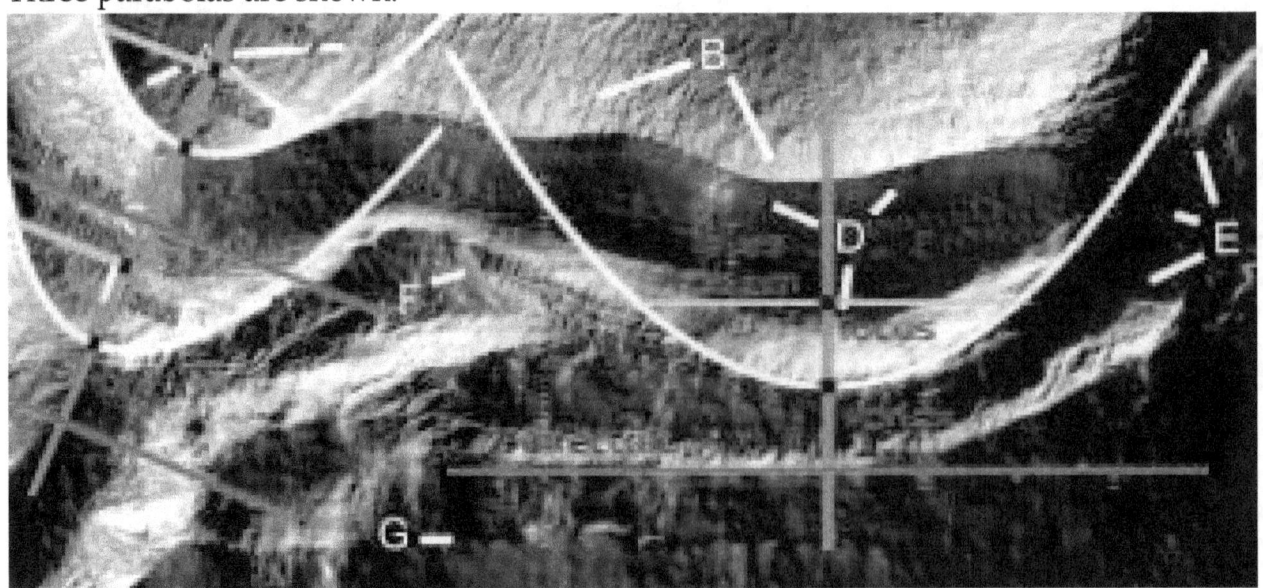

Cymd176c

Hypothesis

A shows how sharp the dam wall top is, with very little erosion. The second line at A at 4 o'clock appears to be a funnel like channel. B shows more of this layer breaking. C shows the edge of a channel.

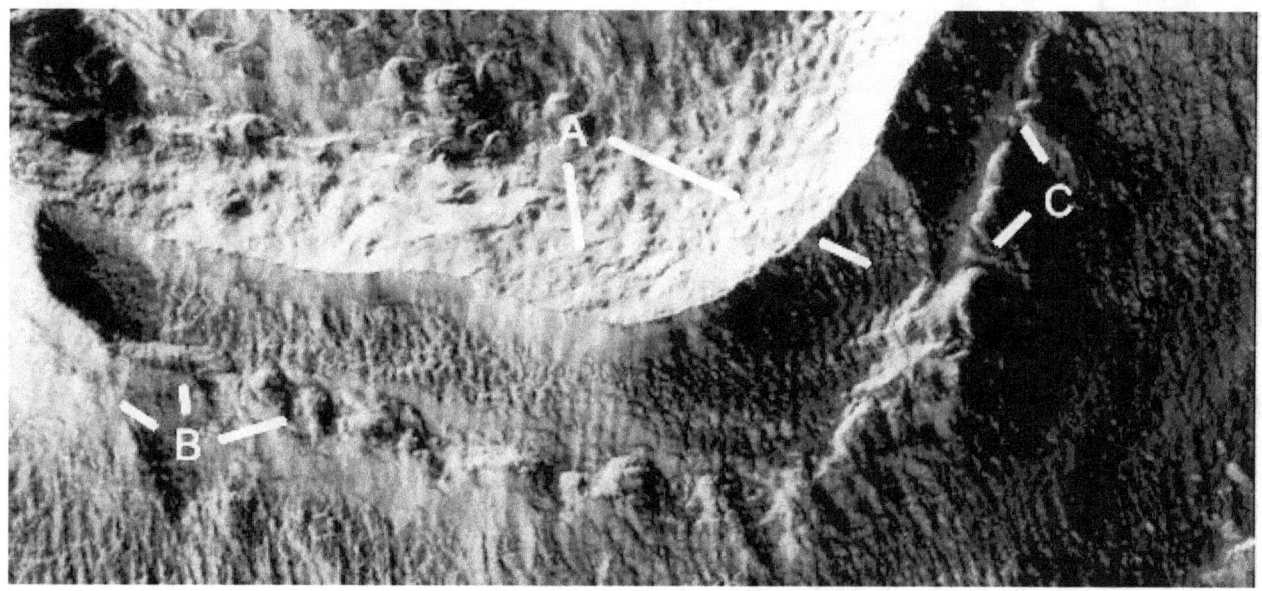

Cymd176c2

Hypothesis

A parabola is shown.

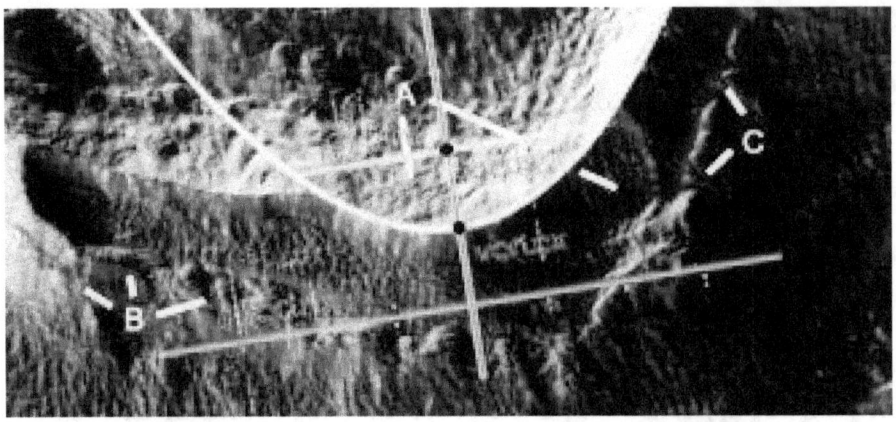

Cymd177

Hypothesis

A shows how the edge of this dam goes up the crater wall, against the tendency for this to be eroded. B shows where the water would pool, F shows the other side of the dam against gravity up the crater wall. C shows another dam, D is also a highly eroded dam.

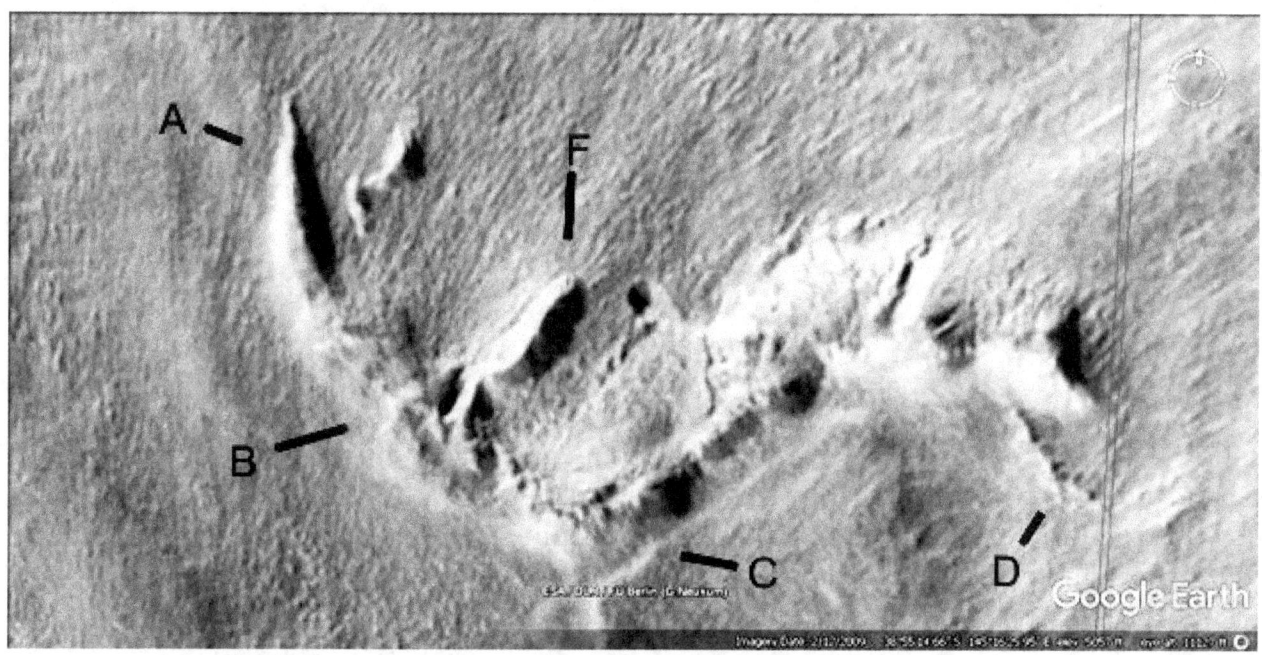

Cymd177a

Hypothesis

A parabola is shown.

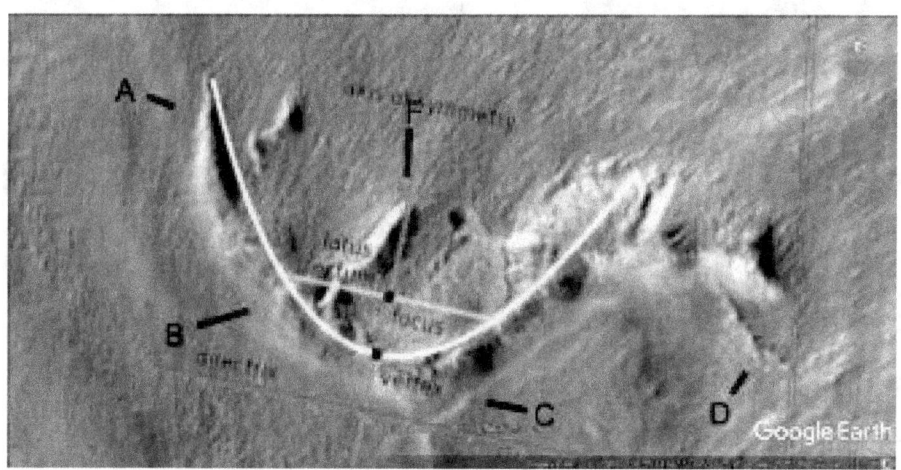

Cymd178f

Hypothesis

A shows a small dam, the cracks may be from the material creeping down the crater wall. B shows another two dams, C shows the left wall of these and D the right. D at 10 o'clock shows possible silt having filled up the dam indicating a lack of maintenance when the water was still flowing. This seems less likely to be dry material from its shape.

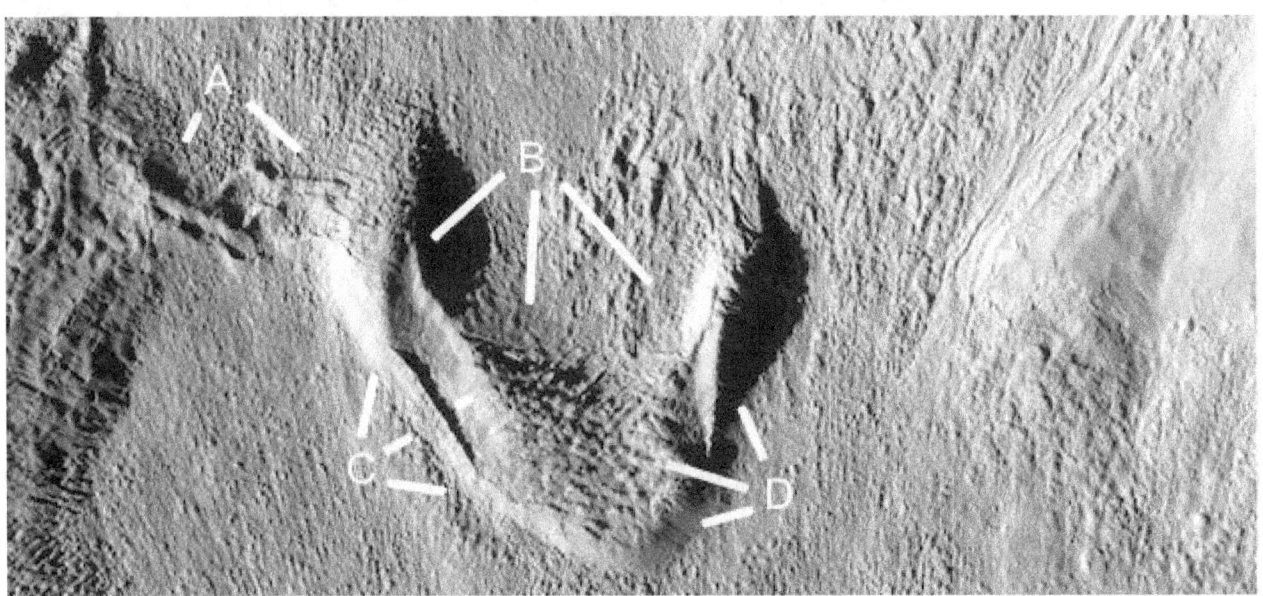

Cymd178f2

Hypothesis

A parabola is shown.

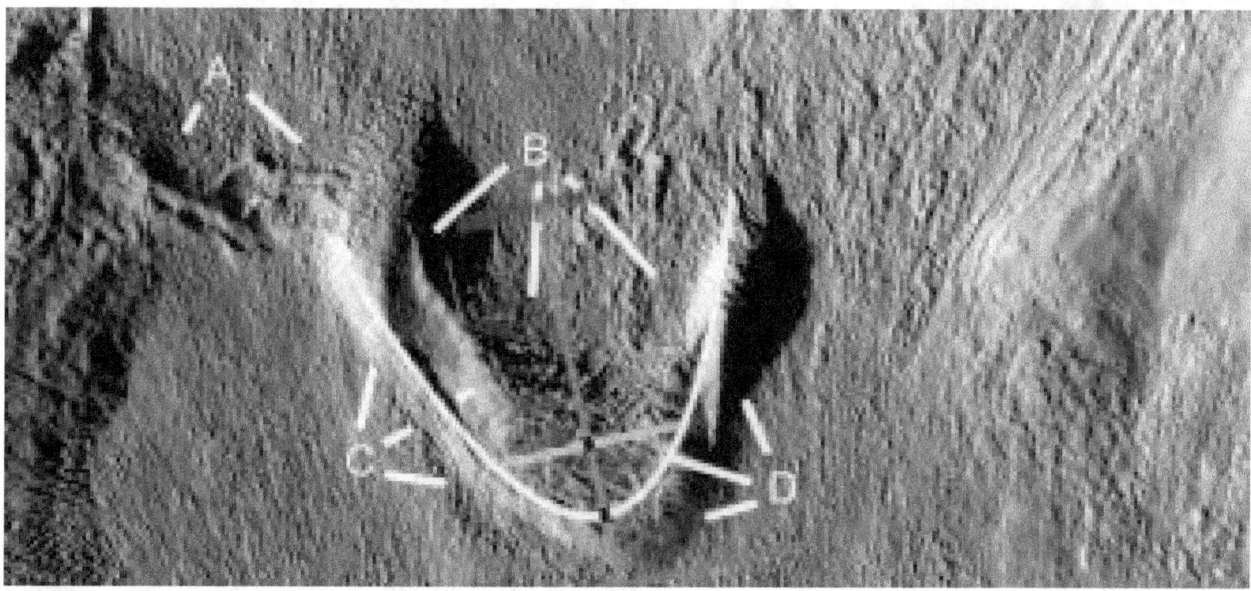

Cymd181

Hypothesis

A, B, and C show excavated dams, D, E, and G show the hill walls which would direct water into them. G appears to be collapsed leaving a cavity in the dam wall, B shows a cavity under the dam wall, this makes the wall stand out from the slope when if natural it should have eroded with the cavity under it.

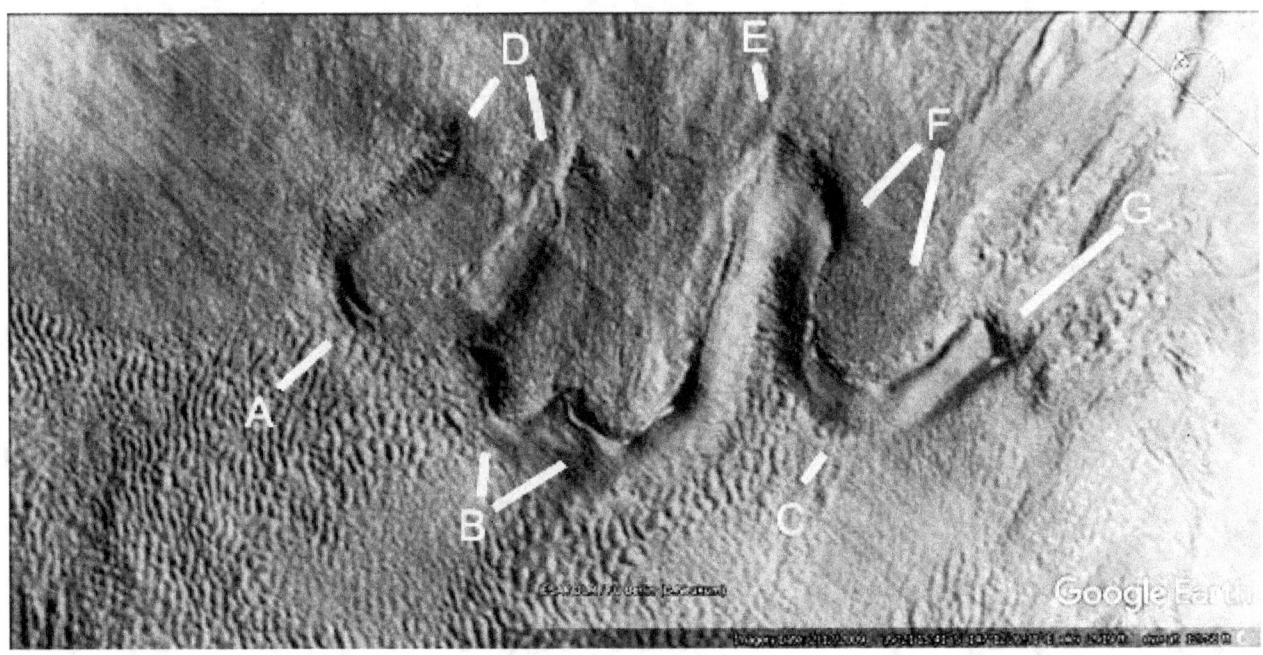

Cymd181a

Hypothesis

Two parabolas are shown. B are probably parabolas too, but too small to overlay.

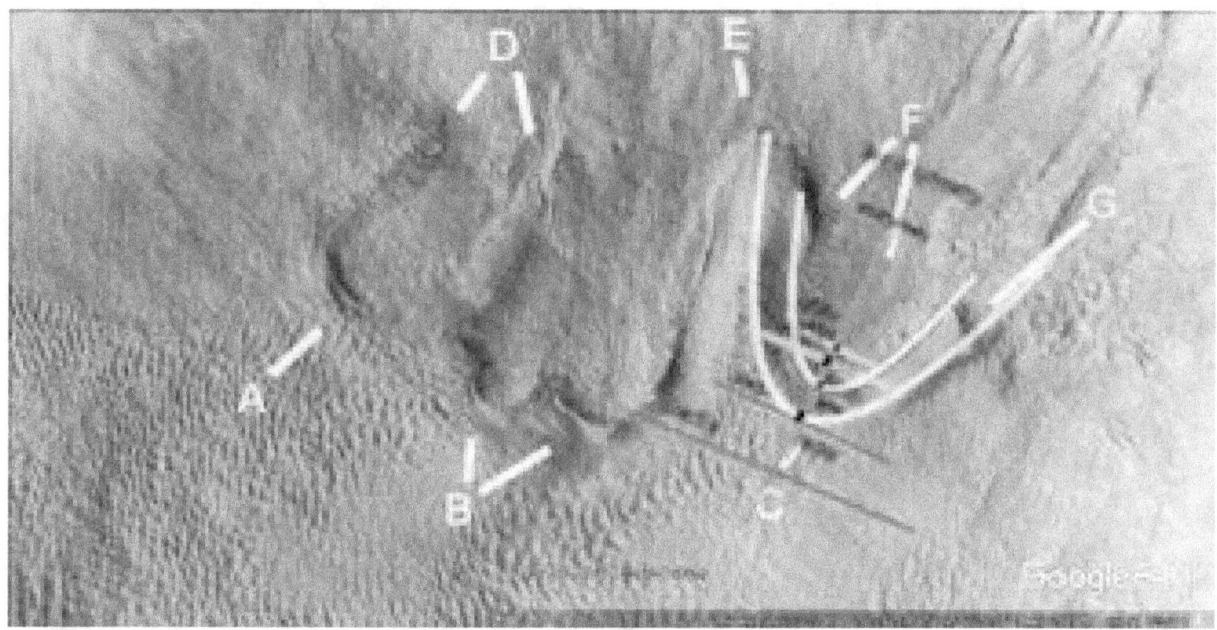

Cymd193

Hypothesis

A may be a small dam, B at 5 o'clock has a smooth wall with a lip. C at 7 o'clock is close to parabolic, D is a faint dam shape. G down to E may be a tube carrying water, F appears to be a support built under the dam.

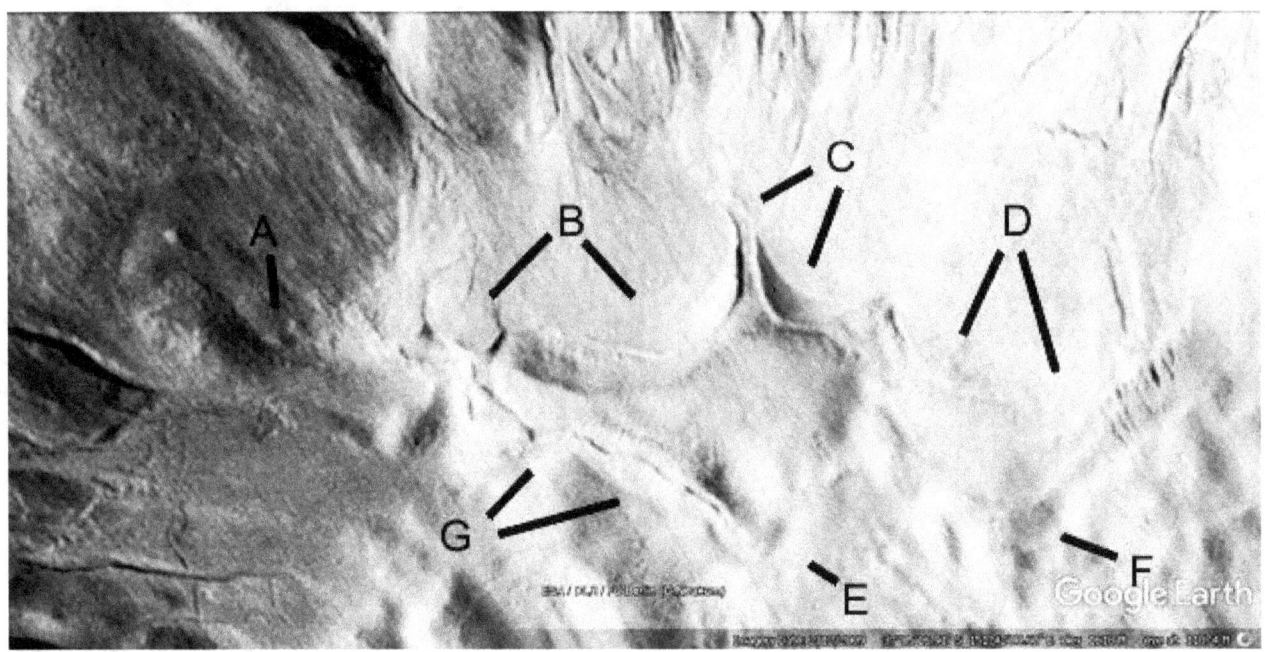

Cymd193a

Hypothesis

A parabola is shown.

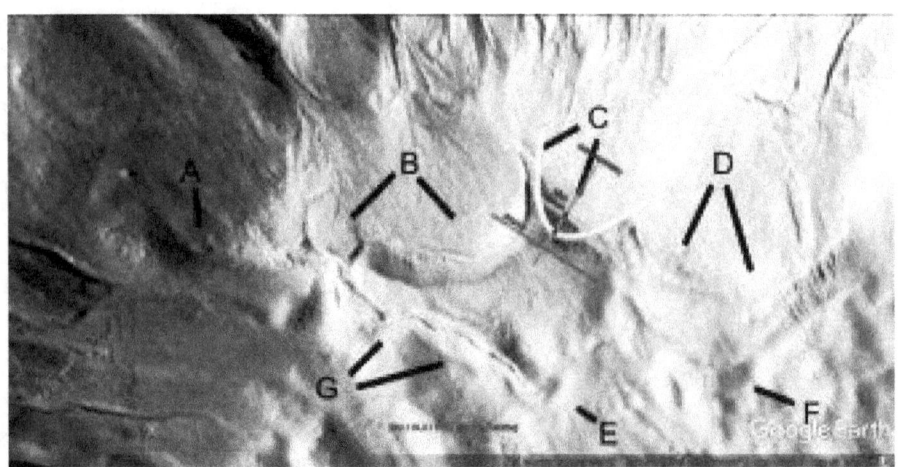

Cymd194

Hypothesis

A, B, and C are also smooth like made from cement. E looks like water has gotten past these dams, D is a parabolic dam.

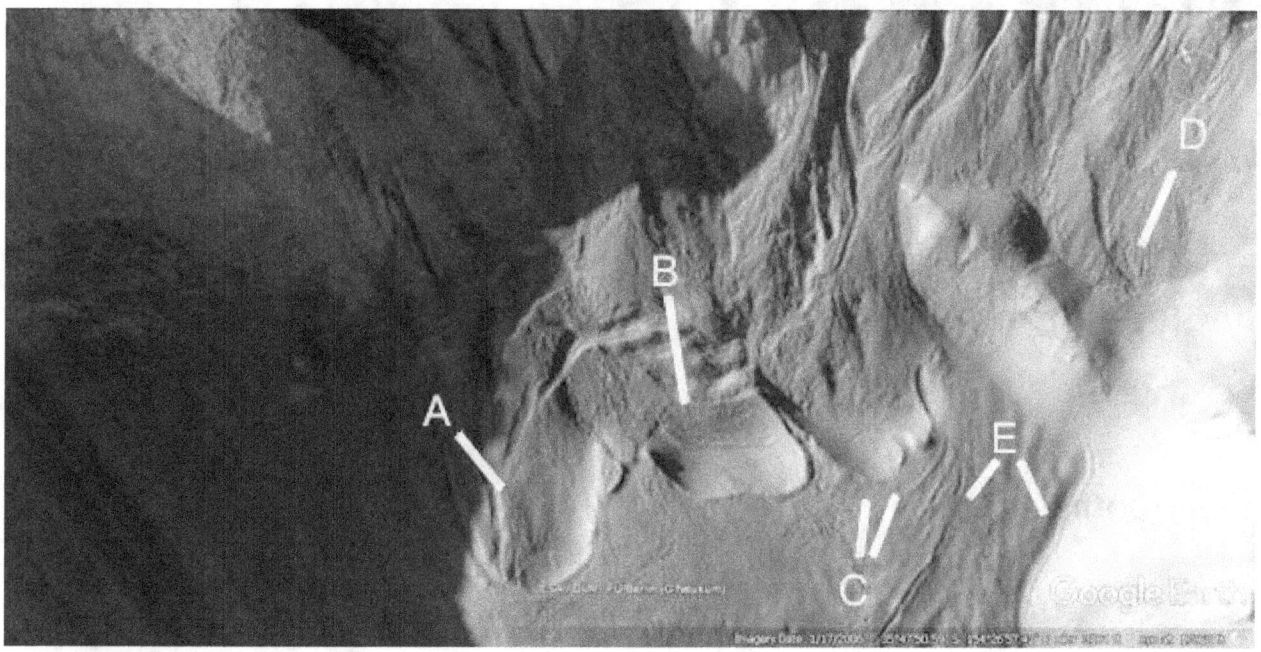

Cymd194a

Hypothesis

This shows a parabola superimposed on D, a very close fit.

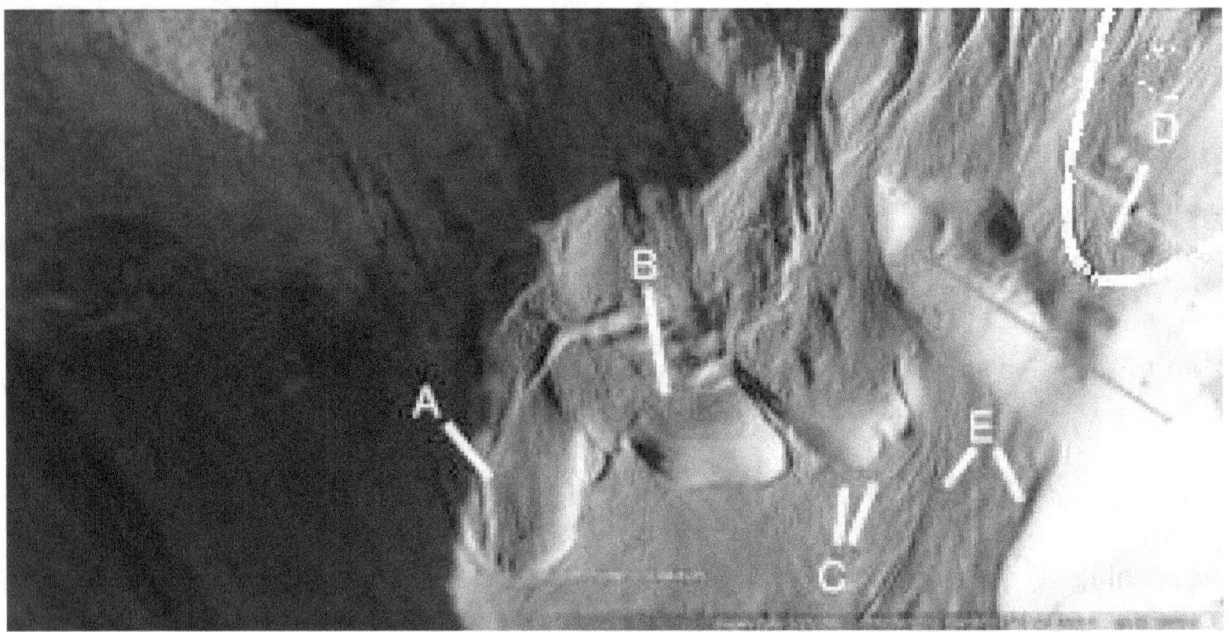

Cymhh200b

Hypothesis

Many hills in this area have room like shapes in them as shown. These may have been for strength like a honeycomb design or for habitation. A shows some rooms to the left of the dividing ridge, B shows them on the other side. Many of these have right angles and similar areas like human buildings. B appears to have some height as if these walls extending up higher in some areas but eroded down. Between A and B there is a ridge, perhaps an interior support for the hollow hill that extends up to E and another ridge. C shows other room shapes with more irregular angles. With uneven terrain these may have also been uneven, using other than right angles. Large areas, as B shows however, are as regular as any human city layout. Many old cities would also have irregular streets like C, these may then have been roads or paths inside the hill not just walls. For example there may have been rooms and then on top of the walls they could have connected to roads or tunnels. We do something similar with roads over underground buildings like bus depots in human cities. While some are very regular with right angles it is not a refutation of the room hypothesis that others are less regular. Some also appear to have gaps in the walls like doors or entrances, however if the top of the wall is intact the doors would not be visible from space above them. D shows a hill at 9 o'clock that appears to be eroding, at 12 o'clock it shows clear wall shapes, and at 1 o'clock more pale material covering them. This may have been a floor or ceiling for the walls, and then another layer of walls. Often this appears to be the case with much higher sections still having walls. E shows these walls with some pale material covering parts of them it implies this material is eroding and showing the wall structures. In some areas, as will be shown, these walls also erode away leaving bare ground and sometimes pits. F appears to be a higher wall casting a shadow on its lower side.

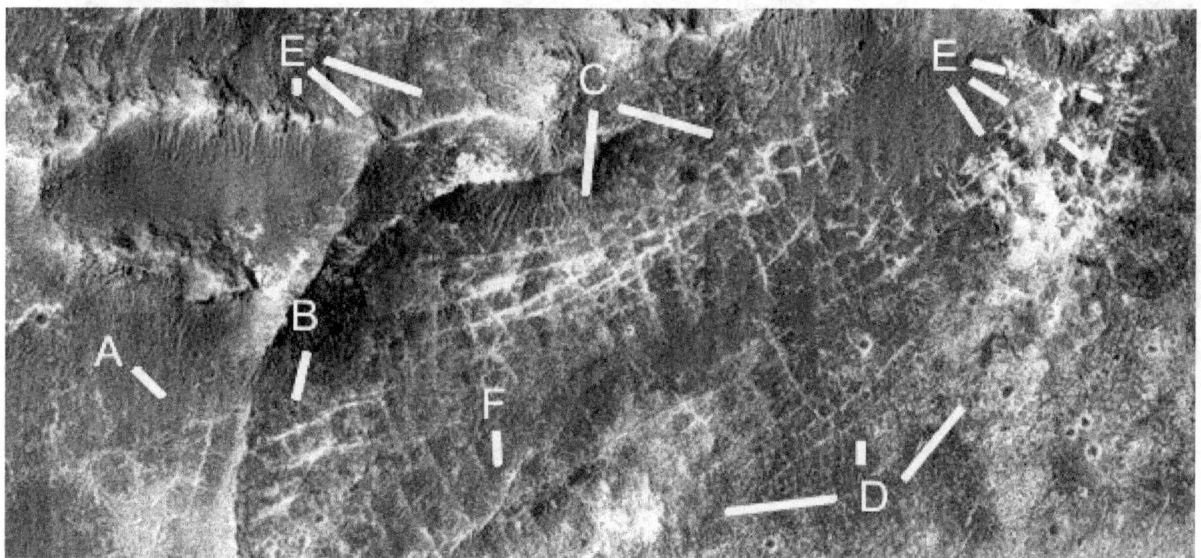

Cymhh200f

Hypothesis

The walls at A are quite high and form a right angle , the pale material in the hollows may be ceiling material covering rooms. These can be seen in some places.
 B at 2 o'clock shows a room shape approximately square, at 10 o'clock there appear to be rooms.

Cymhh200g

Hypothesis

A are interior supports close to another right angled T intersection. B shows an intersection of ridges, one goes down to C and a flat area like a ceiling not collapsed. D at 5 o'clock shows a mesh of walls in a more irregular shape, perhaps because they are partially collapsed. At 10 o'clock the wall shapes are more like standard room shapes with parallel walls. E shows pits that appear to form as these walls erode away. F shows another internal T junction close to a right angle. G looks as if the hill was built around this crater with more rooms on its upper side. Water may have been directed from the roof into this as a dam. It may also have been deep enough to reach the water table and form a permanent internal lake, as could many of these pits.

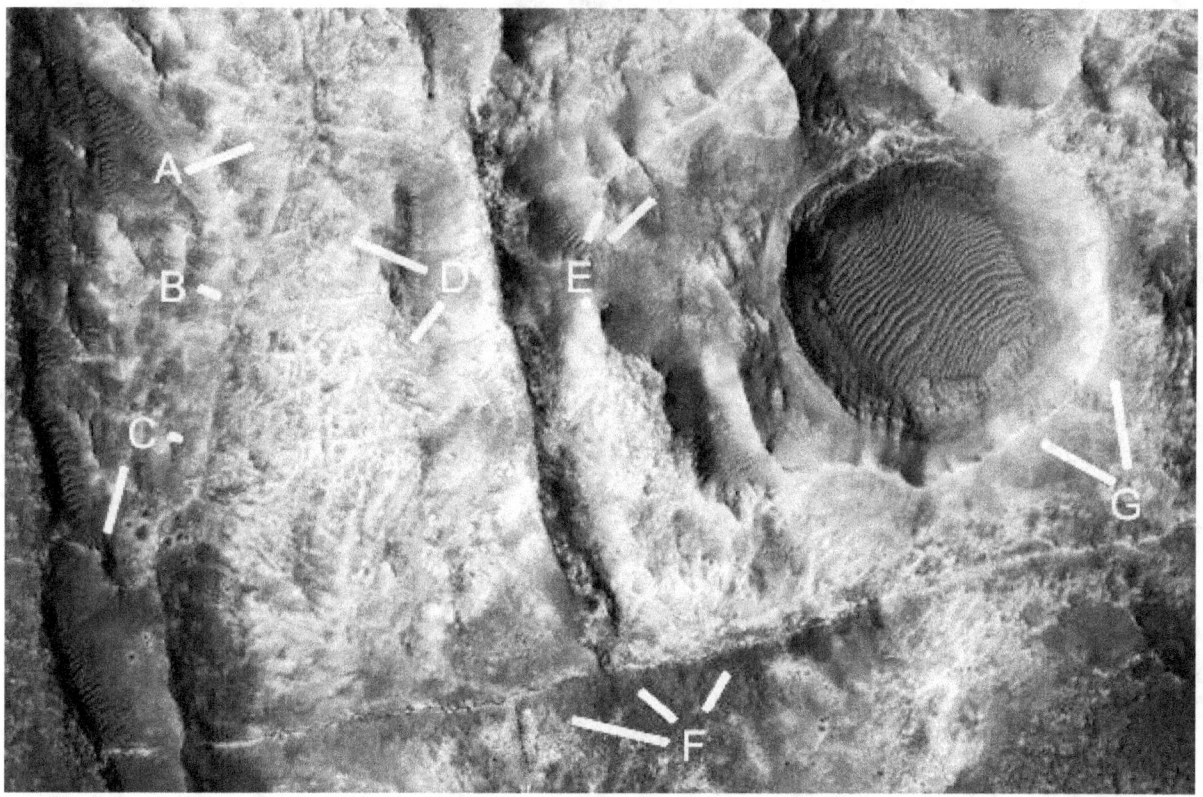

Cymhh200I

Hypothesis

And B show another ridge like an interior support. C and D show more room shapes, at C they appear to be covered more by this ceiling material. D also has some of this ceiling material, above it the pale walls are much clearer without it. Between A and B there is an impression of these walls with the ceiling material draped over them.

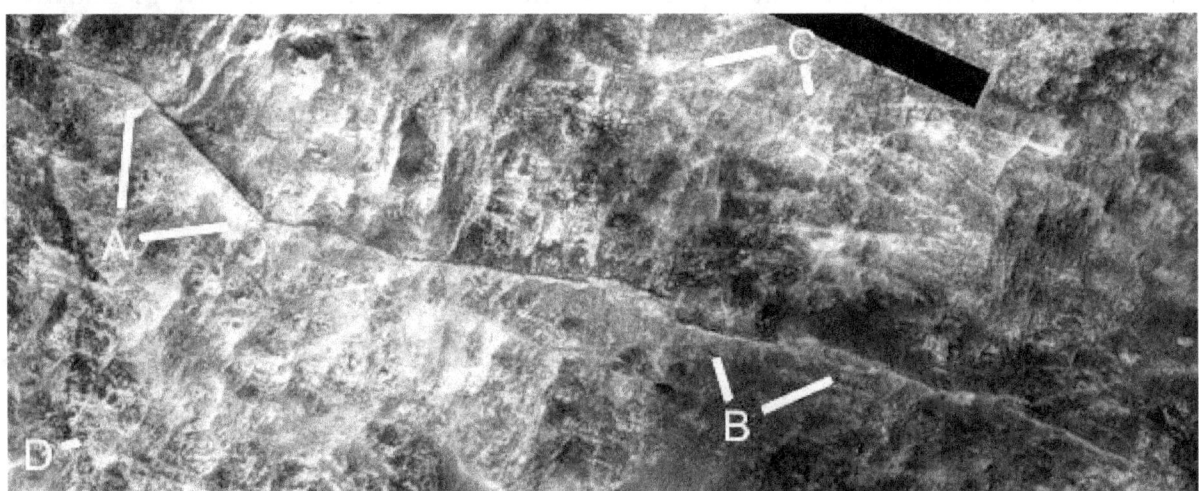

Cymhh203c

Hypothesis

These walls are more like a tetrahedron than cubes, A shows bare walls and a nexus of walls partially covered by ceiling material. B also shows a nexus of walls with hollows around it. Six walls appear to come together at C. D shows a triangle of walls with one bisecting the base. E shows a corner of two walls with a hollow inside, between D and F it is like a rectangular shallow box suspended by some walls.

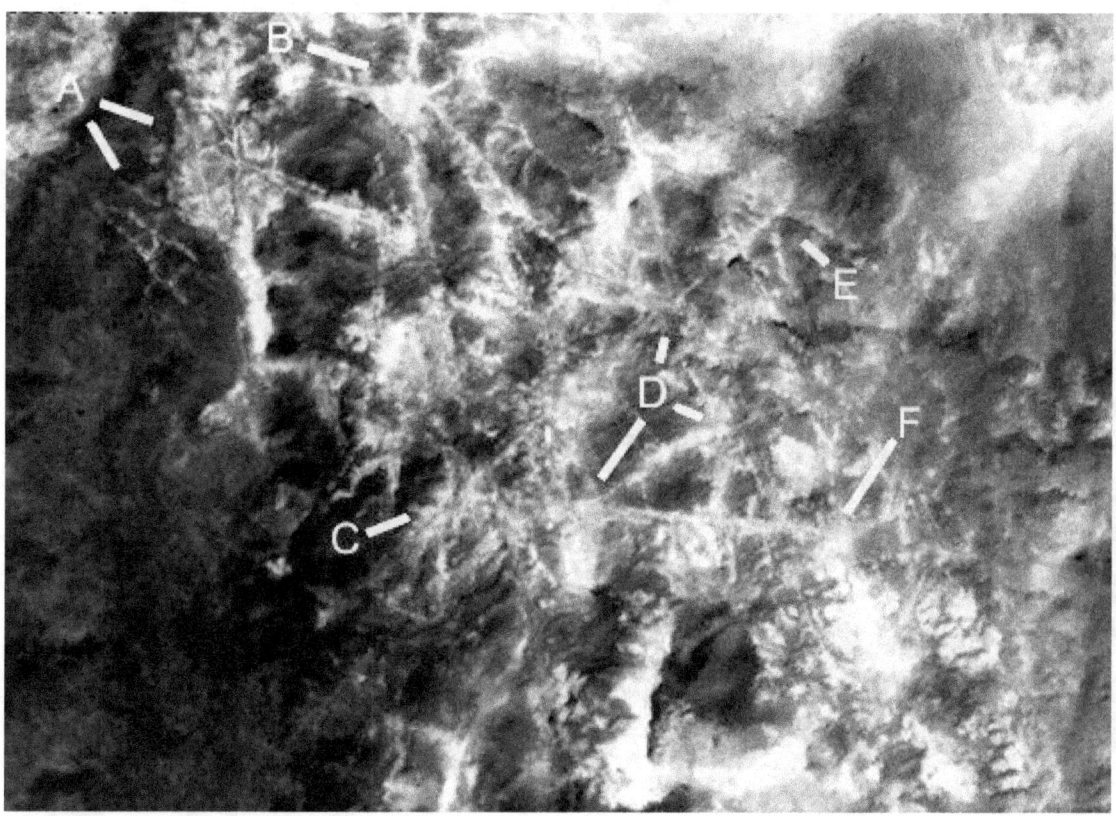

Cymhh203c2

Hypothesis

This shows the possible squarish shape suspended by other walls.

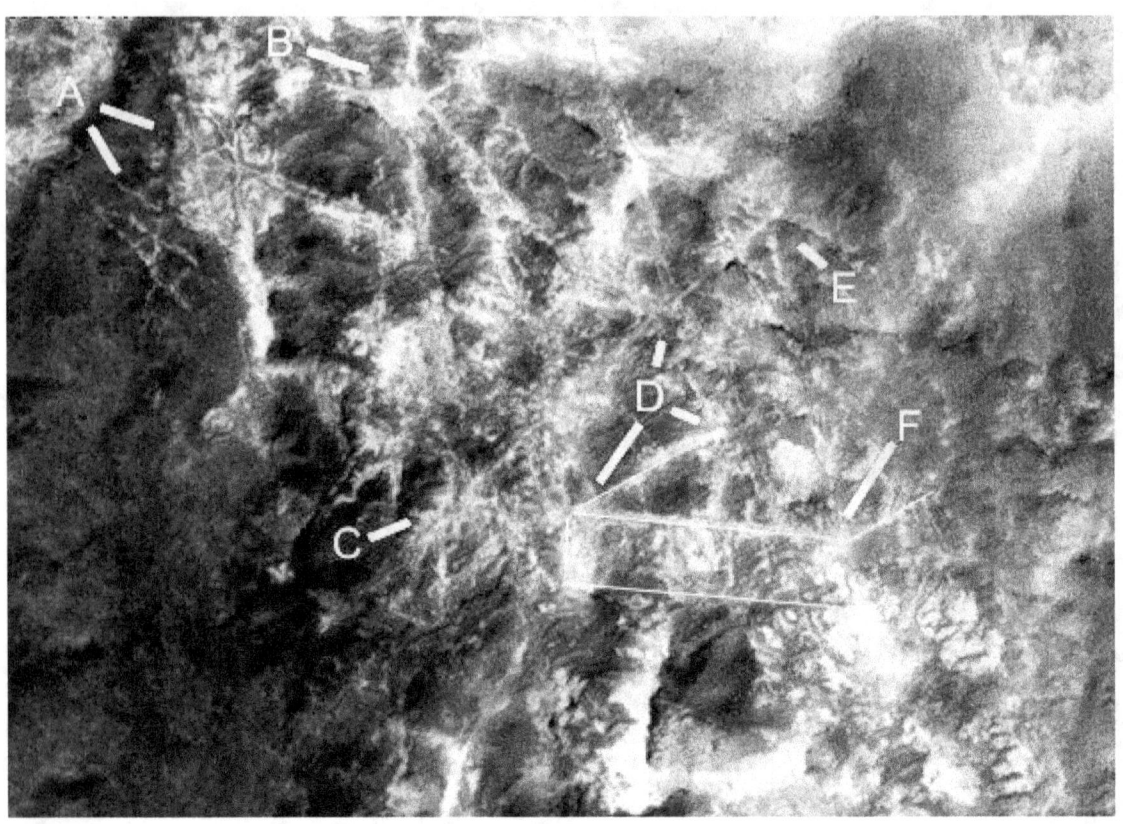

Cymhh203g

Hypothesis

A shows an interior support and T junction, probably a right angle. B shows a wall in good condition yet the hill above it has collapsed, C shows another wall. D at 4 and 10 o'clock shows another wall. D at 8 o'clock shows another wall coming from a T junction. E shows more walls, there are pale squarish shapes like the eroding floors of rooms on the edges of the pit.

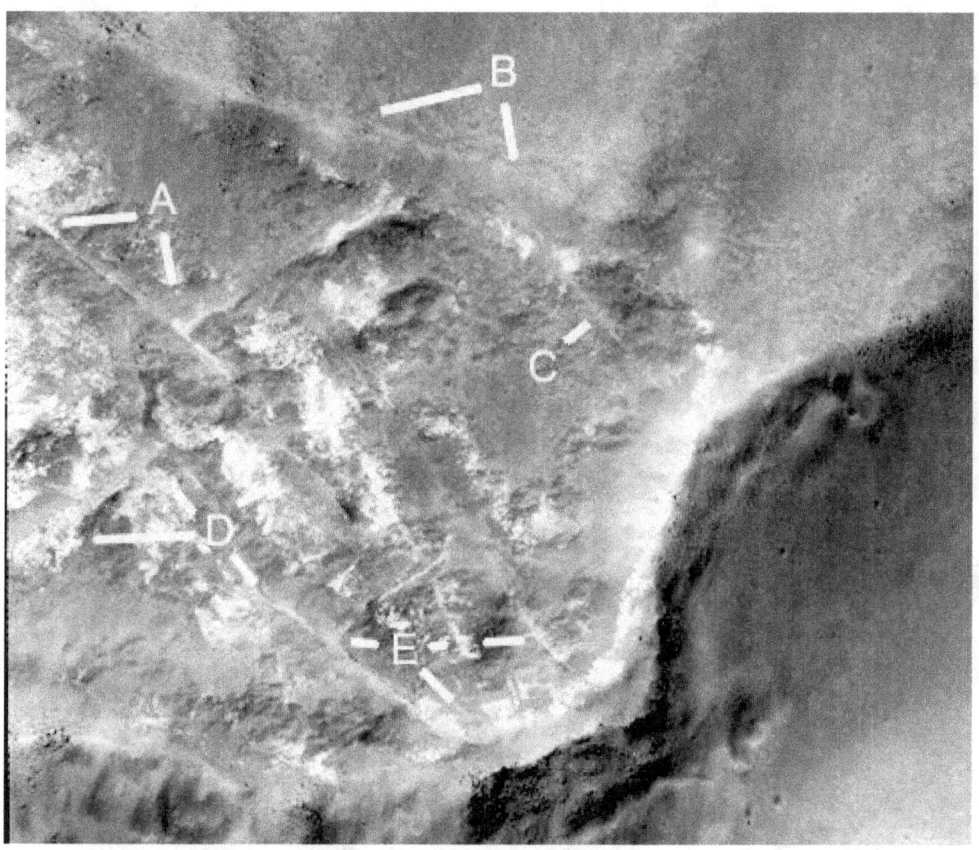

Cymhh209b

Hypothesis

This is a closeup of the previous upper left corner, there is a definite impression of a wall going from A to C and some parts appear to be higher like a cement pad. B also appears higher, the texture of the ground is also different to outside this. If this is cement then rocks would not come from underground, but some might form as the cement broke up.

Cymhh209c

Hypothesis

This is a closeup of the curved wall mentioned earlier, A show this wall is broken at 4 o'clock. B and C show a curve which is close to two parabolas. D shows this part of the wall is breaking up on the right into separate rocks.

Cymdhh209c2

Hypothesis

This shows two parabolas superimposed on the curved wall, the latis rectums are approximately parallel and form a rectangle as they join to the apex of each other. This would be a simple way to build a curved wall so the two curves line up with each other.

Cymhh209f

Hypothesis

A shows another inner and outer wall, or perhaps a road with railings. B shows a curved wall as the corner of the collapsed hill, it has a cavity running outside it. C shows another curved corner, it connects to a wall at E and perhaps a small collapsed hill. This also appears to be a small walled room under the pale material. D shows a wall or tube continuing up the image.

Cymhh209f2

Hypothesis+

This shows a parabola superimposed on the corner for a reasonable fit, the builders could have used other curves but parabolas appear to be the most common. Mathematicians might consider the most efficient method to make a corner like B, perhaps a part circle with a rope fixed to the center, then straight walls connected to the arc from it.

Cymhh209i

Hypothesis

This shows many room shapes, A is a wall on the edge of the collapsed hollow hill. B shows a wall at 1 o'clock and a room at 5 o'clock. C shows a connecting wall to A and B with more rooms. D shows a larger wall that may be a collapsed tube, it has a hollow running down it. E shows more horizontal walls, F, G, and H show many rooms most with right angles. A tube like D might have been used to travel longer distances, there may be tubes like this under H at 2 o'clock. Also there may have been tubes above this area which have already eroded away.

Cymhh209o

Hypothesis

A shows many rooms, also the walls here appear to be doubled or are collapsed tubes. This is important for the room hypothesis, if someone could go to each room in these tubes then each is accessible. If not then how many could be used is problematic. The thicker ridges also appear hollow at some points elsewhere, B shows a main tube that has some collapsed areas along it. C shows an area that may have eroded to the bare ground, there are faint walls here the same as in the other parts. C at 11 o'clock has very high walls as see from the shadows. Engineers could calculate the height of these walls from the shadow knowing the sun angle from HiRise. The higher the wall the longer the shadow would be inside the room. At C at 8 o'clock the walls are lower as if eroding. D at 5 o'clock shows a rounded formation of rooms like a nexus, at 8 o'clock the walls have collapsed apparently leaving some pillars standing in some cases. E shows a zig zag in this wall or tube, as if the access to it gives straight sections for the entrances. F shows areas where the ceiling appears to have either fallen onto the walls or is still secured above them in parts.

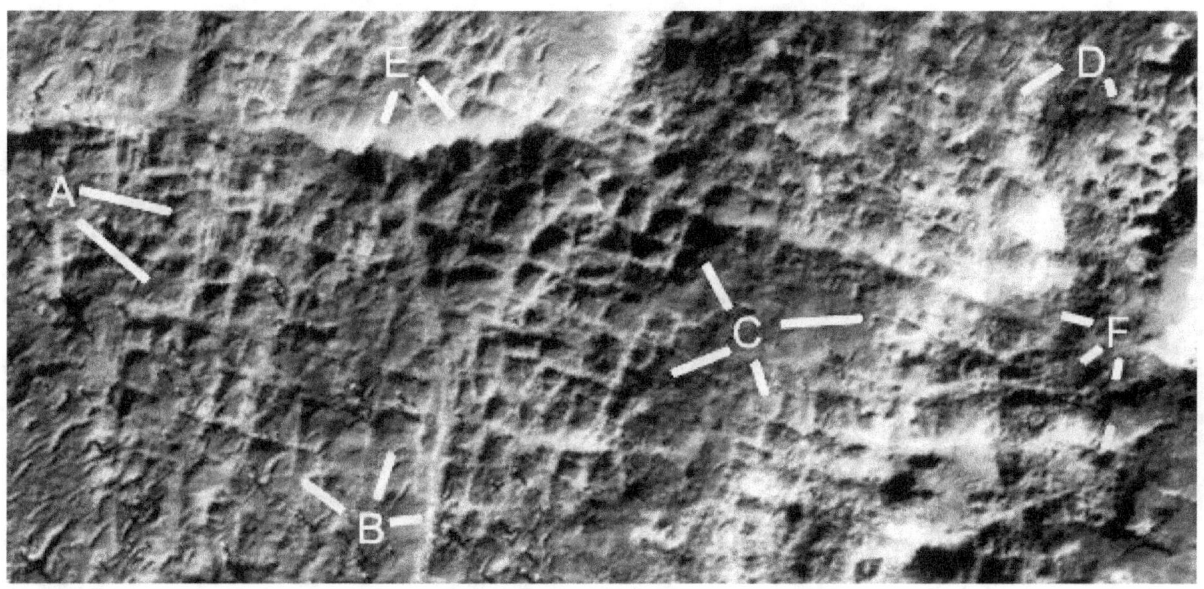

Cymhh209p

Hypothesis

More rooms at right angles are shown at A, B has a dark line along it as if it is a collapsed tube. This would give access to the rooms at A. C shows this tube at 7 o'clock, at 2 o'clock these may also be tubes as they connect to it. At D the shadow in them continues like a collapsed tube. They may also not have been designed to have a roof but be two separate walls to form corridors. E at 11 o'clock shows another tube connected to this nexus , at 8 o'clock is a pit perhaps from erosion or exposing a lower layer of rooms as the ceiling collapses.

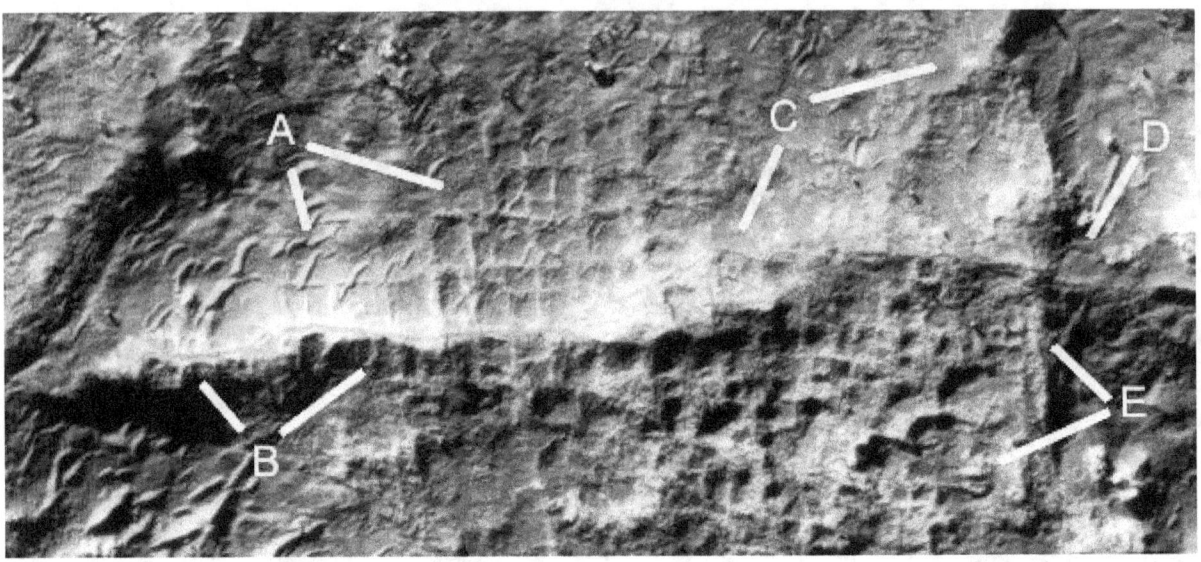

Cymhh211d

Hypothesis

This shows many more rooms and also lower layers of rooms. A shows the ceiling at 5 o'clock is more intact, at 4 and 7 o'clock they connect to the edge of the pit and may also extend to habitats underground. B is the opposite, a pale material with dark lines in similar shapes to the rooms. These may be either the ceiling material has collapsed onto walls or this is a decayed floor. C shows more rooms with the ceiling material still covering them. D shows how there are details in these hollows like lower down walls and rooms, particularly at 12 o'clock. E shows more rooms with deep hollows, at 10 and 11 o'clock it appears to be three dimensional. F shows a ceiling above the hollows at D at 12 o'clock. F at 5 and 6 o'clock appear to show decaying walls as these holes grow. G also gives the impression that these walls are in the hollows and connecting to each other in unseen ways. G at 10 o'clock may appear to be entangled because it is wreckage of walls which have collapsed in these hollows. This is shown from above with H.

Cymhh212b

Hypothesis

This also appears to have a three dimensional shape, the impression at A is like a HiRise being eroded at an angle exposing rooms throughout it. This might then be say 10 stories high with rooms. B appears to show large bridges suspended in the air at 9 o'clock going down to more rooms collapsed at 6 o'clock. At 1 o'clock it may be a ceiling is still on these rooms. Areas like this might be idea for colonization, if these walled areas can be pressurized. Many of these holes in the walls might be at least patched much more easily than building new habitats. C at 6 and D at 8 o'clock also appear to show three dimensional buildings. C at 9 o'clock may be collapsed ceiling material and at 12 o'clock more large tubes. E also appears three dimensional with buildings. Between F and G there are many more walls and hollow three dimensional areas.

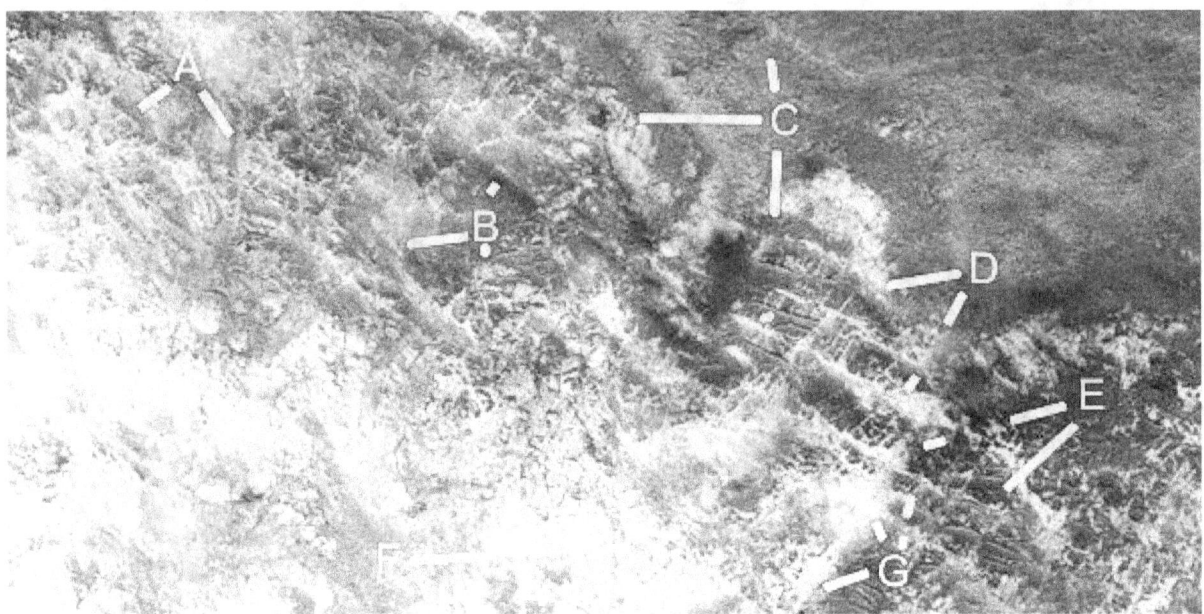

Cymhh212c

Hypothesis

This again appears three dimensional, like a war zone where buildings have collapsed in part. Though a war is unlikely, erosion could have a similar effect. A shows floors with dark lines where the walls probably were. B appears to be a much higher area with the shadows showing how high the walls are. Engineers could determine the dimensions of all of this by the sun angle. C is similar but with the floor material still largely intact. D at 8 o'clock might also be floor material, at 10 o'clock it may be an intact part of the hollow hill.

Cymhh215c

Hypothesis

This shows another tube section that connects to the crater, A at 11 o'clock is there the tube terminates or is broken off. At 1 o'clock there is a groove or platform under the tube. B shows floor like material similar to in the hollow hills. C shows how the tube connects directly to the crater, even though a crater should occur randomly. C at 3 o'clock and D at 7 o'clock show how the tube attaches smoothly to the crater, at 2 and 10 o'clock there is more floor material.

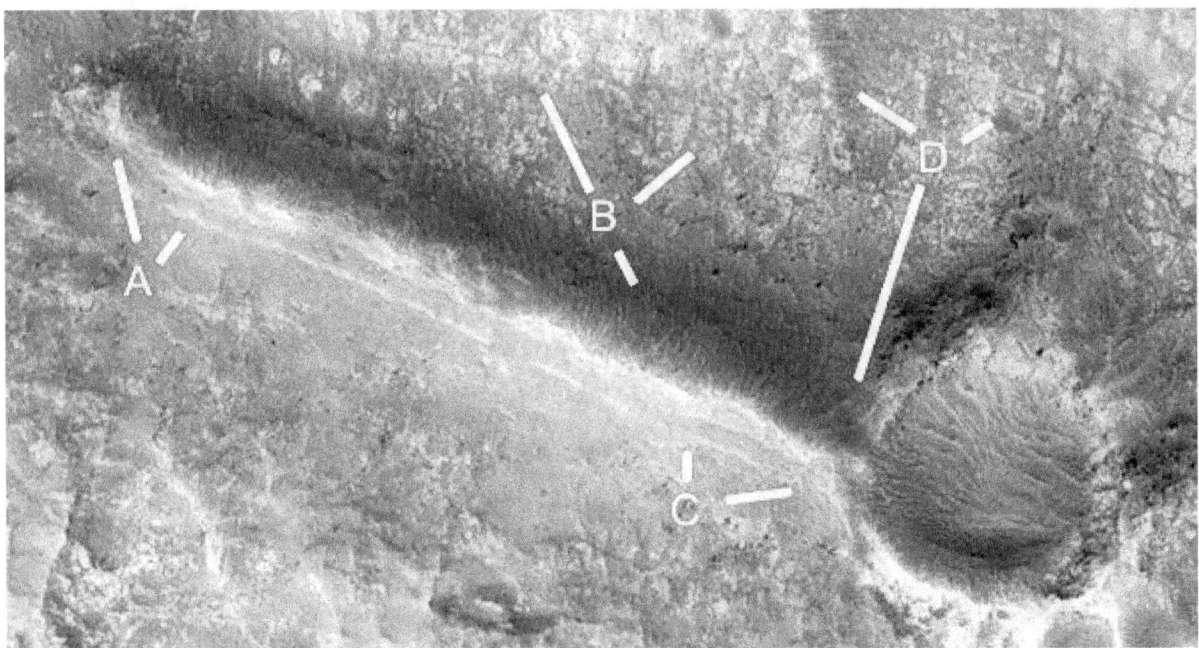

Cymhh216b

Hypothesis

These walls at A may have partially collapsed as they look more broken along them rather than just being joined at corners in irregular shapes. B looks like a dome with the center still intact and shadows of other walls inside it. Between here and E there may be a ceiling largely intact and rooms inside. C may have many floors of rooms in it as this goes up between E and G. Over to F there are many right angles though some walls may still be covered by ceiling material. E may be a tube going to other parts of the formation, it appears to become an edge of the formation under G. D at 10 and 3 o'clock looks like a tube or road, at 5 o'clock these walls are in irregular shapes. G shows an area where the walls may have eroded away. H shows the edge of the pit.

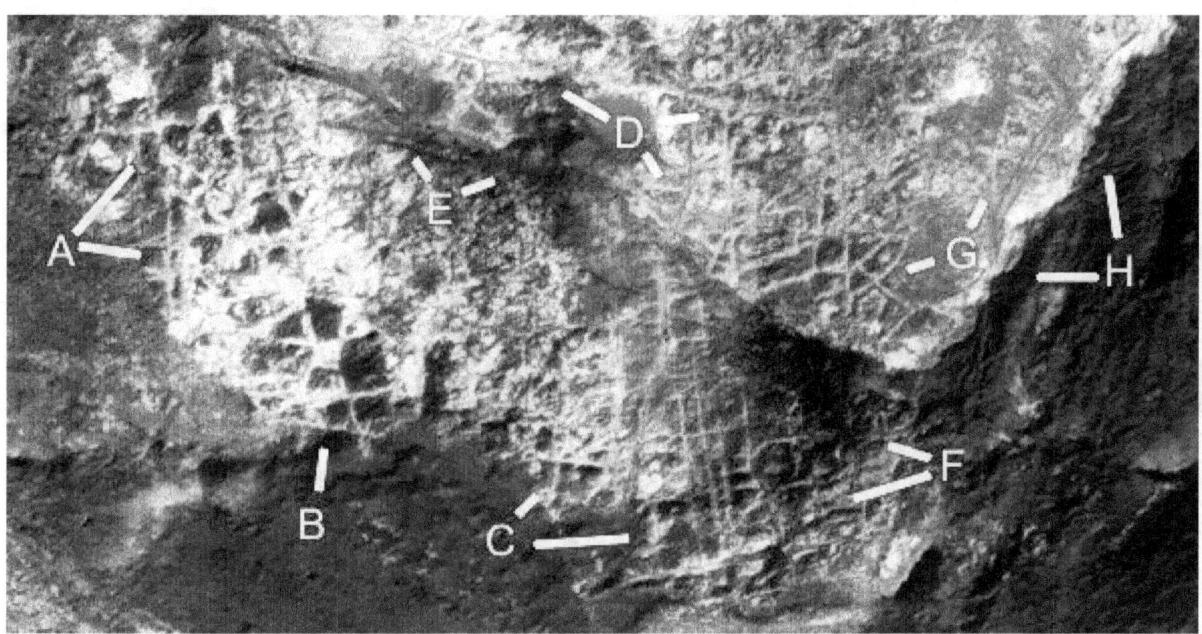

Cymhh217c

Hypothesis

This shows more of the dam shapes, they appear to be three dimensional with shadows and some have become eroded at A and B having collapsed in part.

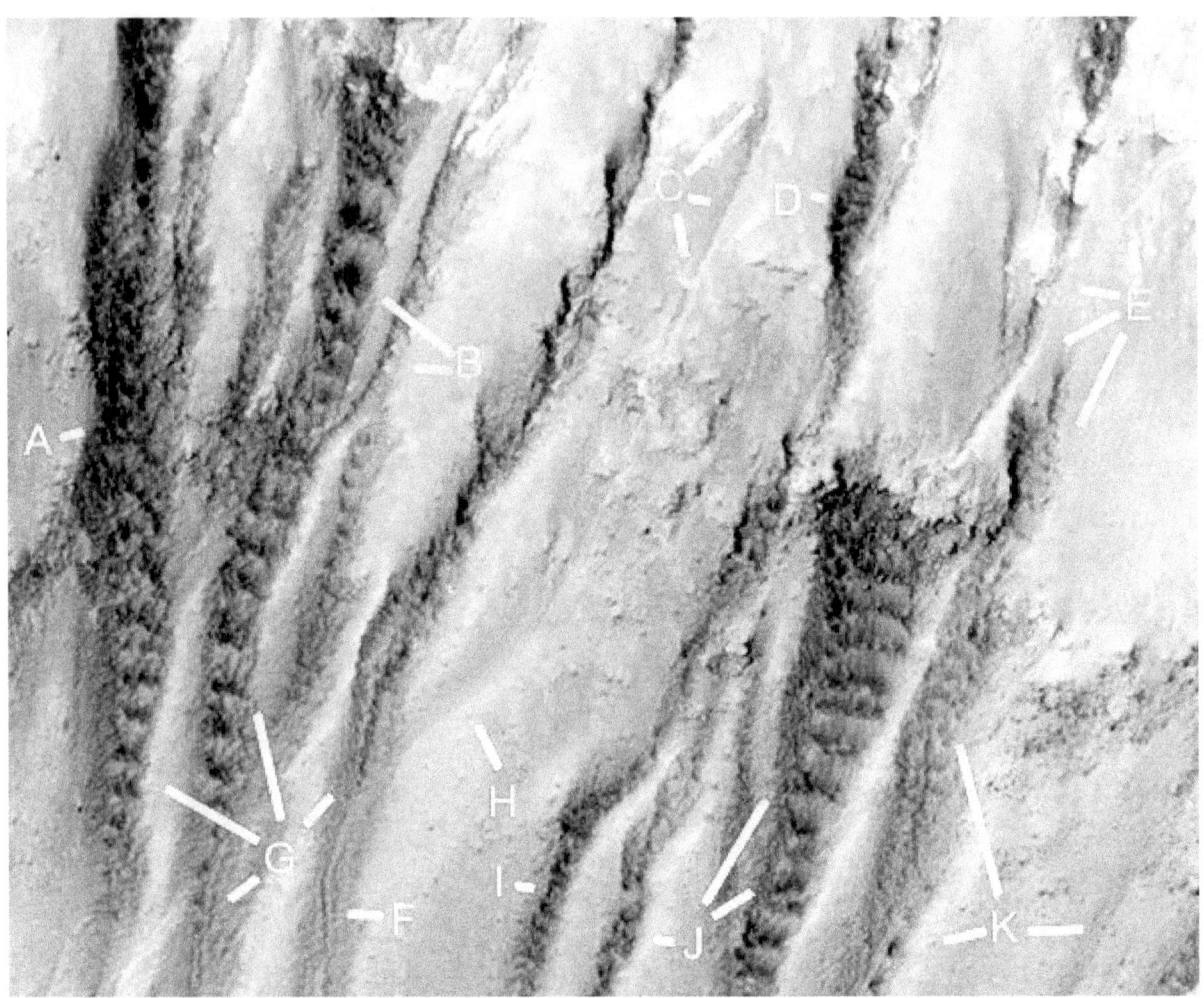

Cymhh217d

Hypothesis

This hill appears to be more like an irregular foam of walls and rooms, engineers might be able to model this to show how strong it would be. A and B show where the ceiling is still intact with some hollows, at C the ceiling is showing the walls as if it is wearing thin. D shows more rooms becoming visible while others retain their ceiling. At E this ceiling has largely collapsed, the rooms are dark which may be shadows or a dark floor. At 2 o'clock there may be a tube which connects this area to one with an intact ceiling. These rooms then might have been suspended in the air in clusters connected by bridges rather than as a HiRise. F also shows longer walls or tubes connecting smaller rooms, the wall at 2 o'clock appears to connect to a less eroded area. G also shows tubes or bridges, the rooms appear to be lower down like in a foam. H shows some hollows developing.

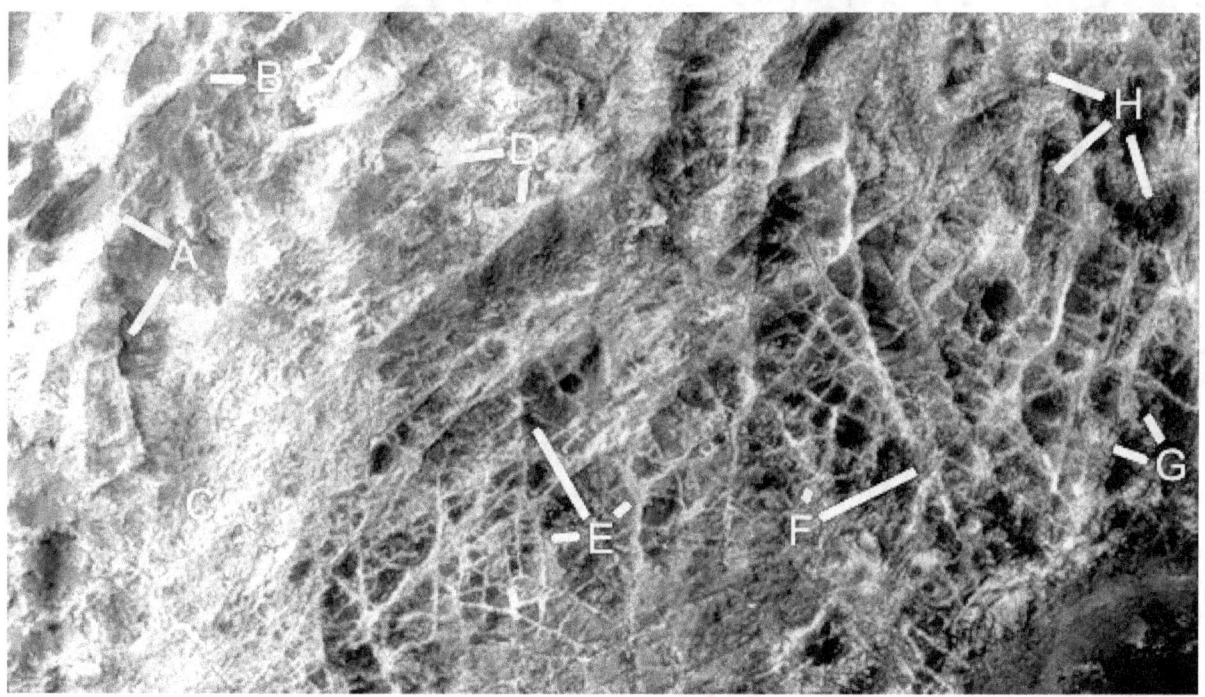

Cymd240c

Hypothesis

This dam also shows signs of collapse at A with cracks along the wall, B may be from creep or cold flow of the dam materials. It may also be mud that accumulated in the dam, engineers might be able to tell the difference by modeling this. By looking at all the differences in these degraded dams with three dimensional modeling the construction techniques might be worked out. C shows another part of the dam wall, intact with the same cracks along the top. D may show a parabolic arch to support this wall, E shows where the smooth part of the dam floor meets the rougher ground above it.

Cymd240c2

Hypothesis

Two parabolas are shown.

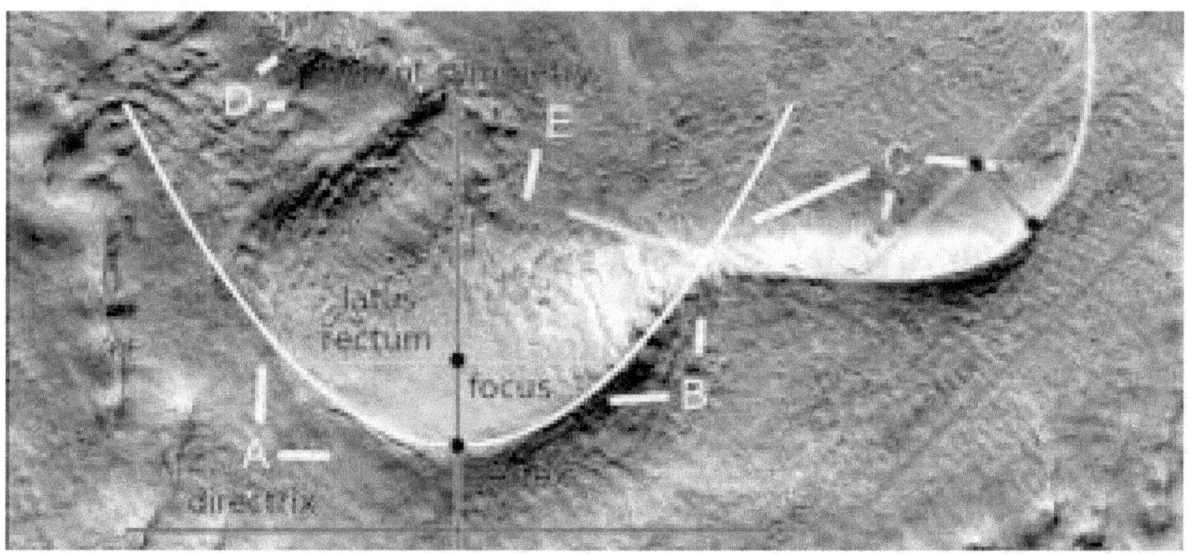

Cymd252a

Hypothesis

A shows a pair of ridges as if the smoother dam wall has broken off. A at 4 o'clock there is a hollow like a parabola. B shows layers in each arch, in other dams this may have had a cement covering. C also shows this step like layers as does G. E and F shows may other hollows where other dams are usually smooth.

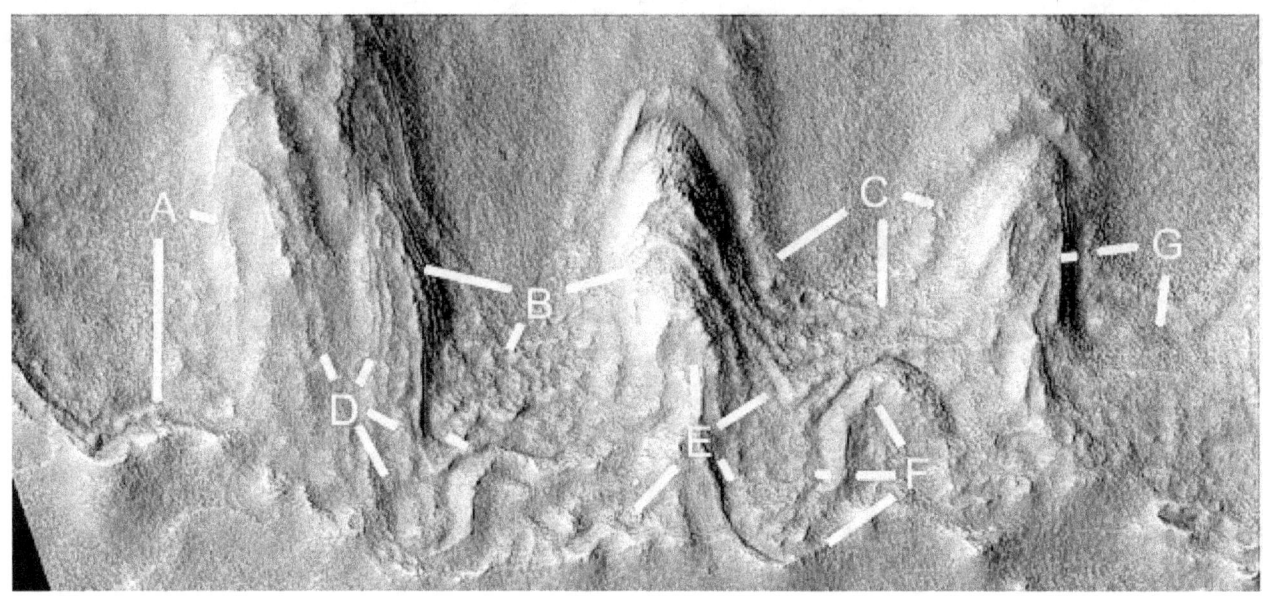

Cymd252b

Hypothesis

A and B show a parabolic arch shape.

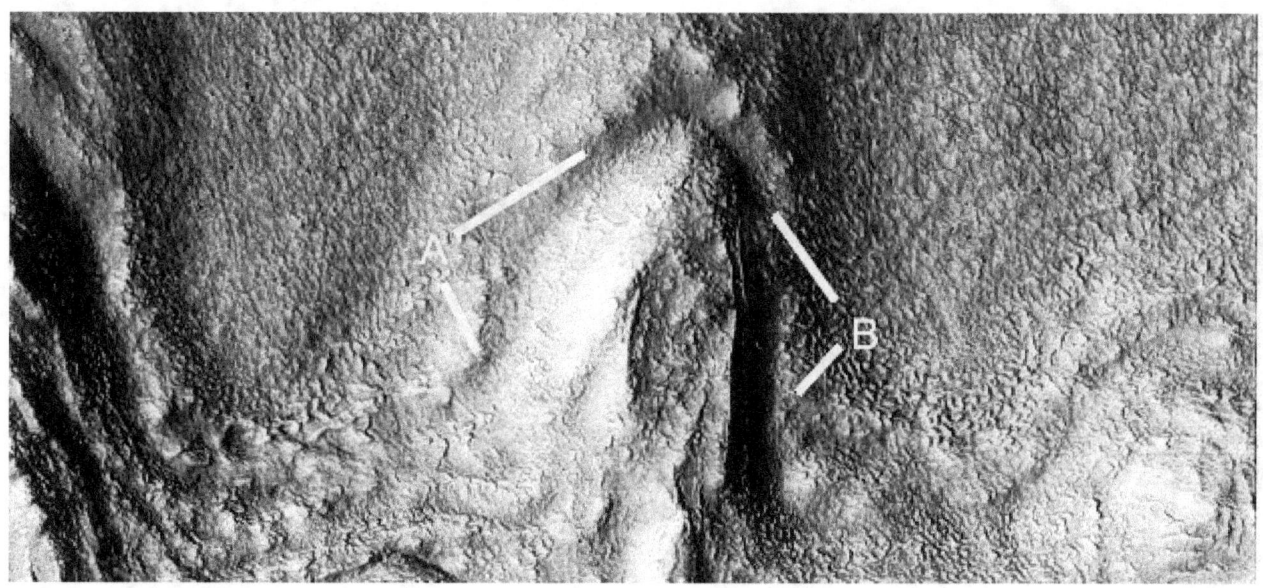

Cymd252b2

Hypothesis

This shows a parabola overlaid onto the formation.

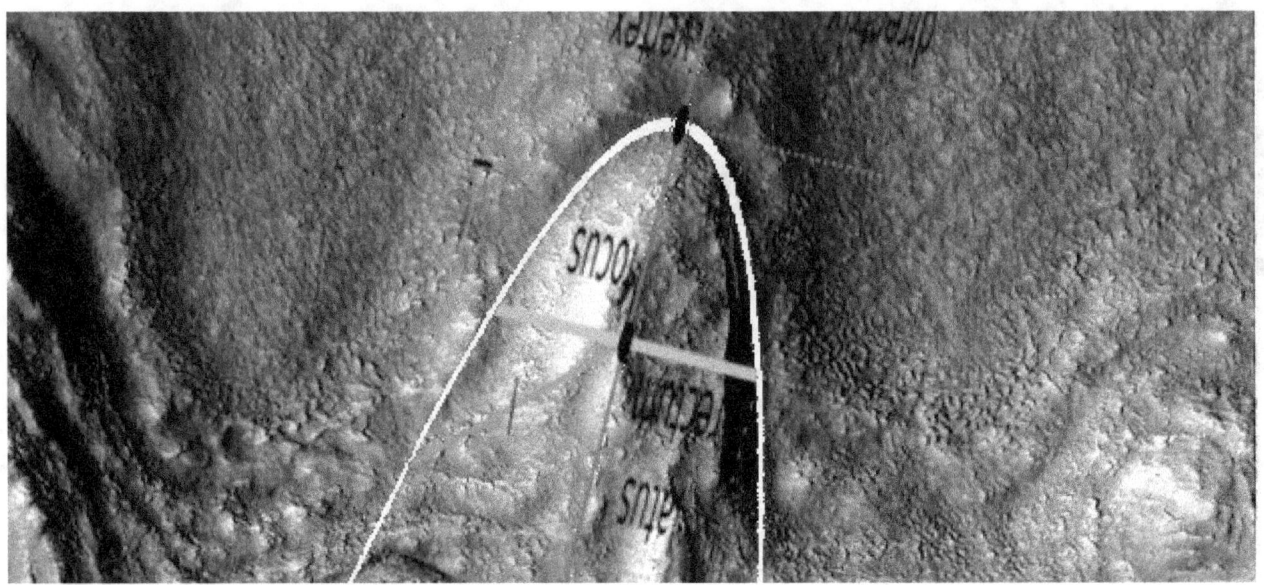

Cymd252c

Hypothesis

A and C show the shape of the parabola, B shows a groove along the inside of this.

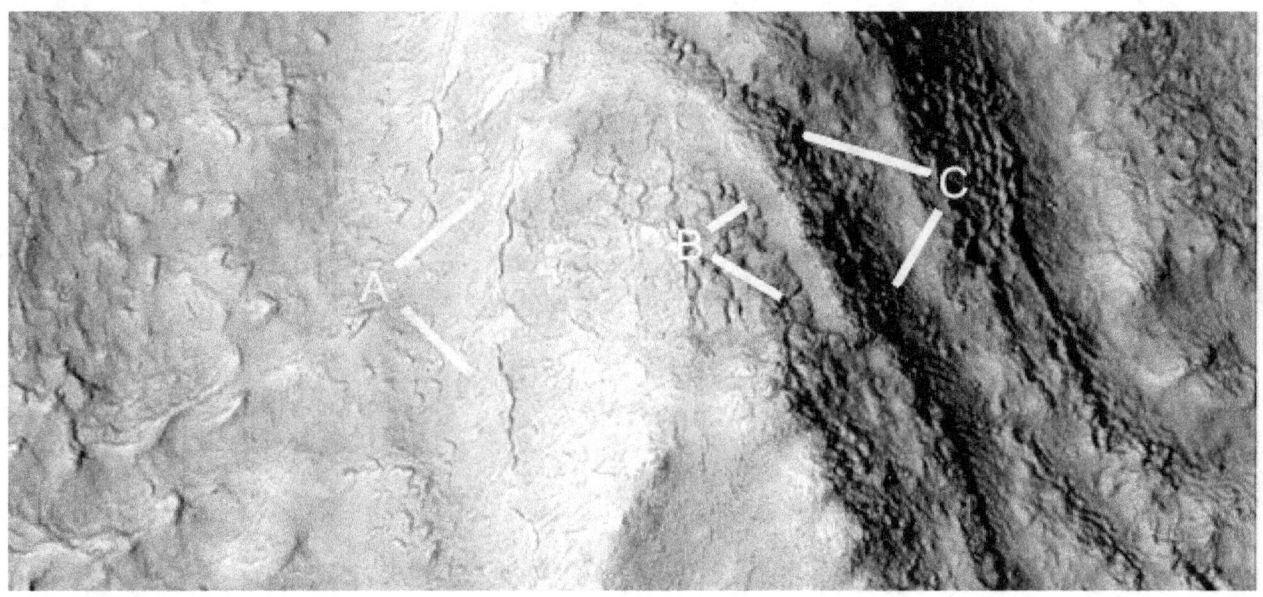

Cymd252c2

Hypothesis

This shows a parabola superimposed on the formation.

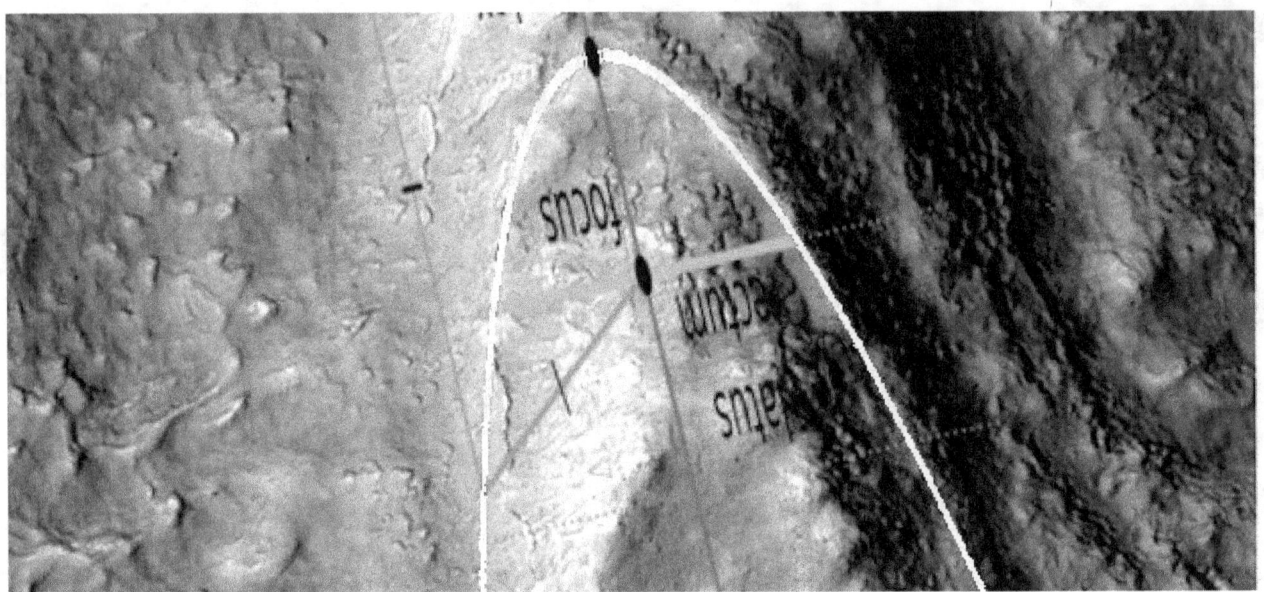

Cymd252d

Hypothesis

All of these could be partial parabolas, A could be extended upward into the full parabola shape, this section is in good condition. B may be three parabolas, C on the outside shows how smooth the lower parts of the wall is. E shows a cavity like the dam wall has eroded away leaving this groove in a parabolic shape. D shows a parabolic shape as well also with a cavity. F shows a more intact part of the dam, engineers might consider how the dams appear to break on the upper walls first moving down to the apex of the parabola.

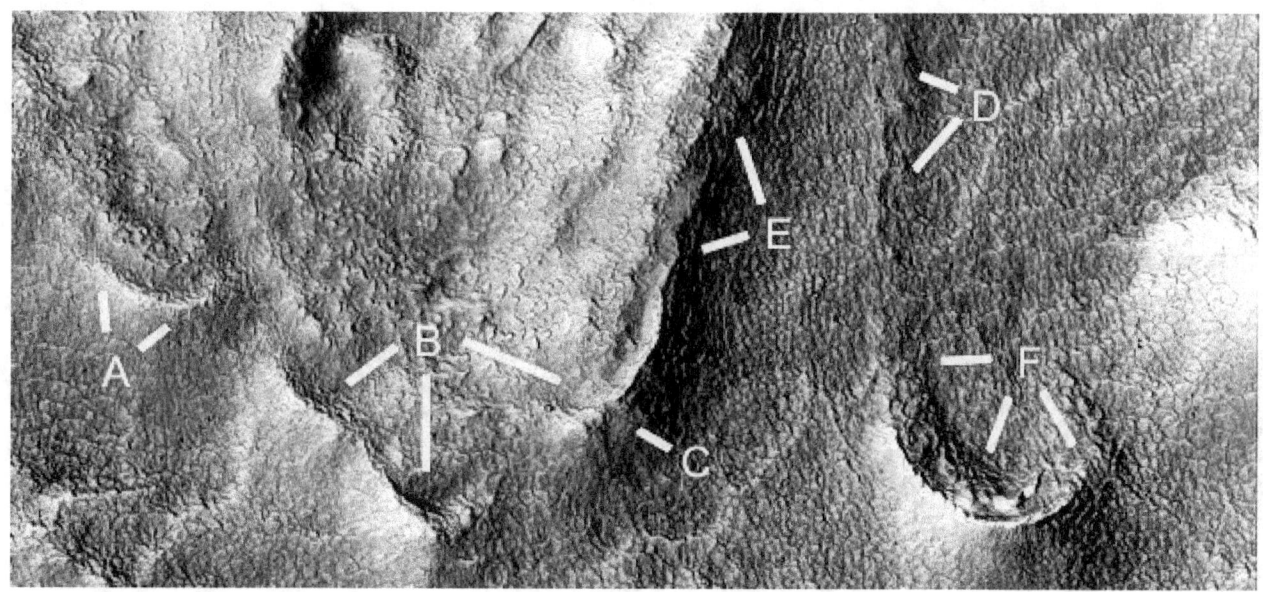

Cymd252d2

Hypothesis

This shows a parabola superimposed on the right dam.

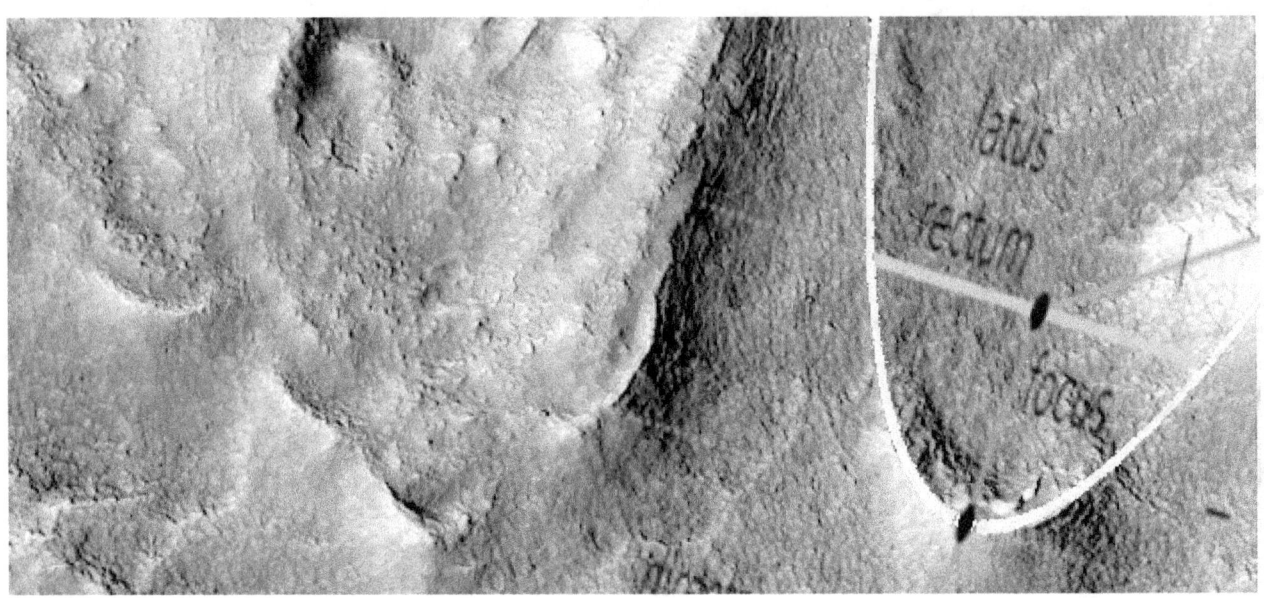

Cymd253

Hypothesis

More examples of excavated dams, here each appears to have an inner and outer wall because the walls have broken off leaving a dam shaped cavity in each case. D at 7 and 8 o'clock appears to be a broken dam and arch.

Cymd253b

Hypothesis

A shows how the arch has an inner and outer wall shape of constant thickness. It looks then like the wall broke off from erosion. B also shows a cavity, these may indicate how deeply set the dam walls are when they are connected to the dam wall as a construction technique.

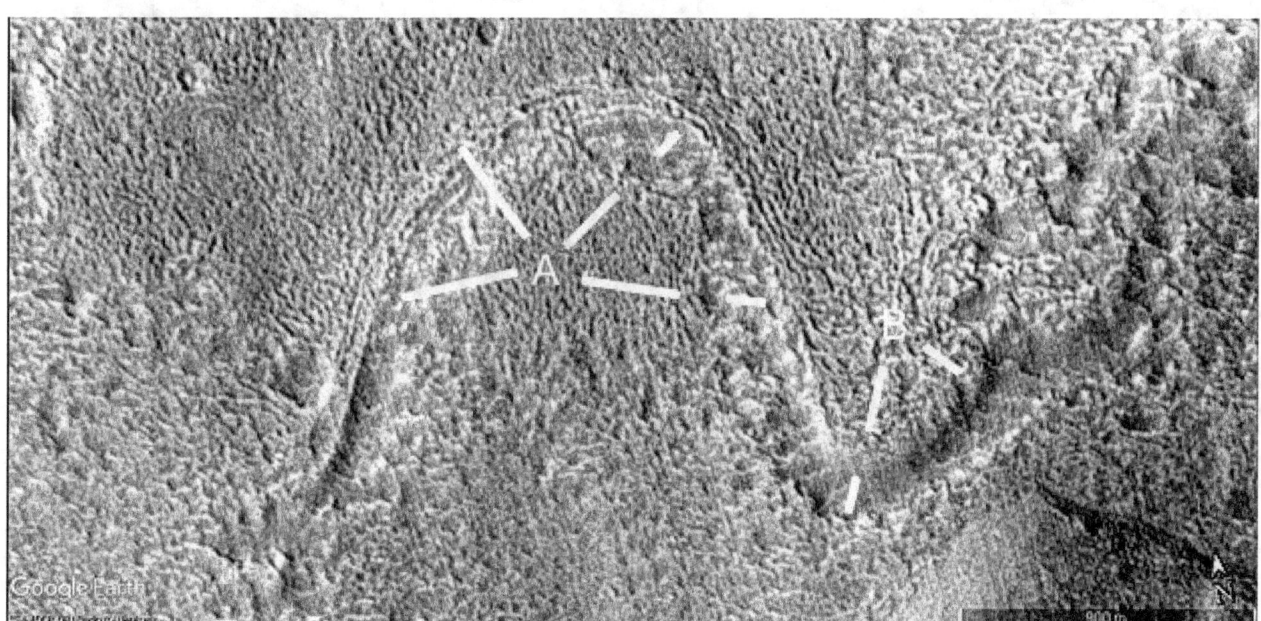

Cymd253b2

Hypothesis

Three parabolas are shown.

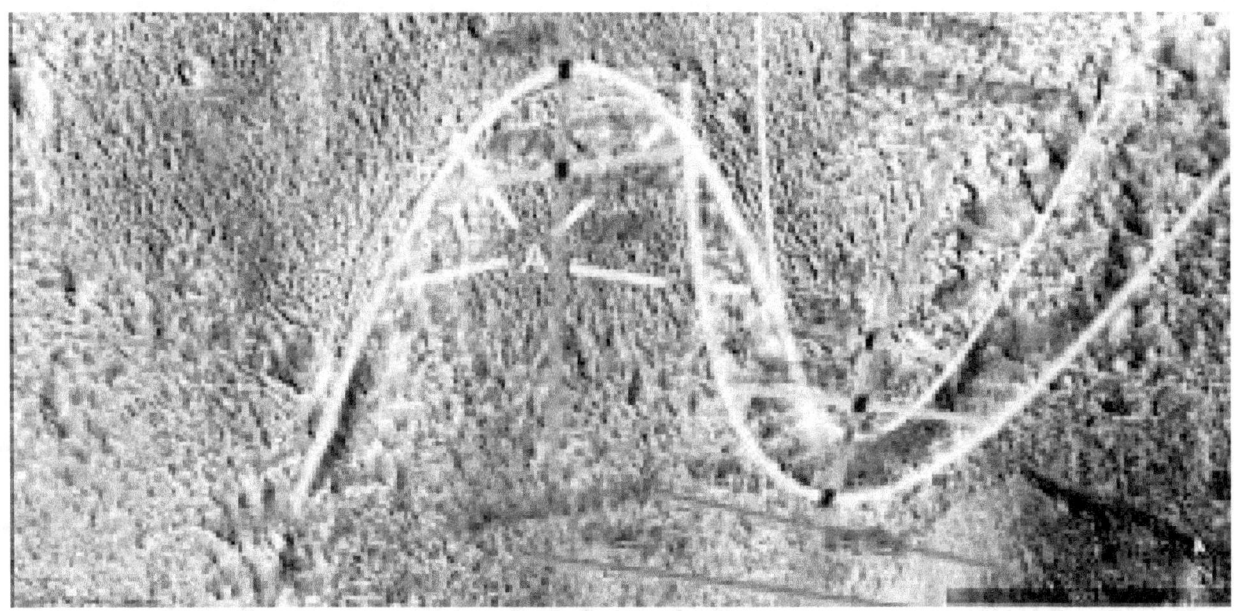

Cymd253c

Hypothesis

A and B show the inner and outer sections of the parabolic dam. C shows the smooth lower wall of this dam, D shows how a large cavity has appeared in this similar to the damage between A and B. E shows the edges of this damage at 7 and 9 o'clock, at 2 o'clock it shows the edge of the dam and the second line shows a ridge similar to a dam wall.

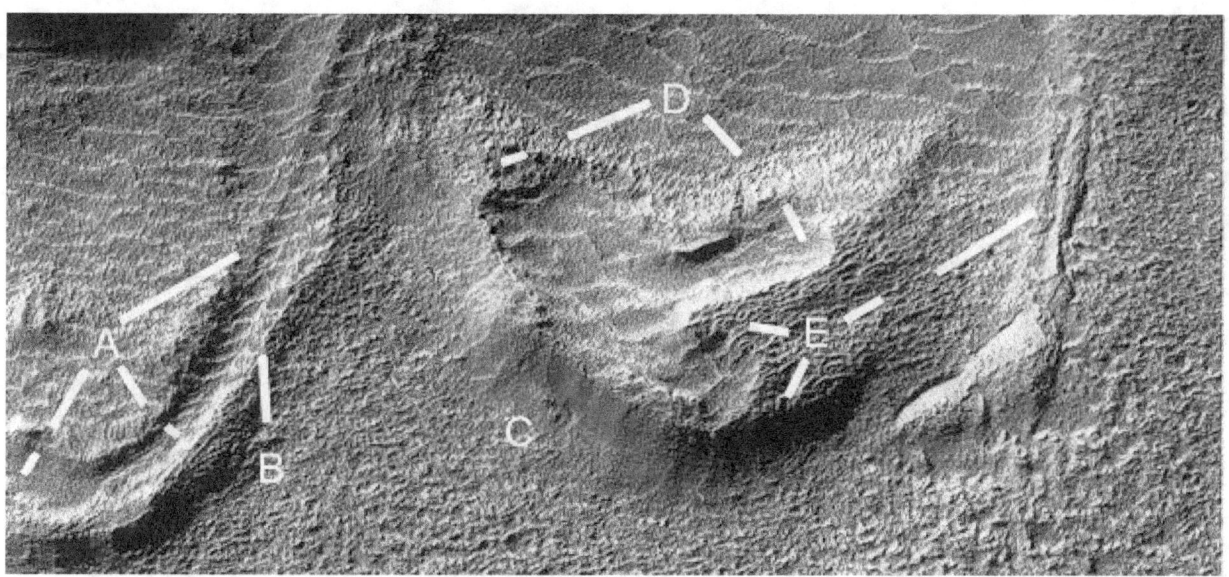

Cymd253c2

Hypothesis

This shows the inner and outer parabolas of the dam, the cavity between them implies there was a dam wall here that broke off.

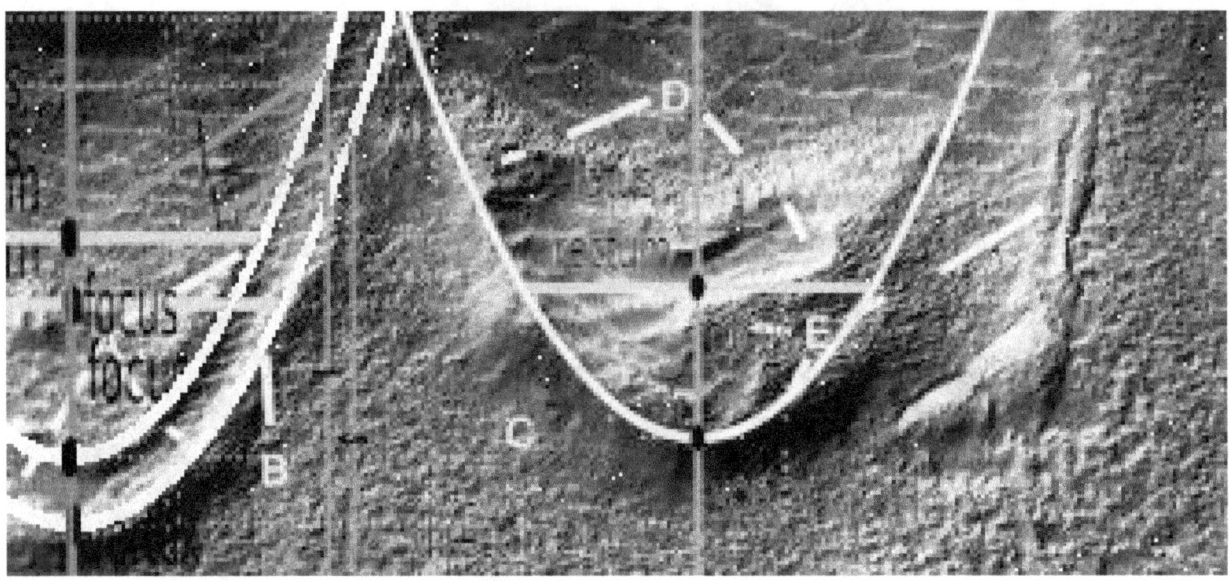

Cymd253d

Hypothesis

This shows another dam wall broken off, inside the cavity there are signs of layers also above B. These may be where the cement flaked off, they do not appear to the left outside the dam. A and B shows the inner and outer wall edges of this cavity, they have a parabolic shape. C at 9 o'clock shows the edge of this wall as if a layer has flaked off as does D.

Cymd253d2

Hypothesis

This shows an inner and outer parabola superimposed on this cavity, it is quite a close fit.

Cymd259b

Hypothesis

A shows the edge of the pit dam, highly eroded. The main evidence here is that water flowed along it, B also appears to have been a smooth dam floor. As will be shown the same crater has dams as well.

Cymd259b2

Hypothesis

A parabola is shown.

Cymd259c

Hypothesis

These dams are in the same crater, A which appears parabolic and B have smooth walls with a few cracks as shown. B at 4 o'clock has a sharp edge to the dam wall in good condition. C at 4 and 6 o'clock show a secondary dam perhaps to catch the overflow, the second line at 6 o'clock shows the base of this wall. D shows another section, perhaps parabolic, with a cracked wall at 5 o'clock. C at 10 o'clock shows a probable parabolic arch. There appear to be faint vertical ridges on the upper part of the dam walls as seen in other dams, these may be for strengthening the wall such as there being pillars inside.

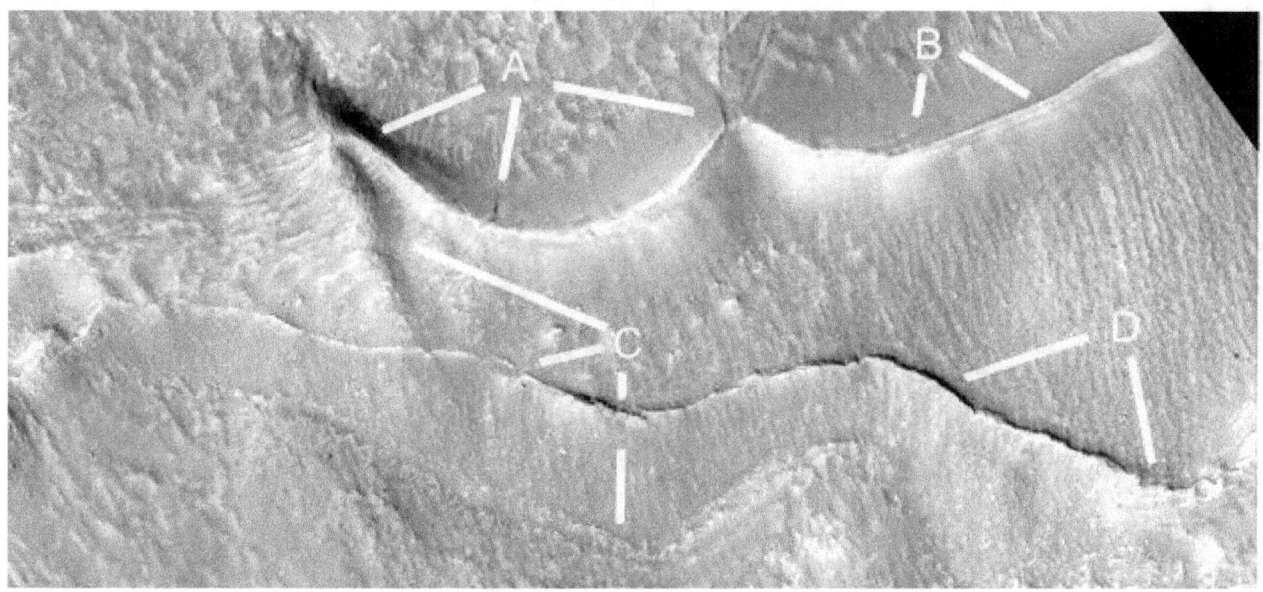

Cymd259c2

Hypothesis

A parabola is shown.

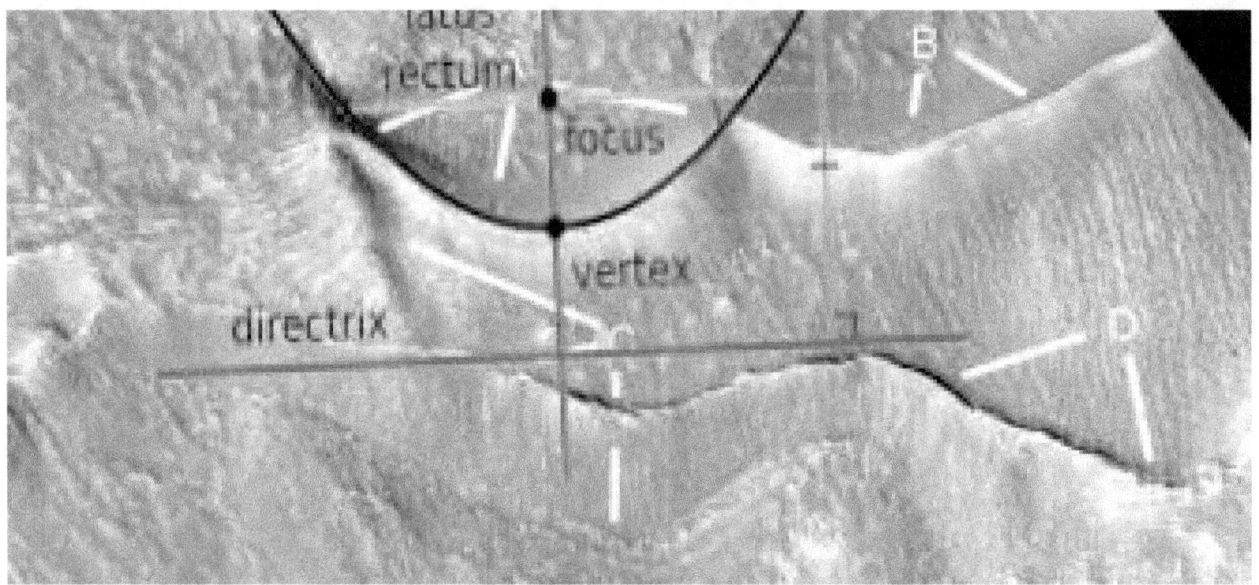

Cymd259b2

Hypothesis

A at 5 o'clock shows another dam, the floor is much rougher here perhaps as the cement has eroded away. Engineers might look at this pattern of how the floor might be prepared before the cement covering. A at 2 o'clock shows a smooth cement layer extending over to B, C shows cracks appear on the edge of the dam and under it.

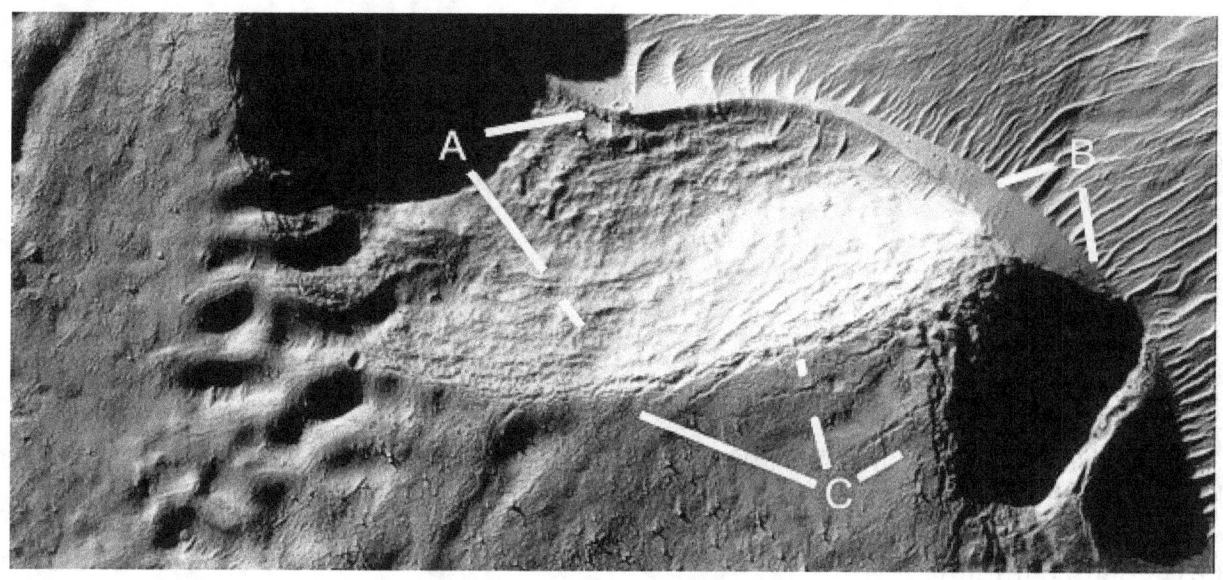

Cymd259b2a

Hypothesis

A parabola is shown.

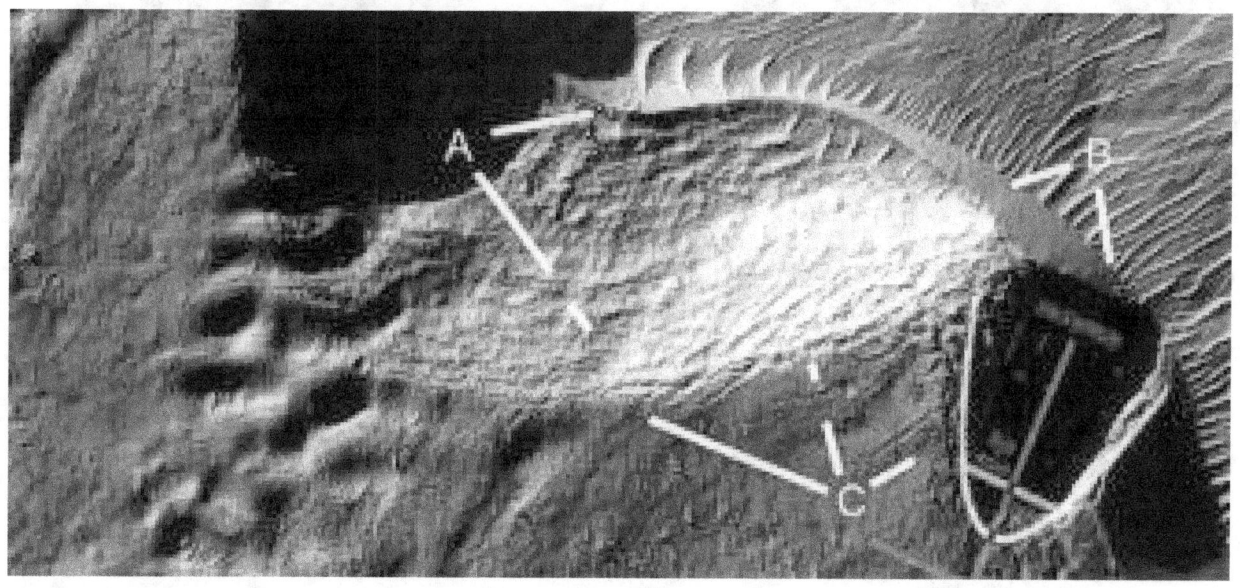

Cymd252c2

Hypothesis

A at 8 o'clock shows some cracks in the dam wall, it is smooth at 5 o'clock with some cracks on the edge. B is also smooth, in both cases it connects to the crater wall even though they are different materials. It appears then to be adhering to the wall which is hard to explain naturally.

Cymd252c2a

Hypothesis

An inner and outer parabola is shown.

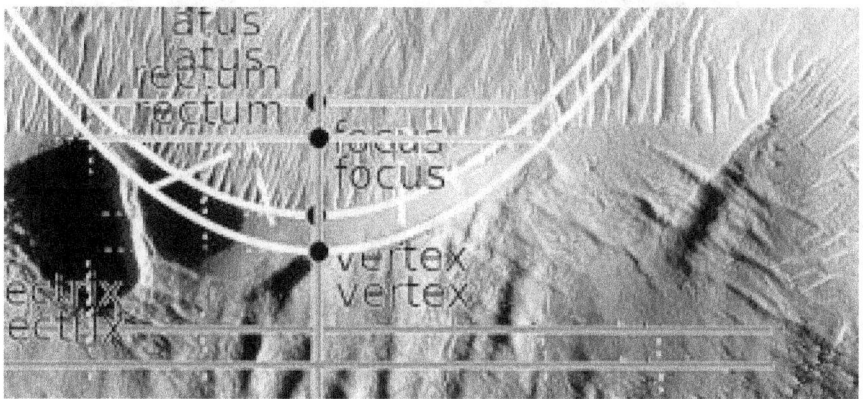

Cymd260a

Hypothesis

A shows parallel grooves on the edge of the dam at 7 and 11 o'clock, at 2 o'clock the dam wall is cracking. Engineers might examine this to see what materials were used, and what would crack this way with the extremes of temperature. B shows vertical grooves under the lip of the dam, in others these are seen as faint lines. It might then indicate what the construction technique is, forming pillars or struts like reinforcing then surrounding it with cement. C shows where the lip of the dam wall is breaking, in some cases it is missing. D shows another dam wall with cracks.

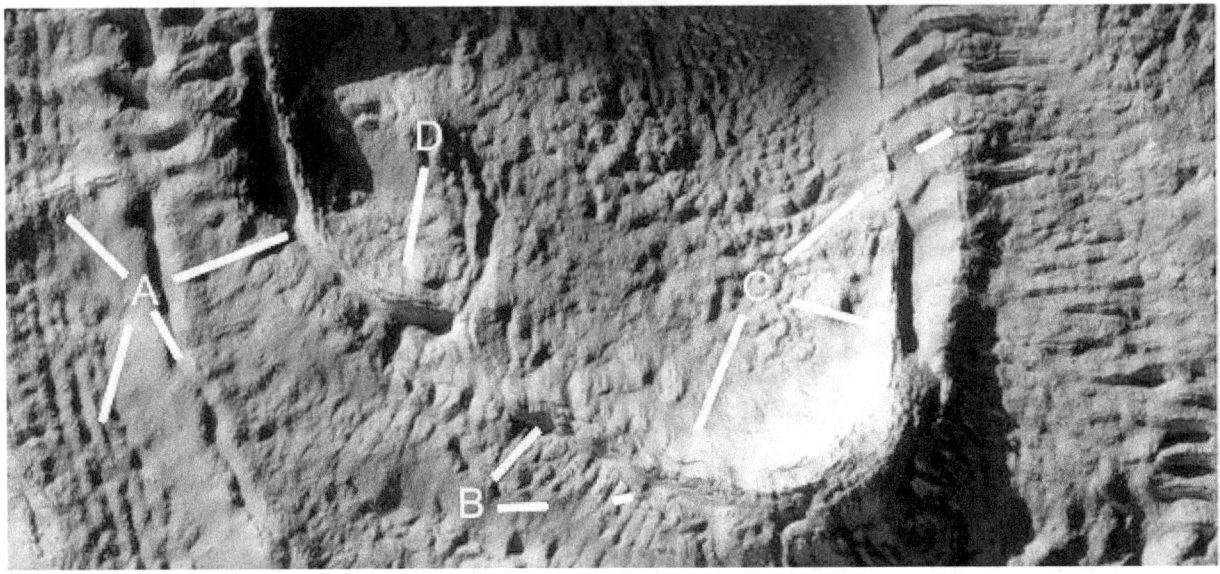

Cymd260a2

Two parabolas are shown, merging together on the left.

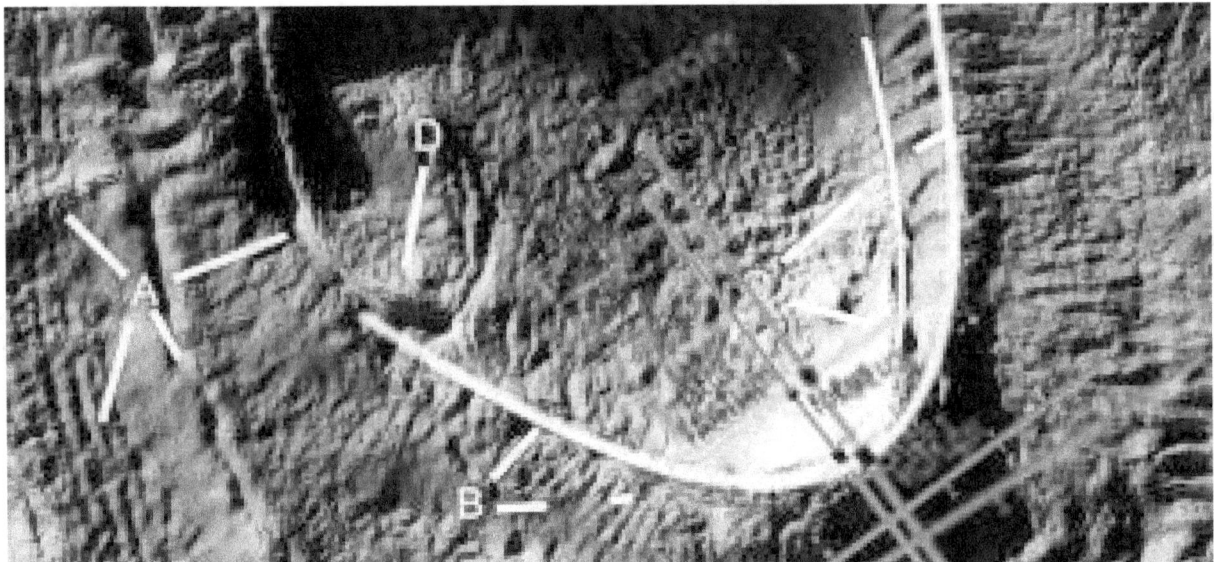

Cymd260c

Hypothesis

A shows at 11 o'clock how the wall is cracking long it, at 6 and 7 o'clock there is an outer skin that has broken off. At 4 o'clock there is a broken layer on the inside. B at 11 and 1 o'clock shows how the wall is smooth below the break. At 3 and 10 o'clock there are pillars which may usually be covered with cement. C at 5 o'clock appears to show these pillars more covered in cement. At 1 o'clock the wall appears to be cracking again longitudinally as if it would break off. At 11 o'clock the dam wall may also be breaking. D at 10 o'clock shows cracks perhaps from creep, at 7 o'clock is the top of the arch supporting this area.

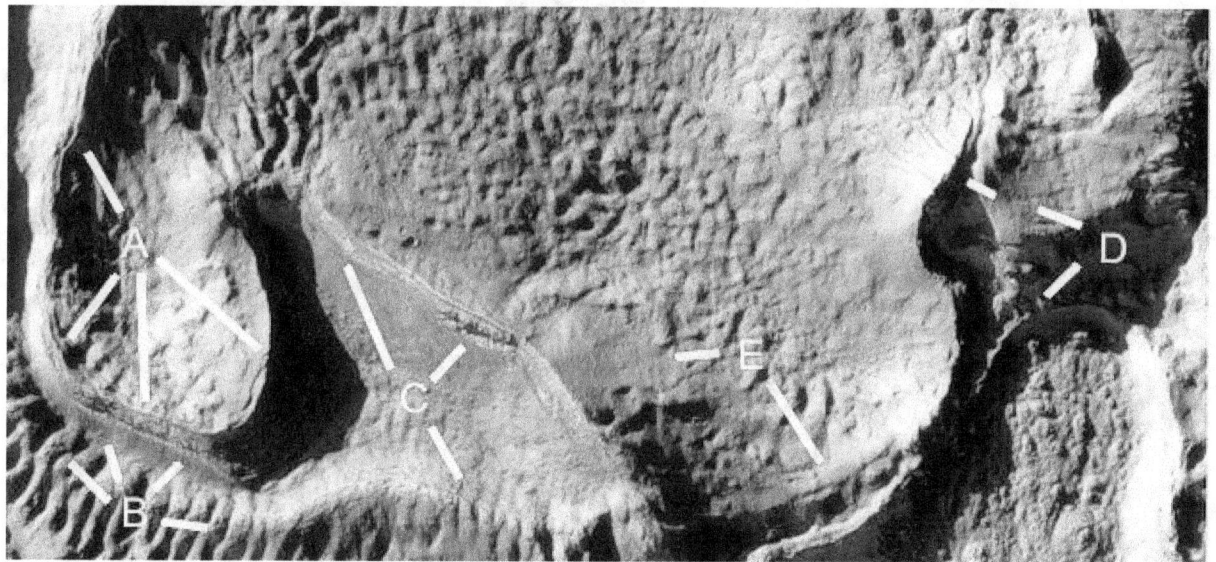

Cymd260c2

Hypothesis

Two parabolas are shown.

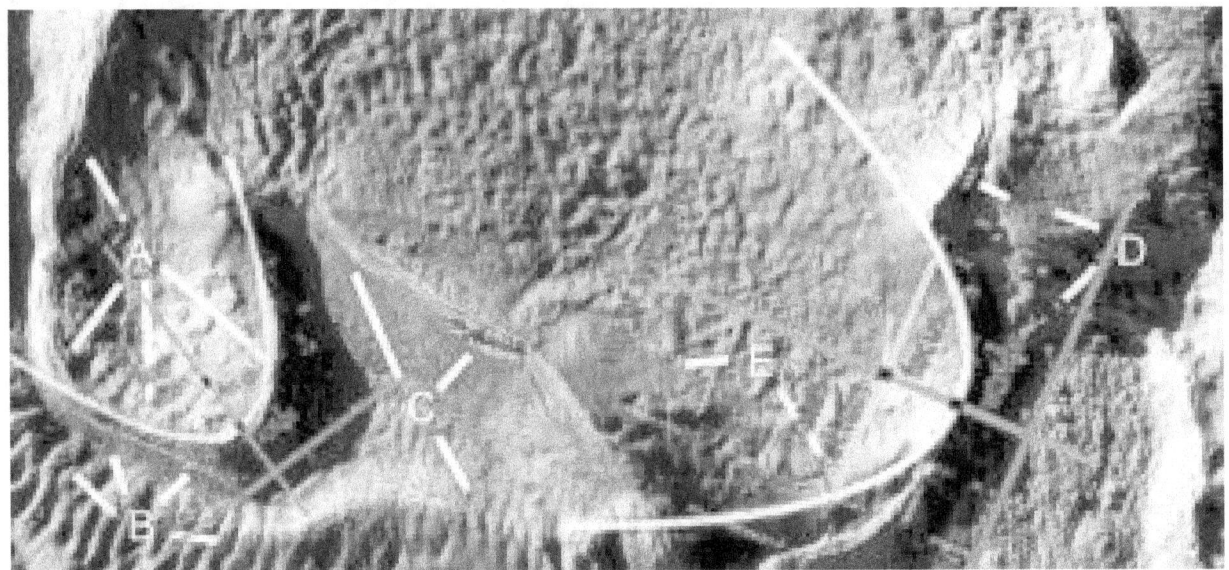

Cymd260e

Hypothesis

A shows a crack at 4 o'clock at a 45 degree angle, this runs into horizontal cracks across to B at 8 o'clock perhaps from creep. Engineers might consider whether the age of these materials can be determined from this creep. A at 4 o'clock on the second line is a small dam with a continuation of this crack, the dam is smooth at a 5 o'clock. B also shows a crack at 2 o'clock. C shows either cracks or horizontal pillars at 7 o'clock, either cracks or the thin cement over rocks is shown at 10 o'clock and 11 o'clock. D at 8 o'clock shows a wide dark layer in the wall, perhaps this indicates the wall is built up in two layers, the dam is in good condition at 7 o'clock. E shows the lip of the dam cracking at 4, 7 and 8 o'clock. At 10 o'clock there is again this dark longitudinal layer in the wall. F also shows cracks perhaps from creep.

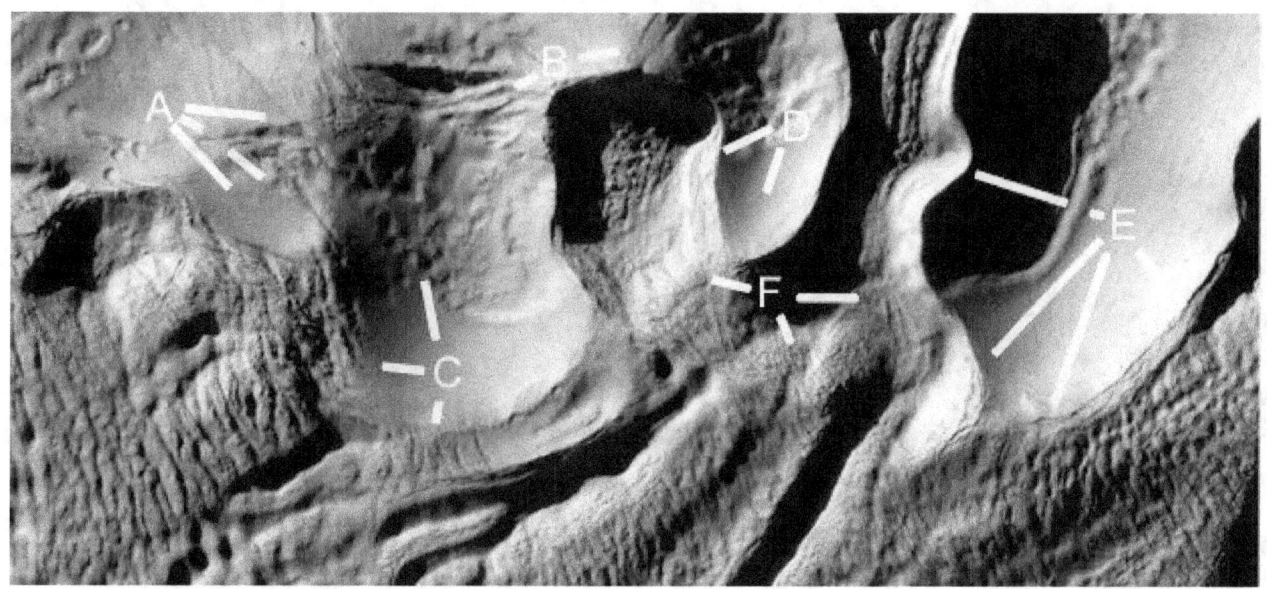

Cymd260e2

Hypothesis

Four parabolas are shown.

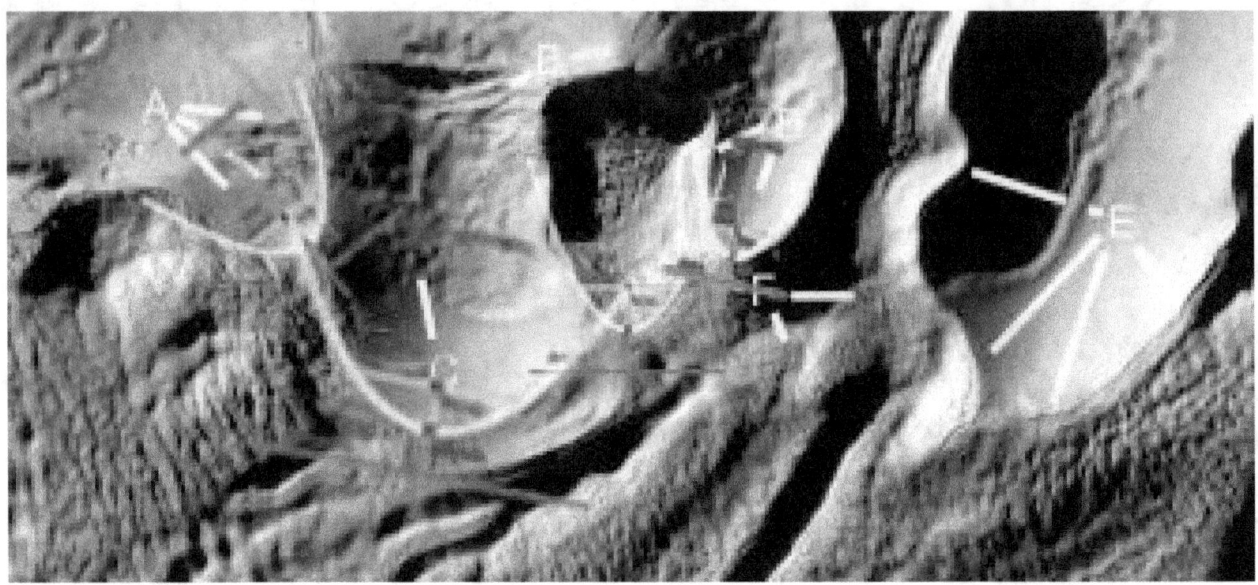

Cymd269b

Hypothesis

A shows a dam with the lip at 4 and 7 o'clock. B shows another dam edge at 1 o'clock, perhaps some cold flow caused these cracks. C shows some cracks on the dam walls, D has the rounded arch shape underneath the wall.

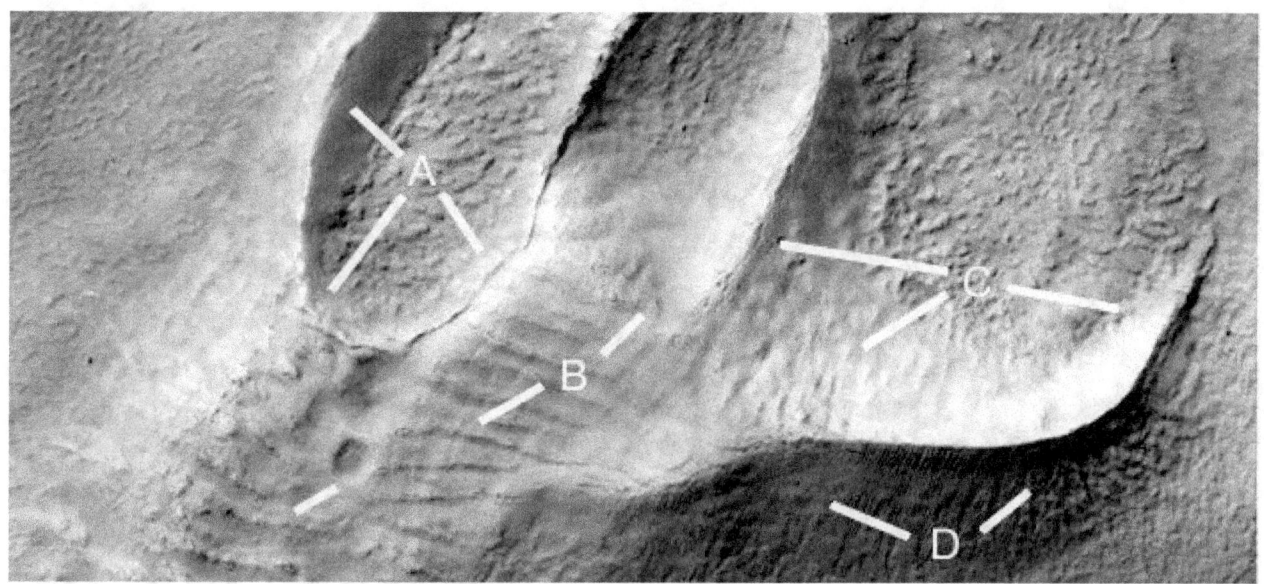

Cymd269b2

Hypothesis

Two parabolas are shown.

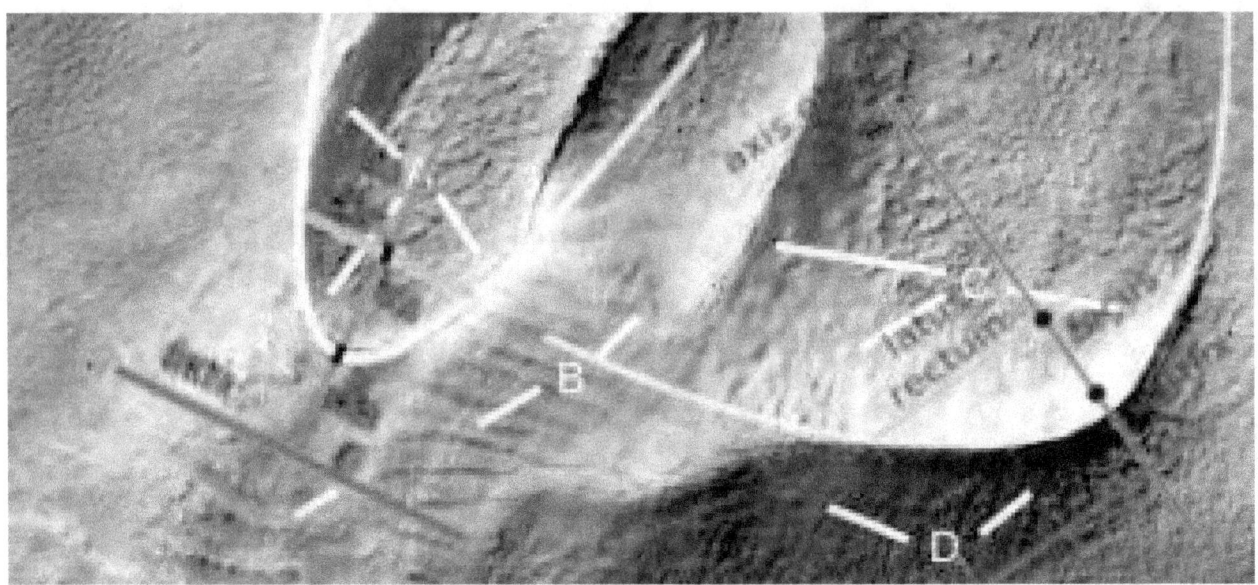

Cymd278a

Hypothesis

A and B are small dams, C would be an arch to support the wall above it. At 9 and 10 o'clock there is a water channel going down to B, it appears to zig zag to reduce the speed of the water before going into the B dam. D appears to be a larger dam, three dimensional modeling would work out how the water flowed in these. E shows the typical transverse ridges around the dam wall.

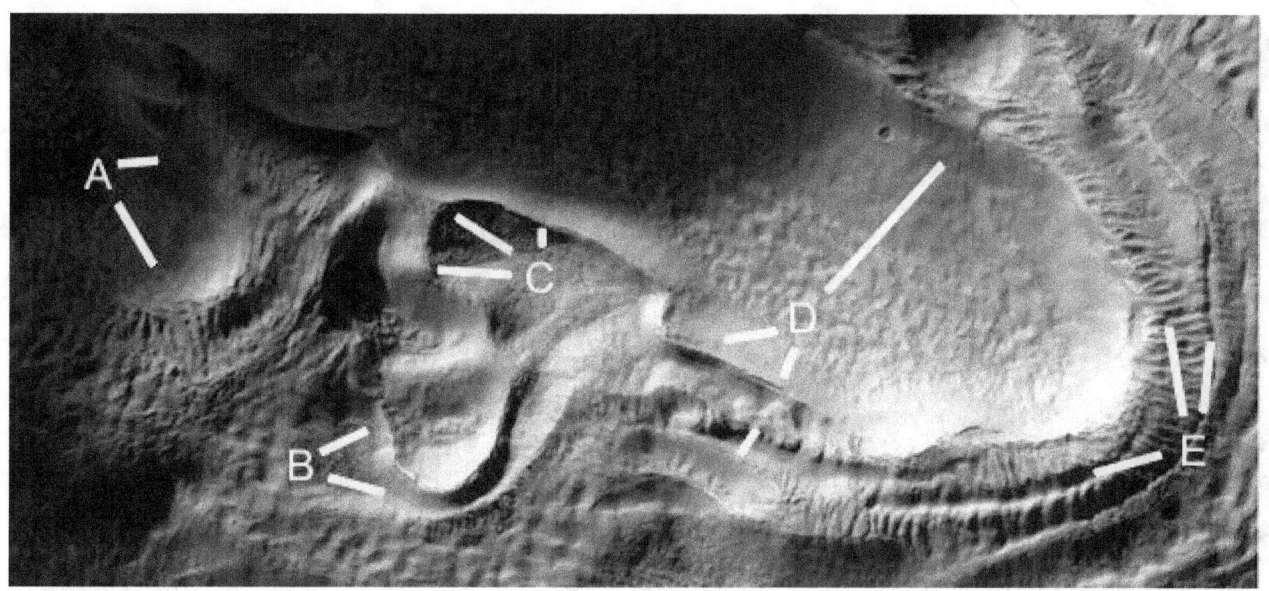

Cymd278a2

Hypothesis

Five parabolas are shown.

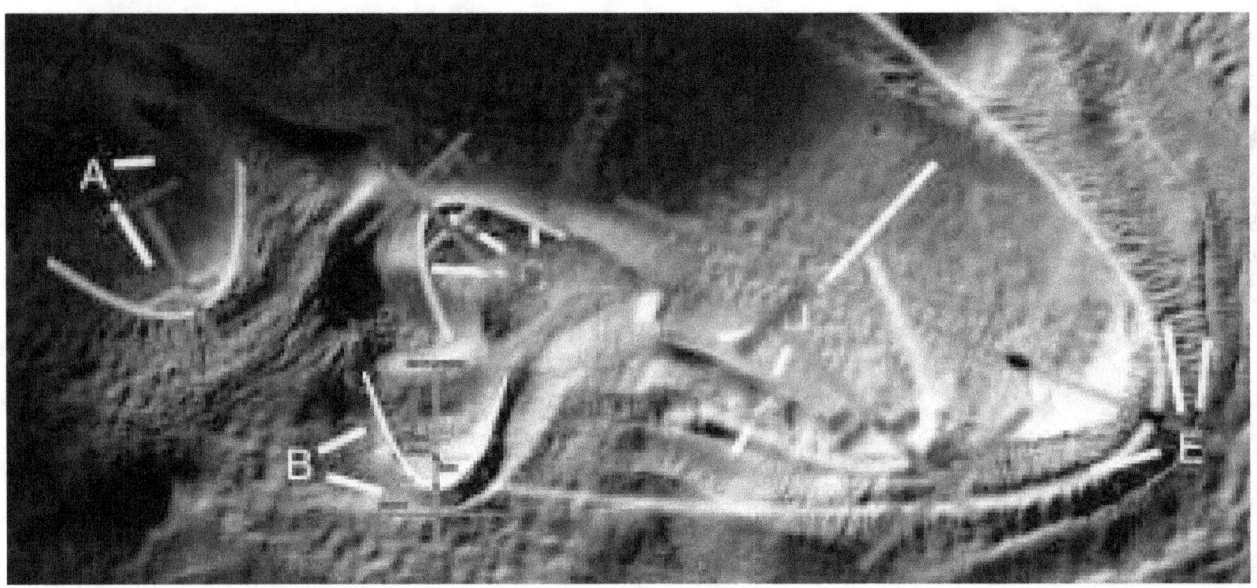

Cymd278e

Hypothesis

A is a small dam, as is B at 5 o'clock. At 3 o'clock may be another small dam, three dimensional modeling would determine if water would collect here. C appears to be a water channel for lower dams.

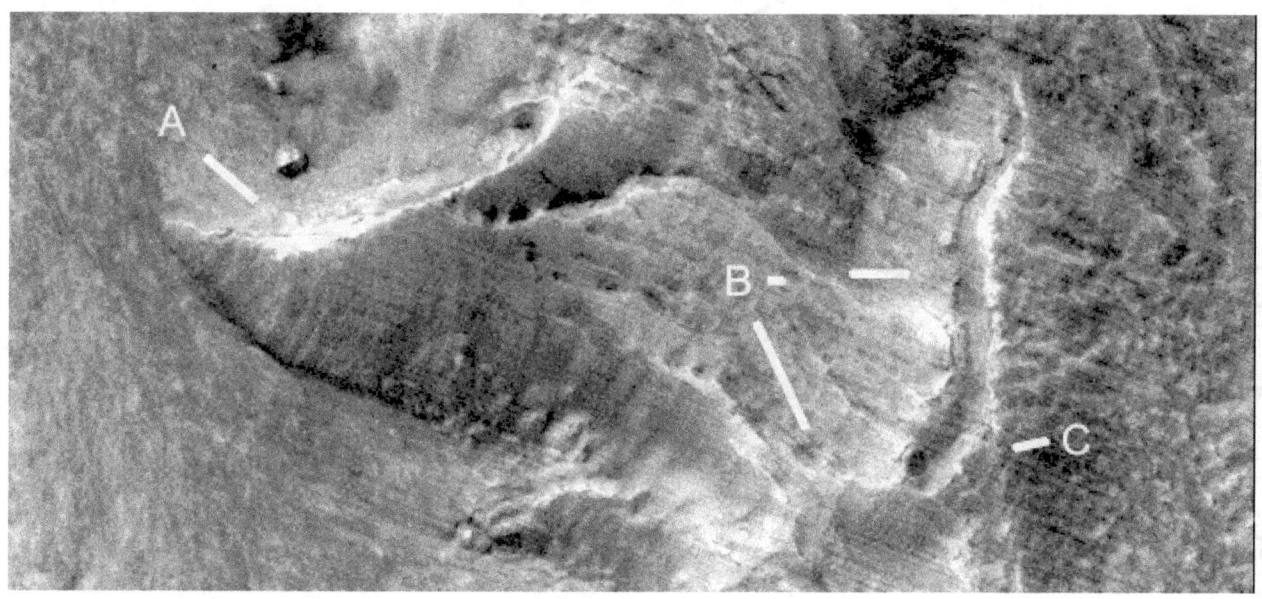

Cymd278e2

Hypothesis

Two parabolas are shown.

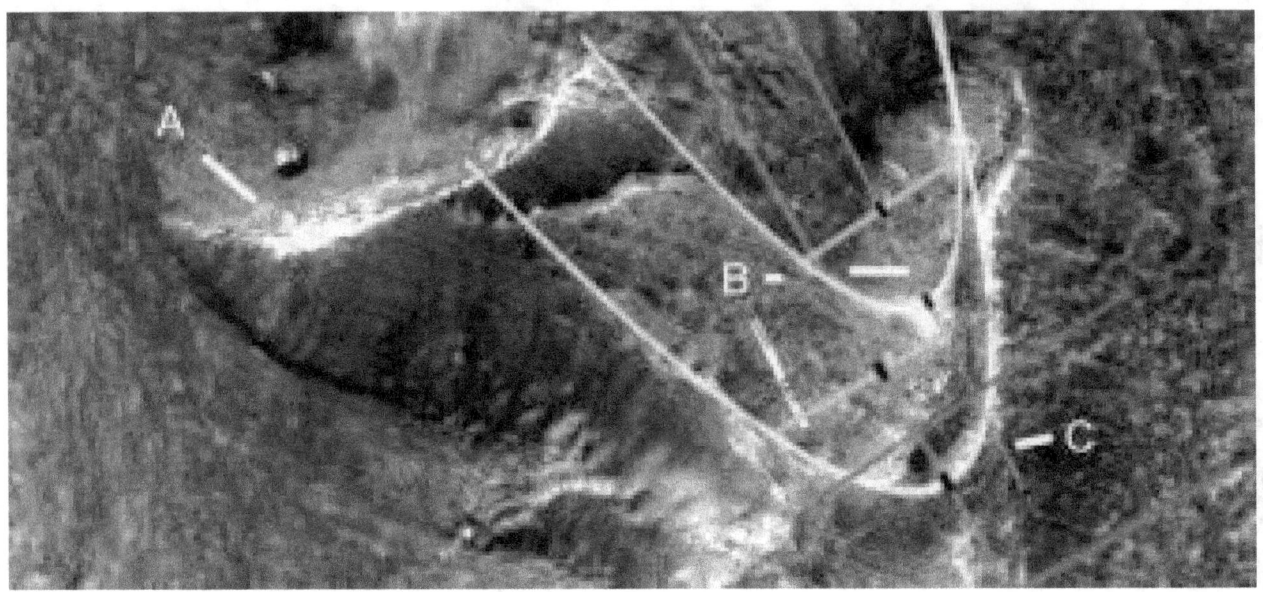

Cymd279

Hypothesis

A and B show double parabolas, like the dam wall has broken off leaving a hollow.

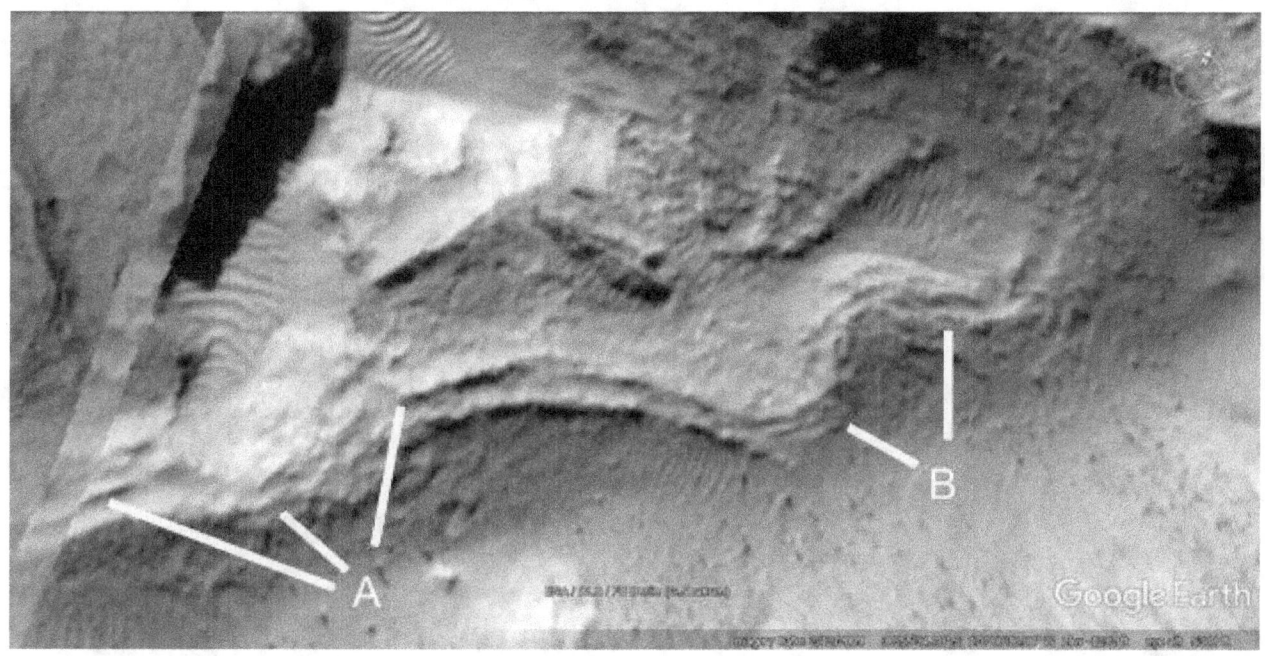

Cymd279a

Hypothesis

Six parabolas are shown.

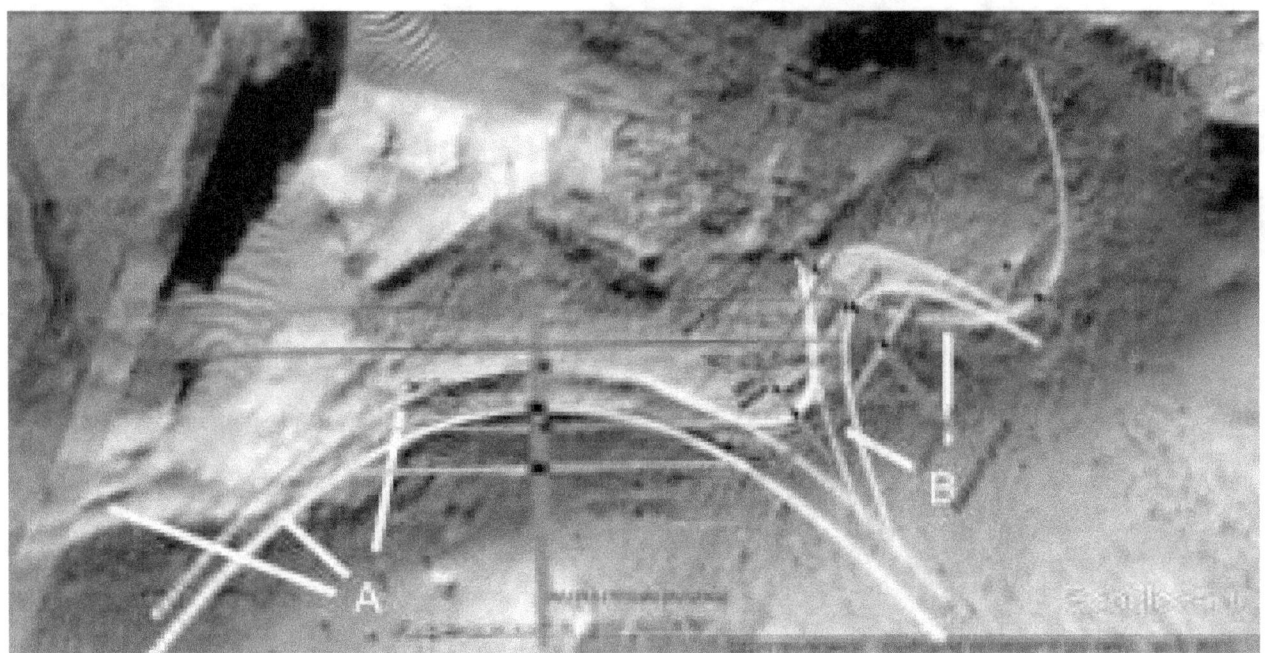

Cymd280a

Hypothesis

A shows how the skin on the dam wall is peeled off, at 3 o'clock is has many pits like on the skin of hollow hills. At 4 o'clock this rough interior is exposed but just below it the skin is smooth. At 6 o'clock is another edge of the smooth skin. B shows at 8 o'clock. How it is peeling off, at 5 o'clock it is more stable. At 10 o'clock there are many pits as it degrades, at 2 o'clock it shows the lip of the dam has broken off. C shows a smooth area that goes up to the broken lip of the dam wall like an external layer, perhaps a patch.

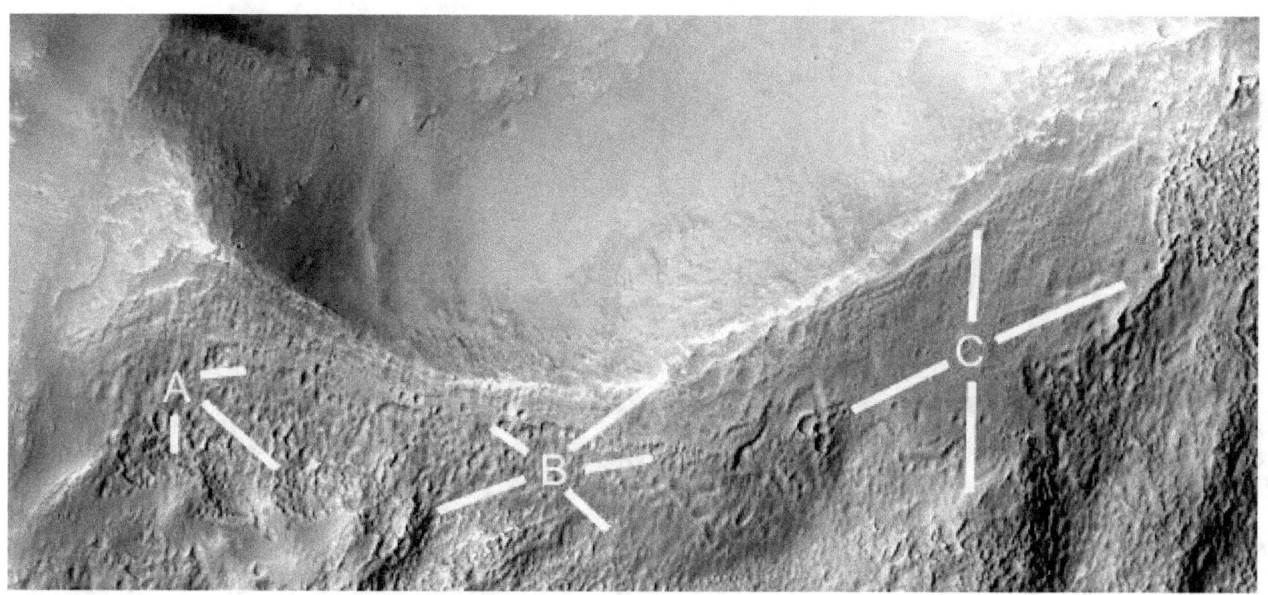

Cymd280a2

Hypothesis

A parabola is shown.

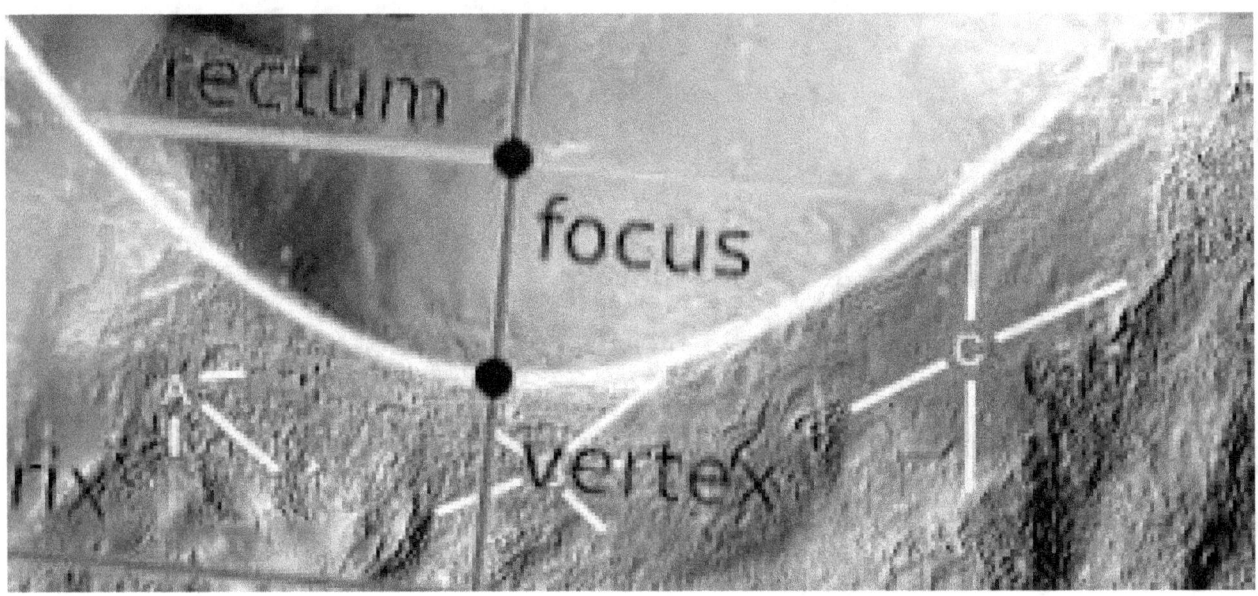

Cymd280d

Hypothesis

A looks like a dam wall attached to the crater, the construction technique is probably a cavity in this shape and the wall inserted into it. This is showing signs of breaking off, B shows the layers in it that are separating. C also shows a separation from the crater, usually these dams fit on very smoothly here.

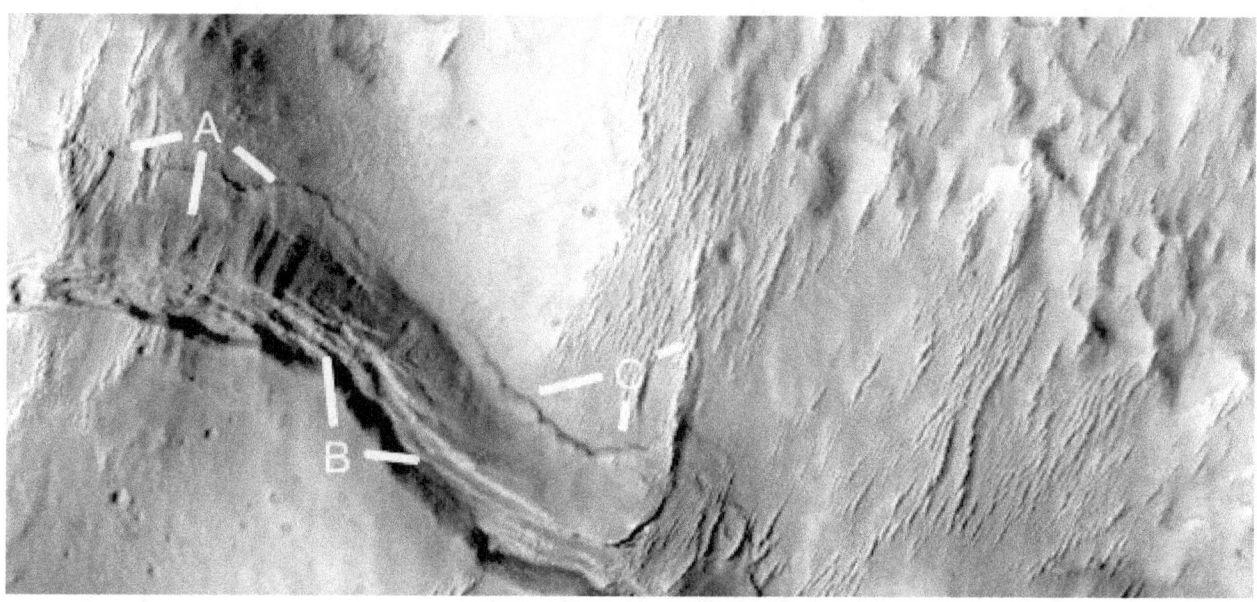

Cymd280d2

Hypothesis

Two parabolas are shown.

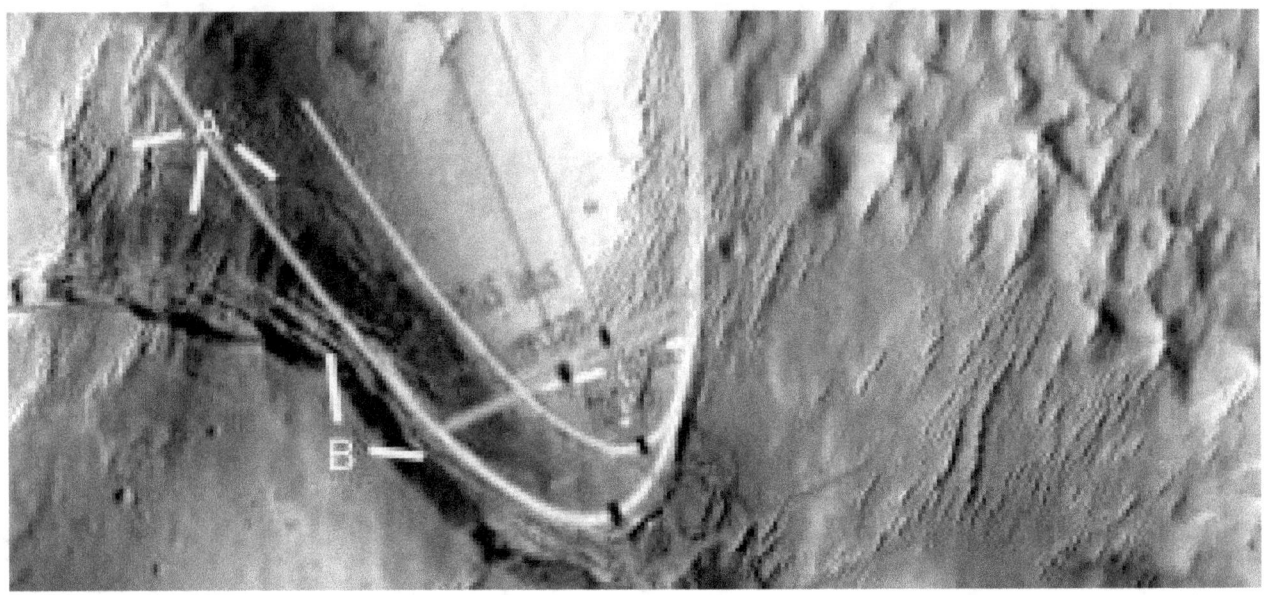

Cymd280i

Hypothesis

Engineers might examine how this wall is fracturing at A to D, Also D at 2 o'clock shows the thicker base holding the dam wall in place.

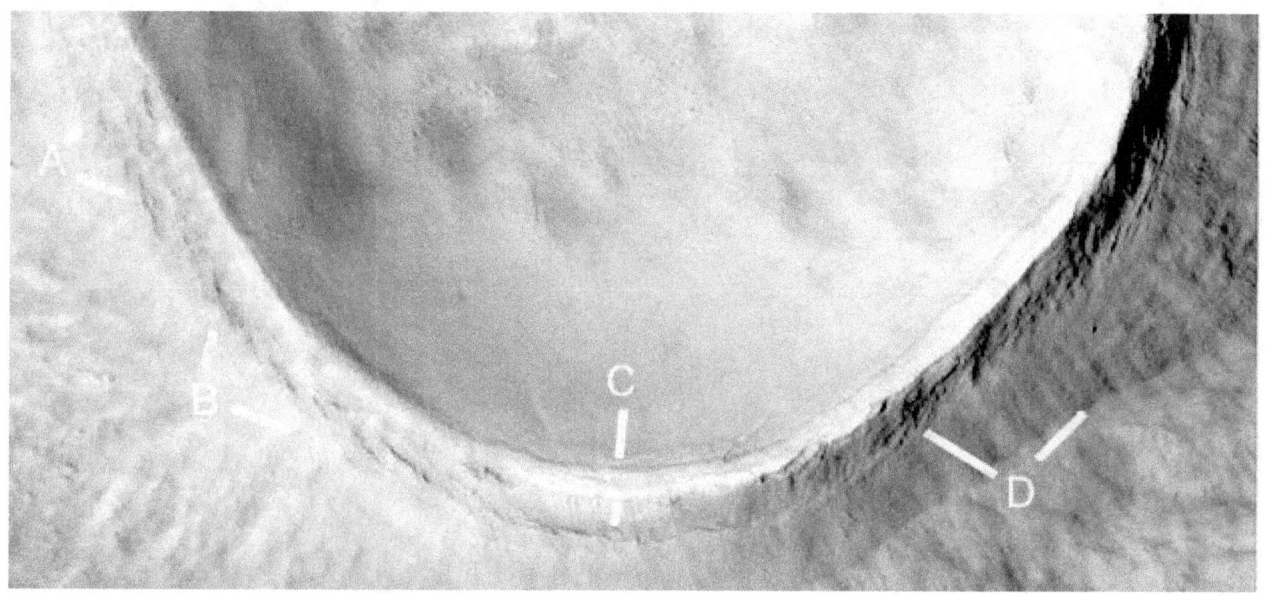

Cymd280i2

Hypothesis

A parabola is shown.

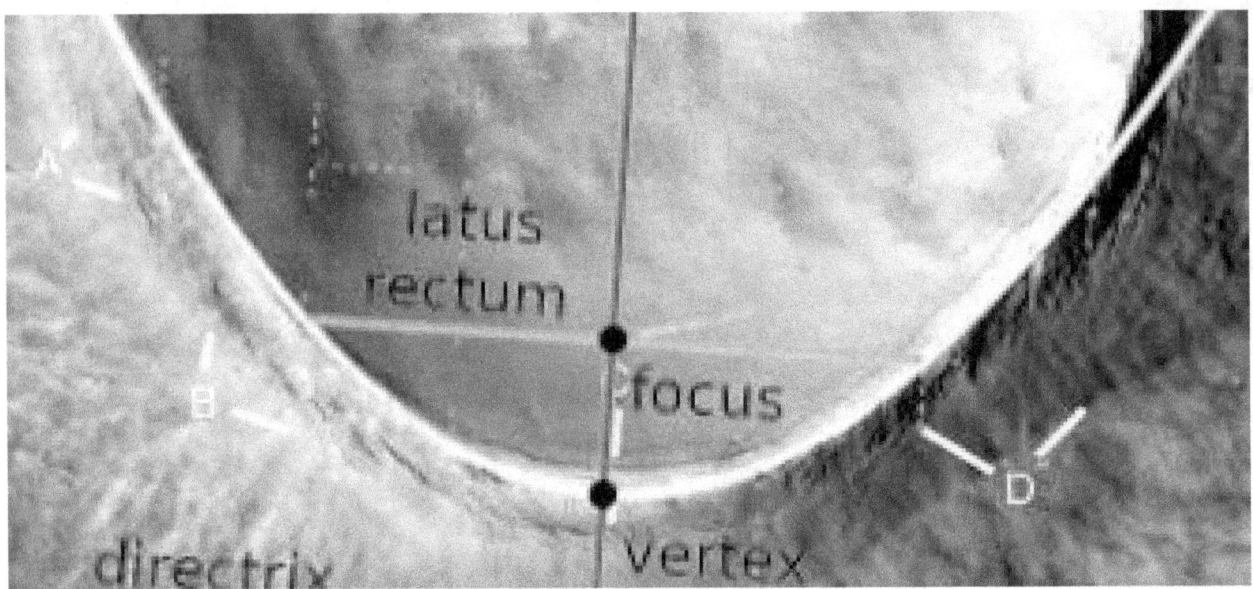

Cymch281

A shows a wall above a parabola, there may have been a dam wall here that eroded away or it could have been a water channel. B shows a small dam and how this channel continues on. C and D show a narrow wall on the side of the channel. E shows signs of a double wall implying a higher wall broke off. F shows another double wall at 11 o'clock, At 9 and 12 o'clock are the boundaries of a dark area, seemingly excavated between F and C to add a row of hills. G shows more walls.

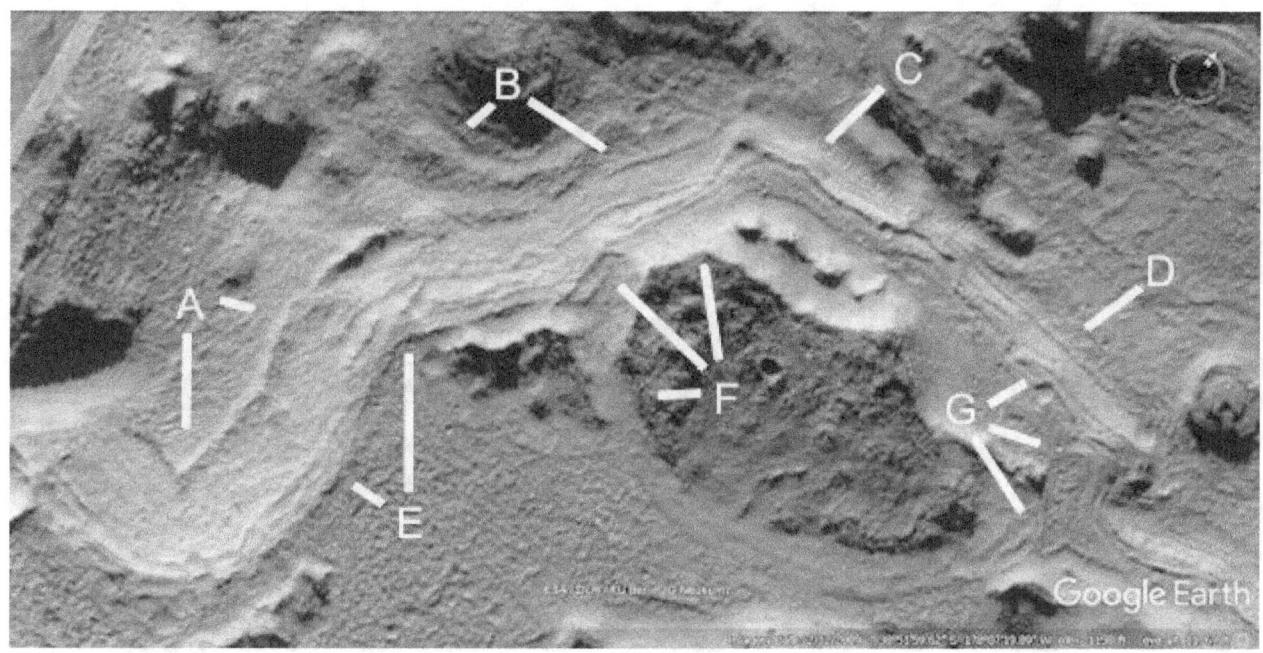

Cymch281a

Hypothesis

A parabola is shown.

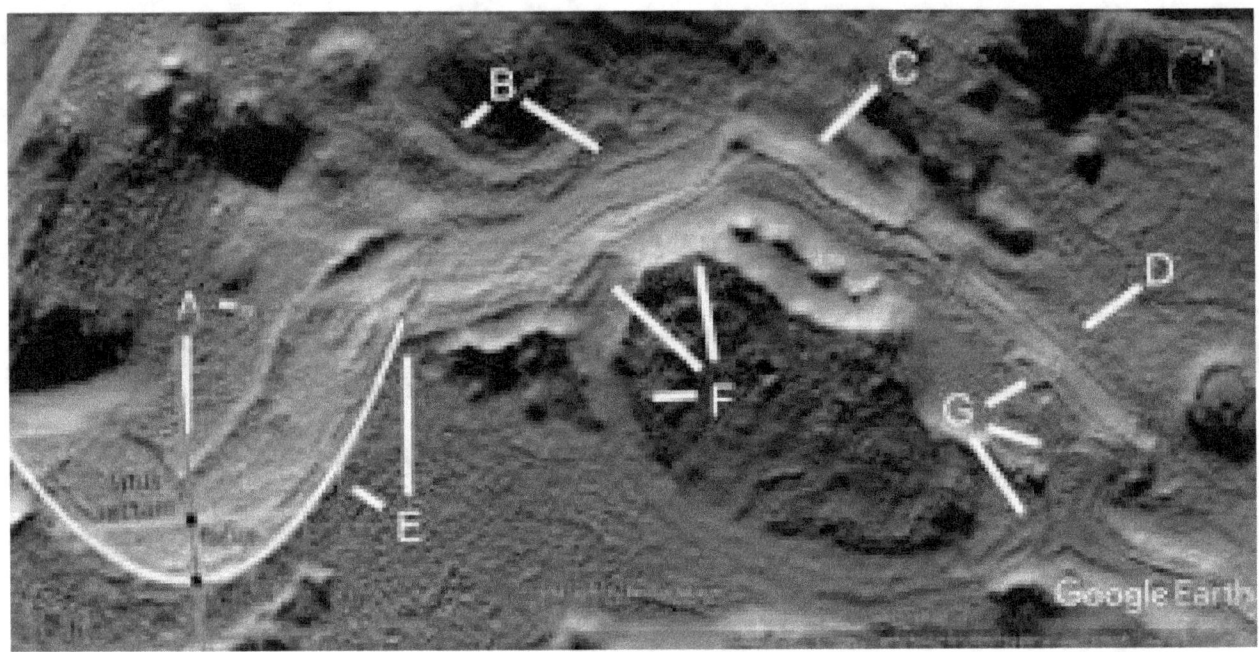

Cymd282a

Hypothesis

A shows a smaller dam with cracks at 6 and 8 o'clock, also a crack right along the dam floor at 2 o'clock. A at 11 o'clock may show how this cement has completely peeled off right along past E at 4 and 8 o'clock. B shows this part of the wall has also fractured. E at 10 o'clock may be a water channel. D at 1 o'clock is in good condition, at 7 and 8 o'clock either the dam lip has broken off or the inner and outer walls are separating. D at 2 o'clock shocks a small basin perhaps for storing water separately. F shows more examples of cracked walls as does G. H at 10 o'clock may show a water channel, at 7 o'clock is the edge of the cement dam floor.

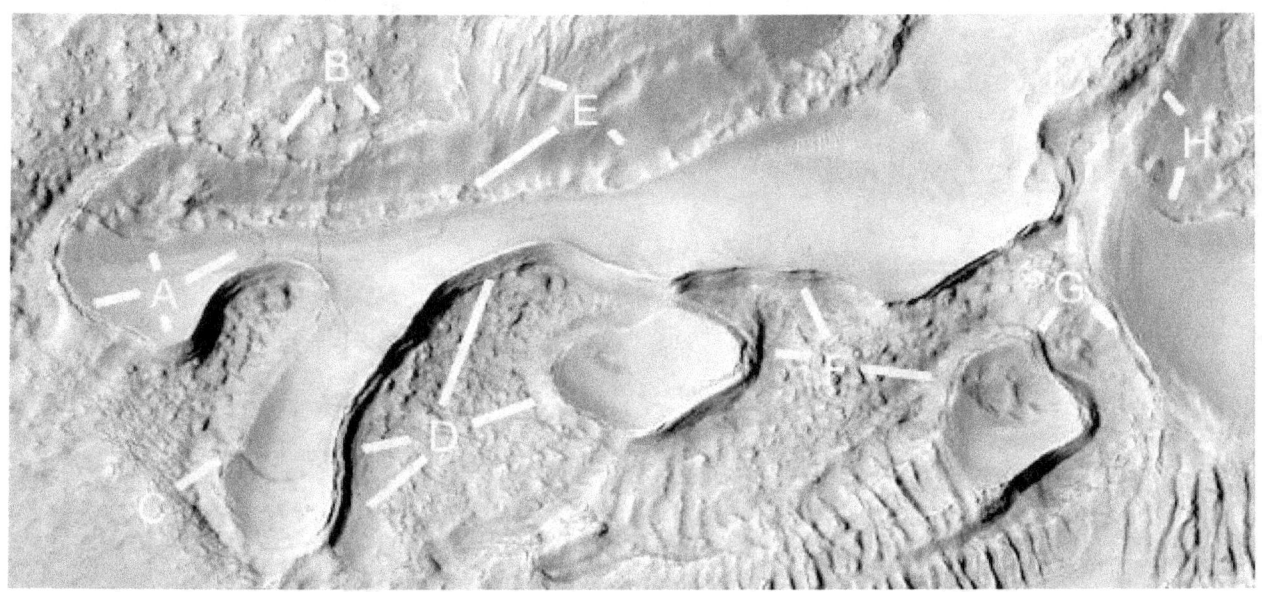

Cymd282a2

Hypothesis

Three parabolas are shown.

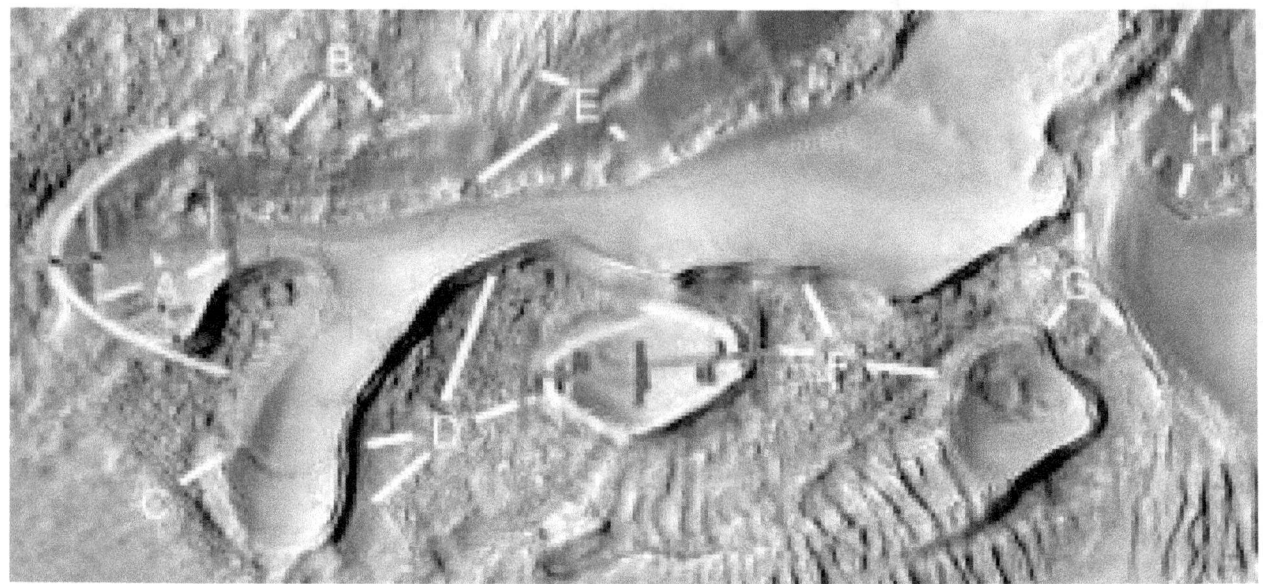

Cymd282c

Hypothesis

A shows a crater that apparently has been cemented over on both sides, this would appear to be impossible to occur naturally. It has cracked more on the lower side as of from creep. It could even be the exit hole for a water tunnel from above. Between A and D there is a water channel, D at 2 o'clock appears to show another channel. C shows the other side of the crater at 11 and 12 o'clock, like the area is another water channel going down to C at 8 o'clock. At 7 o'clock is a thick layer to guide this water, at 2 o'clock it appears to be a small water channel. E also looks like a water channel, F shows signs of the wall cracking and the second line at 12 o'clock appears to be the cement floor breaking up.

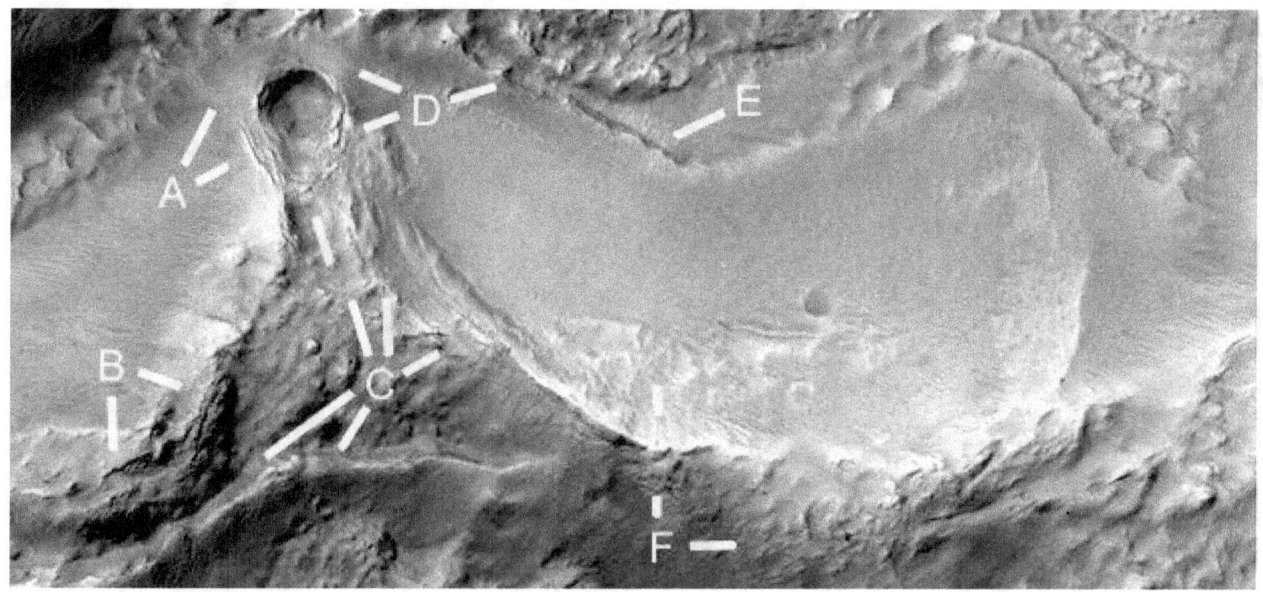

Cymd282c2

Hypothesis

A parabola is shown.

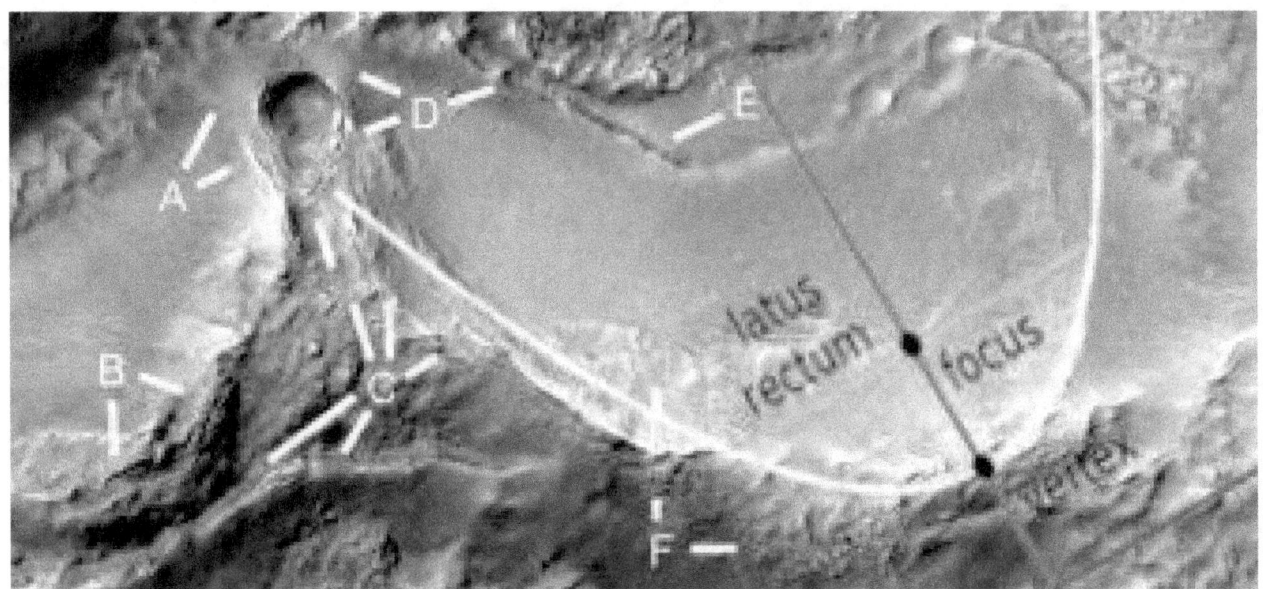

Cymhh291a

Hypothesis

A appears to be a collapsed hollow hill with an internal support at B at 8 o'clock. At 5 o'clock there appears to an entrance to the other hollow hill, D and E show the walls. C and F have a mottled appearance, this may be collapsed corridors or tunnels. G shows a third hollow hill with several internal supports or tubes crossing the hill from the top to the bottom.

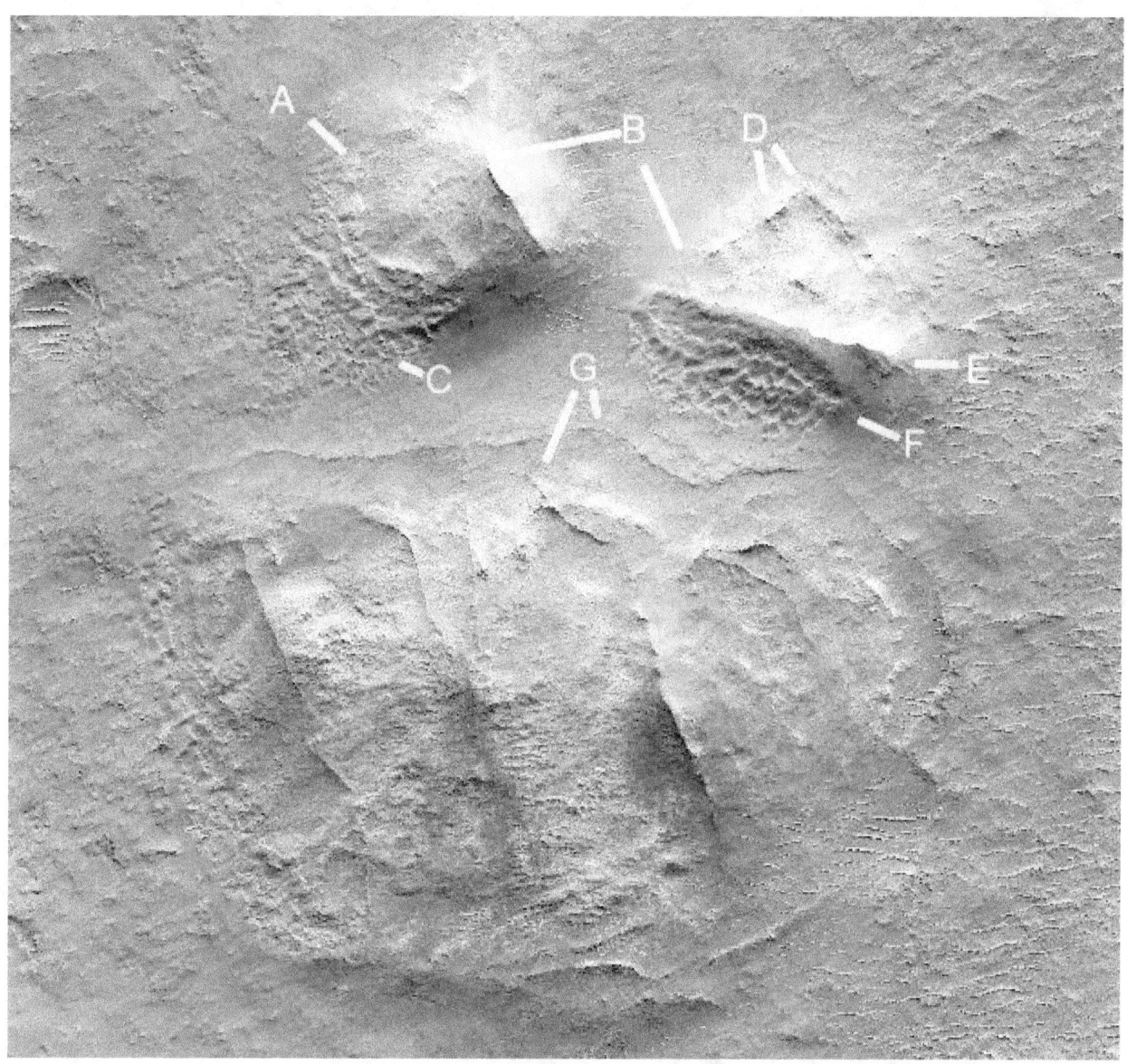

Cymhh291e

Hypothesis

A probably shows ceiling material which is exposing some rooms. B shows some room walls perhaps full of the ceiling material. At C the rooms appear to have less material in them and are clearer. D at 4 and 5 o'clock may be an internal tube that has collapsed showing a corridor inside it, this goes along to E back under a more intact ceiling. D at 7 o'clock shows a more intact ceiling with the walls showing around it, at 2 o'clock there is a well defined walled area like a house with internal rooms. F may have had the internal walls also collapse to bare ground, G may be ceiling or hill material the collapsed tube goes under. This implies there are many corridors like this connecting the rooms.

Cymhh291g

Hypothesis

The ceiling material here is thicker, A shows a small set of rooms in a circle and under the letter A there may be more rooms. B shows large walls or tubes, there may be plenty of space between these to allow for ventilation with multiple layers of rooms. C appears to be the lowest layer as if the rest have eroded away, this is almost gone to bare ground. D is less eroded, the walls are much higher though most of the ceiling material is gone. From this the height of the walls might be worked out by engineers, when the ceiling material is just gone the walls should be at their maximum height. E and F may contain rooms under the ground, some of the mottled areas may be like this where a habitat can extend into the ground around a hill. G appears to show a dome with walls coming to a point, perhaps this was a meeting place. H is also a curved area partially exposed. I shows a pit or dam at 7 o'clock and a large enclosure connected to G, there may have been internal areas for livestock or fish farms. Some areas like J and K appear to have larger enclosures, these might have been bigger housing or more industrial sized buildings.

Cymhh291i

Hypothesis

Here there appears to be some rooms at A and B in a squarish shape, perhaps residential with curved tubes above them perhaps for traveling. We see this with Earth roads how they can be more curved. A at 10 o'clock shows squarish rooms but at 6 o'clock it is laid out more in a circular pattern. D may be ceiling material, at 5, 7, and 9 the rooms are becoming exposed while at 3 o'clock there is a wider tube perhaps to handle more traffic. E appears to be more rooms clustered around this like a main road on Earth. F shows more ceiling material while G at 8 and 10 o'clock shows long walls parallel to each other. H is probably similar but still has the ceiling material. I has more curves but many residential areas have curved roads like this.

Cymt300

Hypothesis

This shows tubes coming out of a hollow hill at A, B shows how this goes into a square shaped hollow and then onwards to the right. C appears to be another tube or wall as is D.

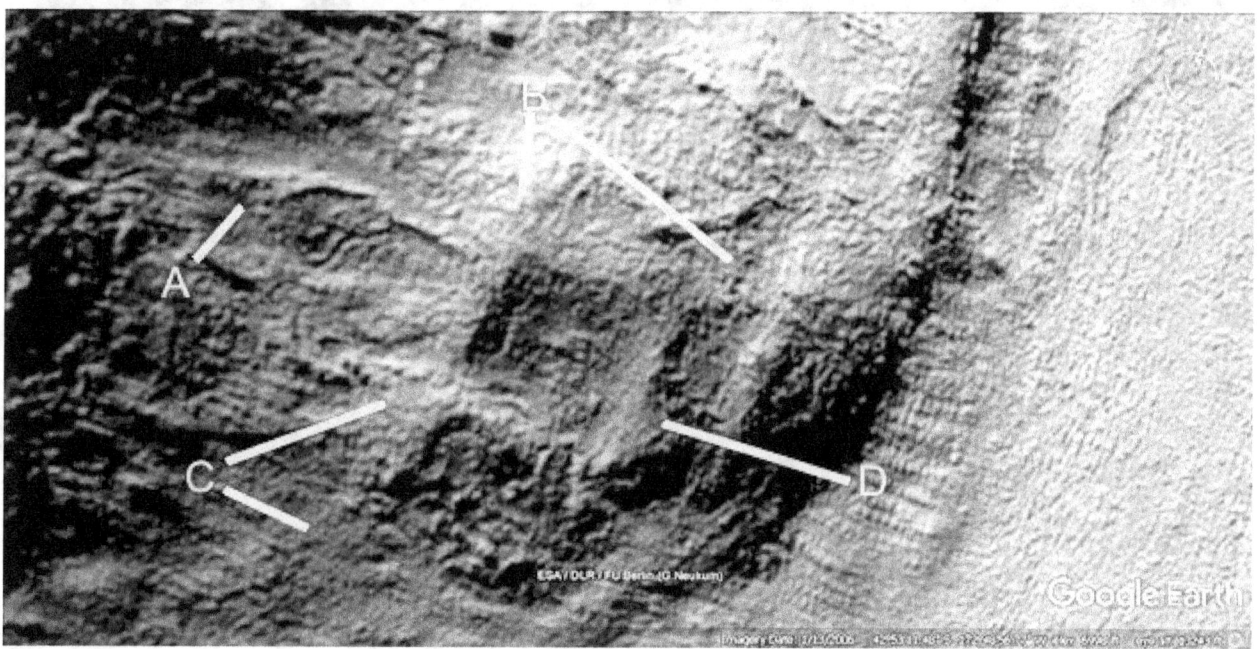

Cymt310e

Hypothesis

A may also be a water channel, B may be where the cement has eroded. C also can be a water channel or pit dam.

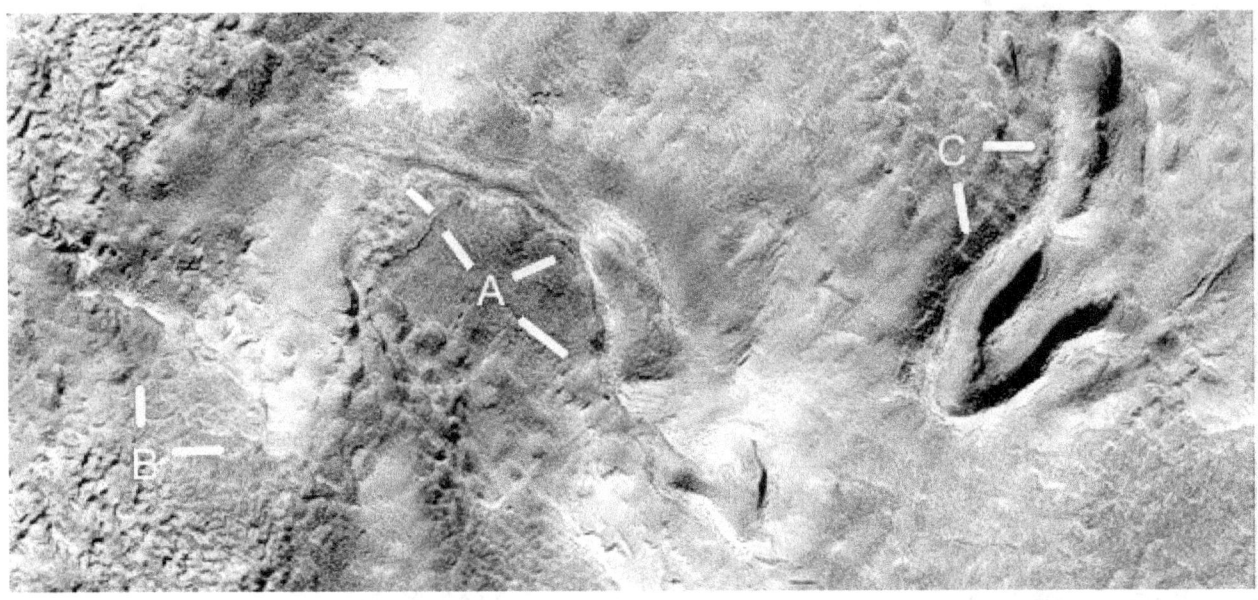

Cymt310e2

Hypothesis

This shows how C is partially parabolic in shape. Often the curved wall deviates from the parabola where it is no longer load bearing.

Cymt316

Hypothesis

A shows a collapsed hill at 1 o'clock, at 2 and 4 o'clock there is a groove like a road or collapsed tube. At 6 o'clock there is a tube going from a crater to this groove. B shows several tubes going to craters then cross the groove over to C and more tubes. D shows many more tubes, E shows a larger tube coming from the hollow hill at the left through a crater to F where it flattens as if collapsed. At G these tubes have become grooves going into the crater, implies the two are the same construction technique. H and I show more tubes.

Cymd341a

Hypothesis

The dam wall at A, C, and D is in good condition, with a sharp top of the wall. A shows some cracks of the cement at 4 o'clock. B shows the two smooth and curved sides of the inner dam, E shows an unusual shape the groove at 1 o'clock goes around.

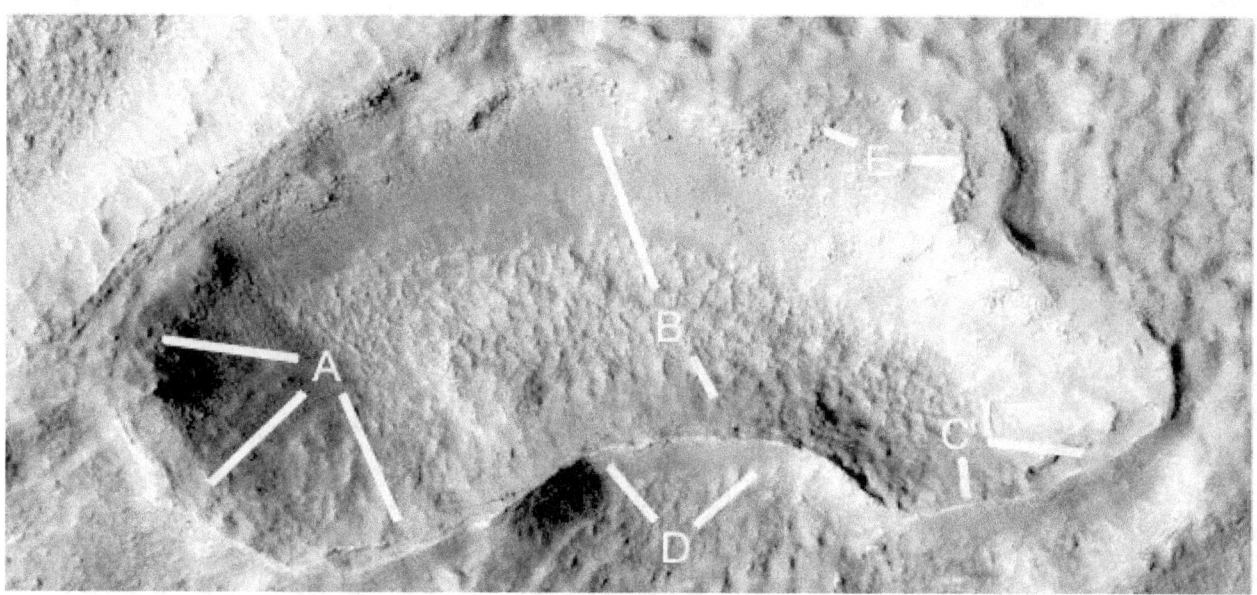

Cymd341a2

Hypothesis

This shows how two curves in the pit dam are parabolic.

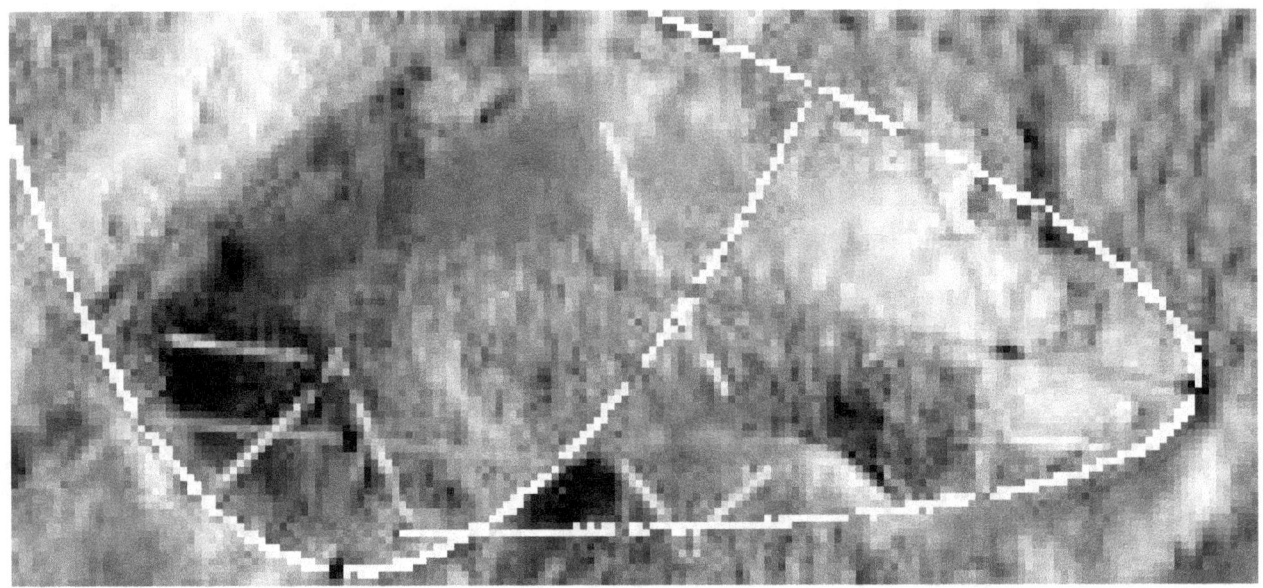

Cymd341c

Hypothesis

A is an unusual shape, it may have had a dam wall which has broken off. B may be a water channel and a pit dam under it.

Cymd341c2

Hypothesis

This shows how the shape closely follows two parabolas.

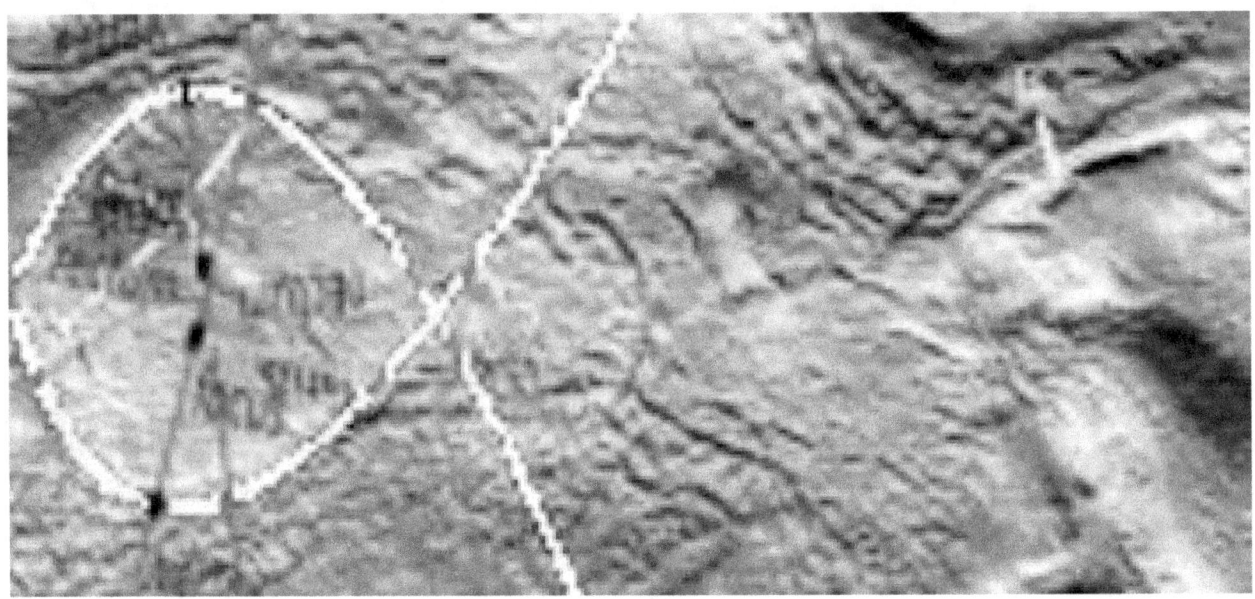

Cymd346d

Hypothesis

These dams also appear to be cracked from creep, A and B may also show how this material has cold flowed into wave shapes perhaps before it cracks. C may be for overflows, D shows some erosion in the lip of the dam.

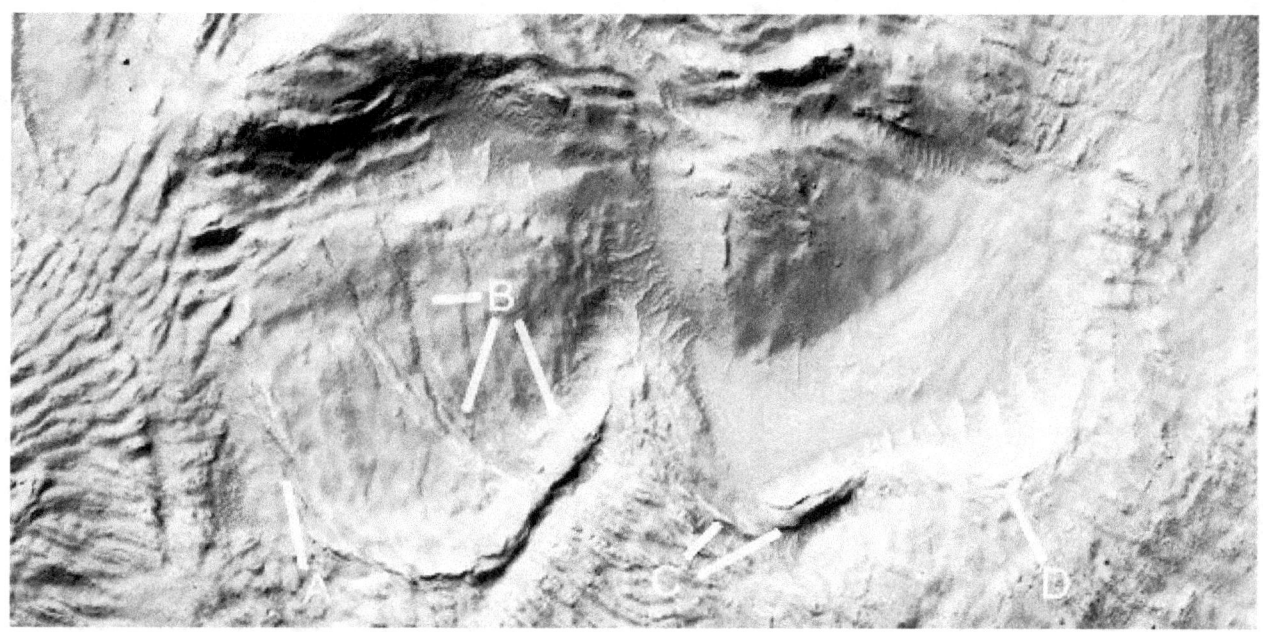

Cymd346d2

Hypothesis

This shows two parabolas in the dams.

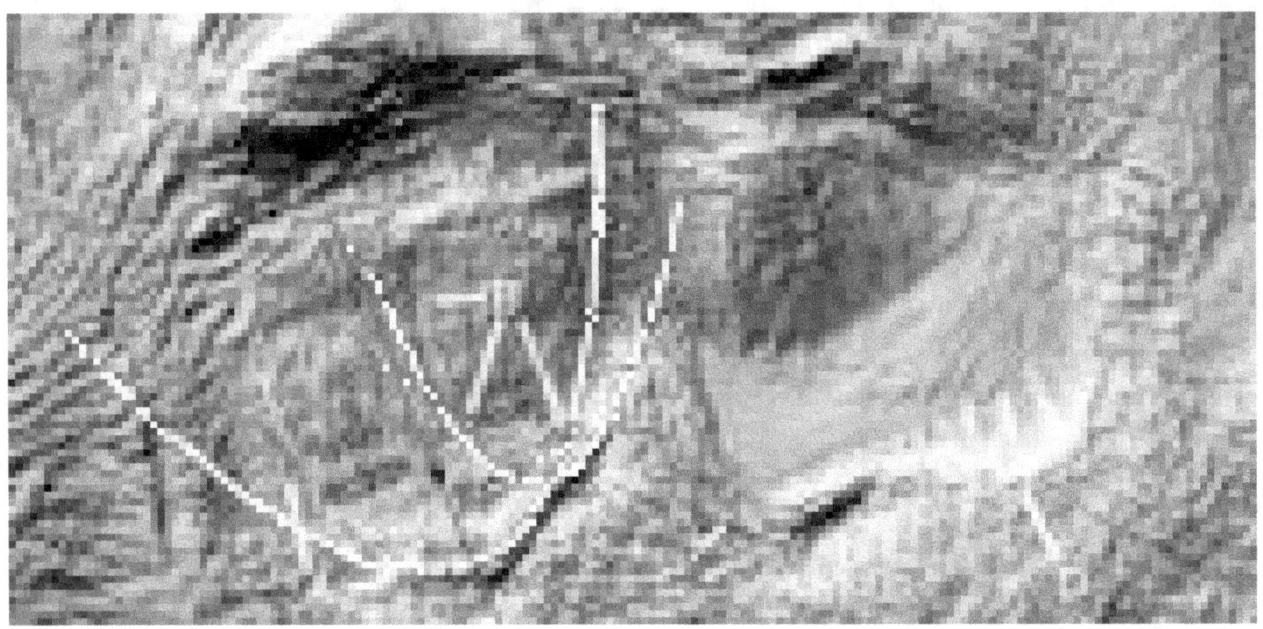

Cymd346e

Hypothesis

A shows more cracks in this skin, B may have been another parabola.

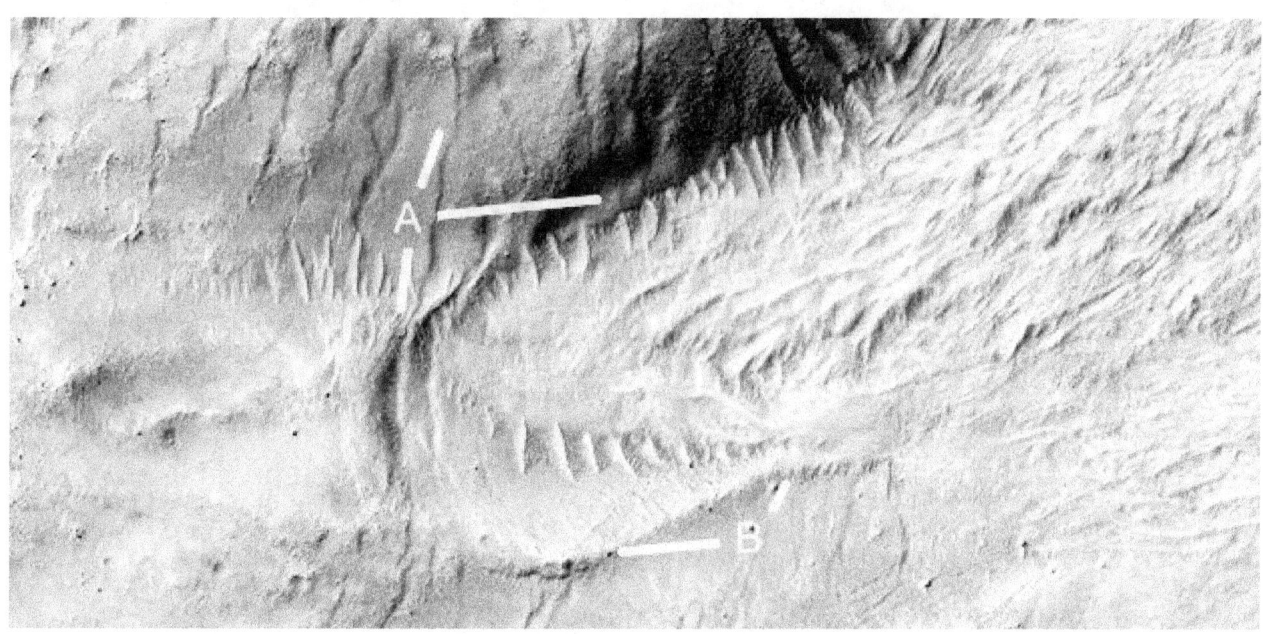

Cymd346e2

Hypothesis

A parabola is shown.

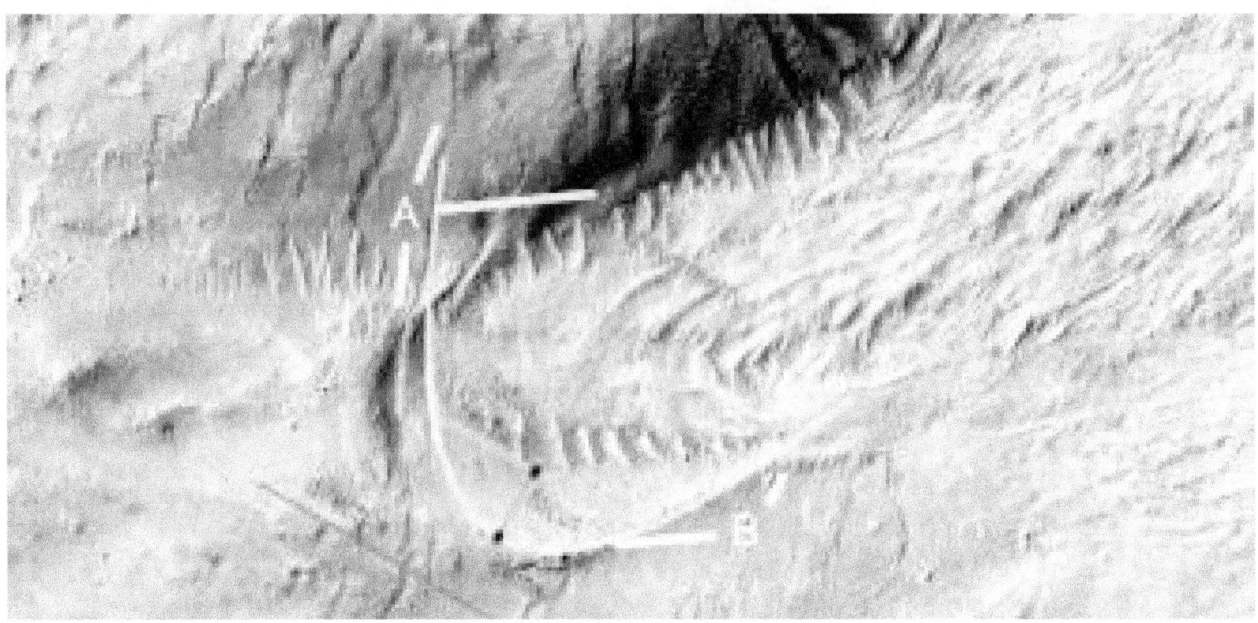

Cymd346g

Hypothesis

A shows several layers exposed in the lip of the dam wall, these can be estimated in how thick they are from the resolution of the image. B shows a small hollow under the lip at B at 5 and 7 o'clock, going along to C at 7 o'clock. The upper side of this hollow is in better condition, the water would have flowed down C at 4 o'clock. D shows signs of creep causing cracks at 1 and 2 o'clock, E is in better condition. F shows more cracks in the layers in the dam floor. G and H show smaller pit dams with these layers exposed, as does I. J shows another water channel.

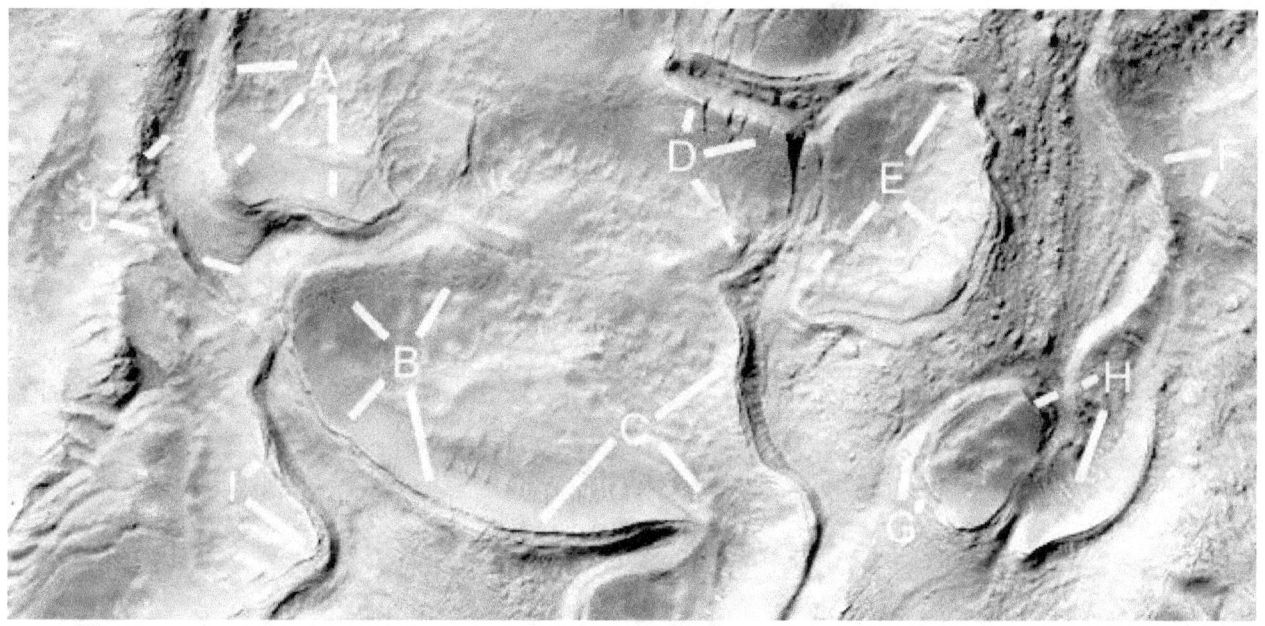

Cymd346g2

Hypothesis

This shows another parabolic dam.

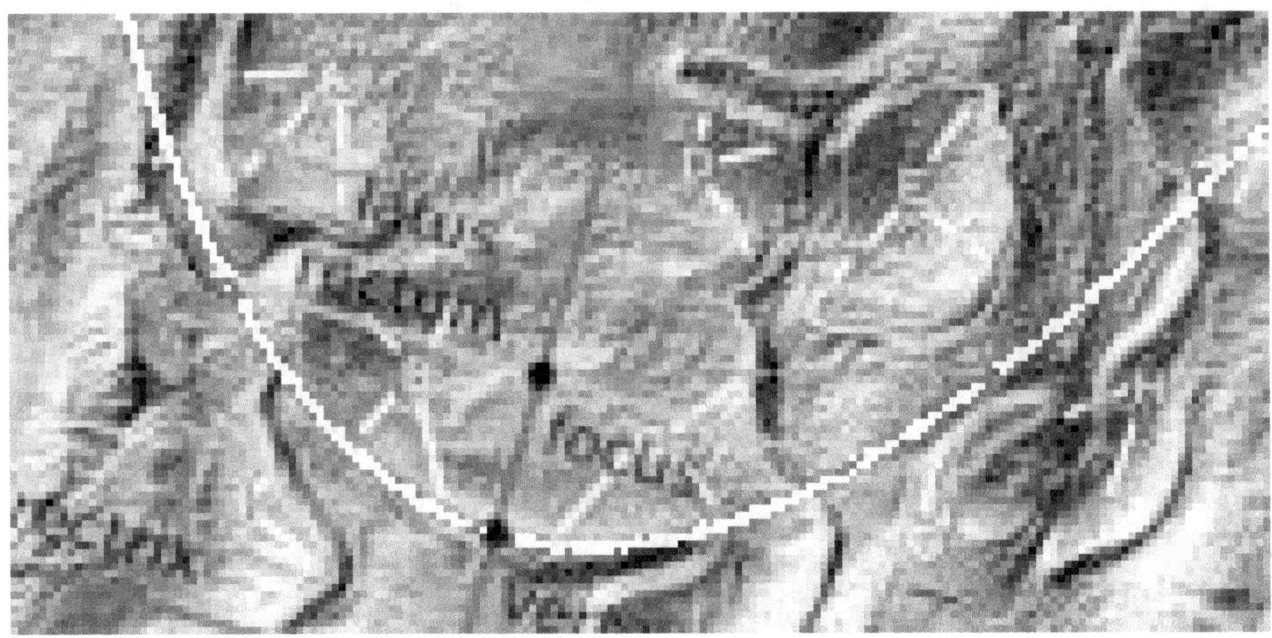

Cymd346h

Hypothesis

More cracks in the dam floors and walls are shown here.

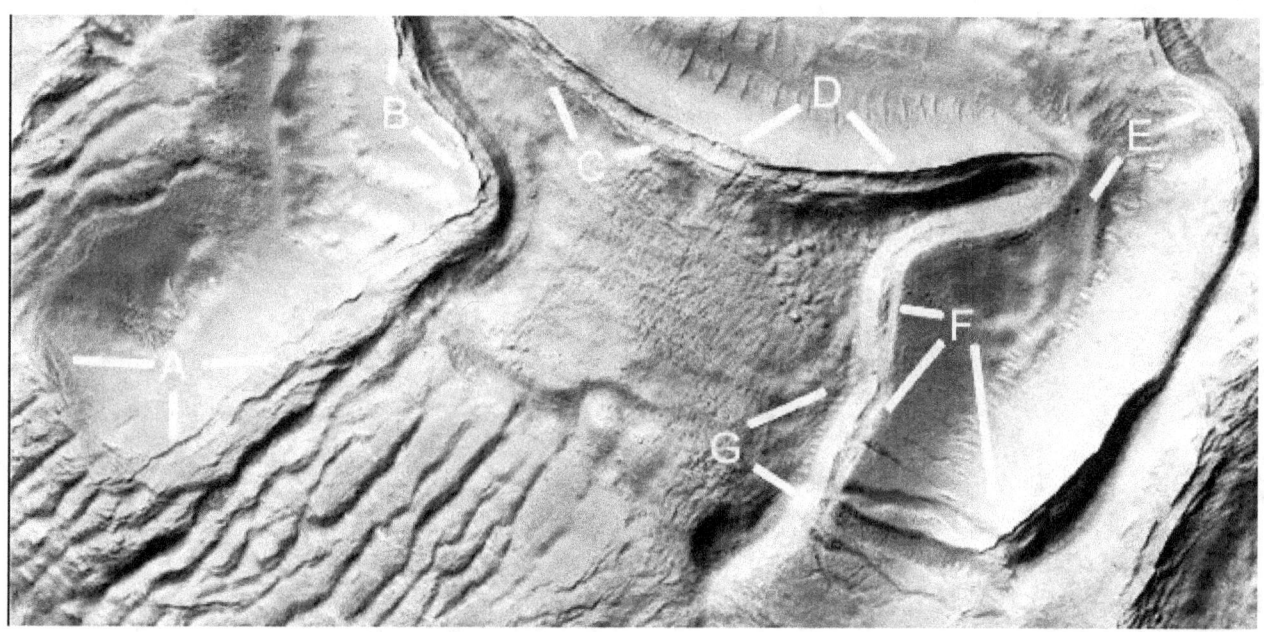

Cymd346h2

Hypothesis

This shows four parabolas were used here.

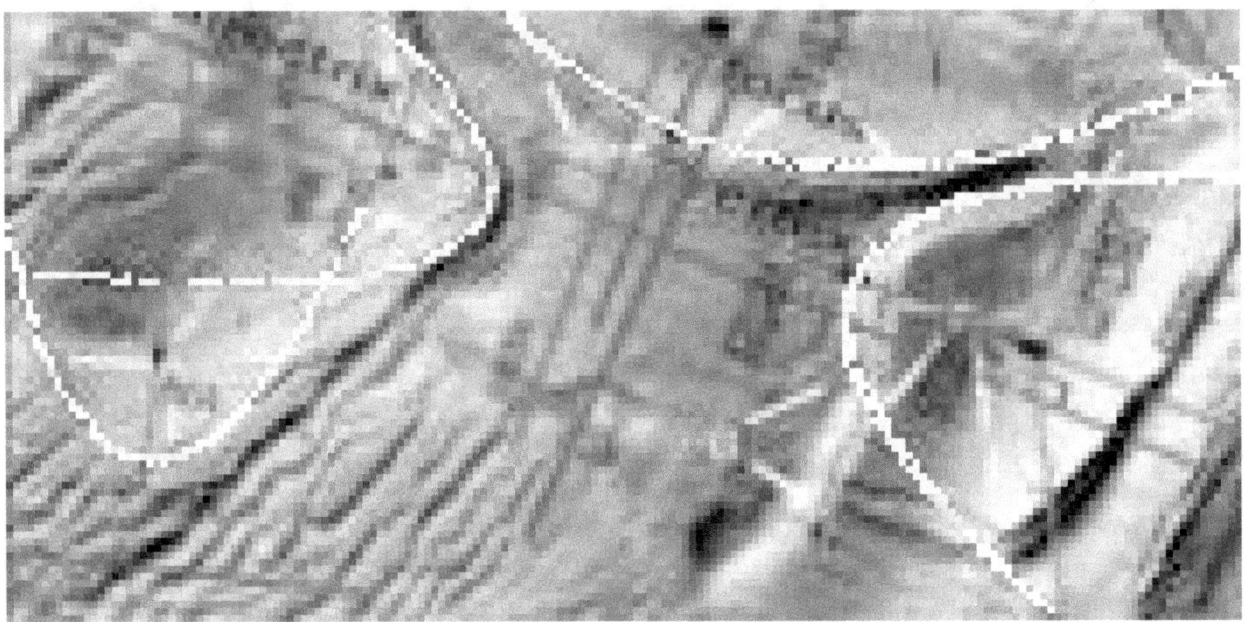

Cymd346i

Hypothesis

A shows a smooth area like cement, the dam walls around it may have broken off. B shows cracks in the dam wall at 6 and 9 o'clock, at 3 o'clock the wall is in good condition. C shows a double wall, it may be expanding and cracking, or the top of the wall may have broken off. D and E show more cracks developing.

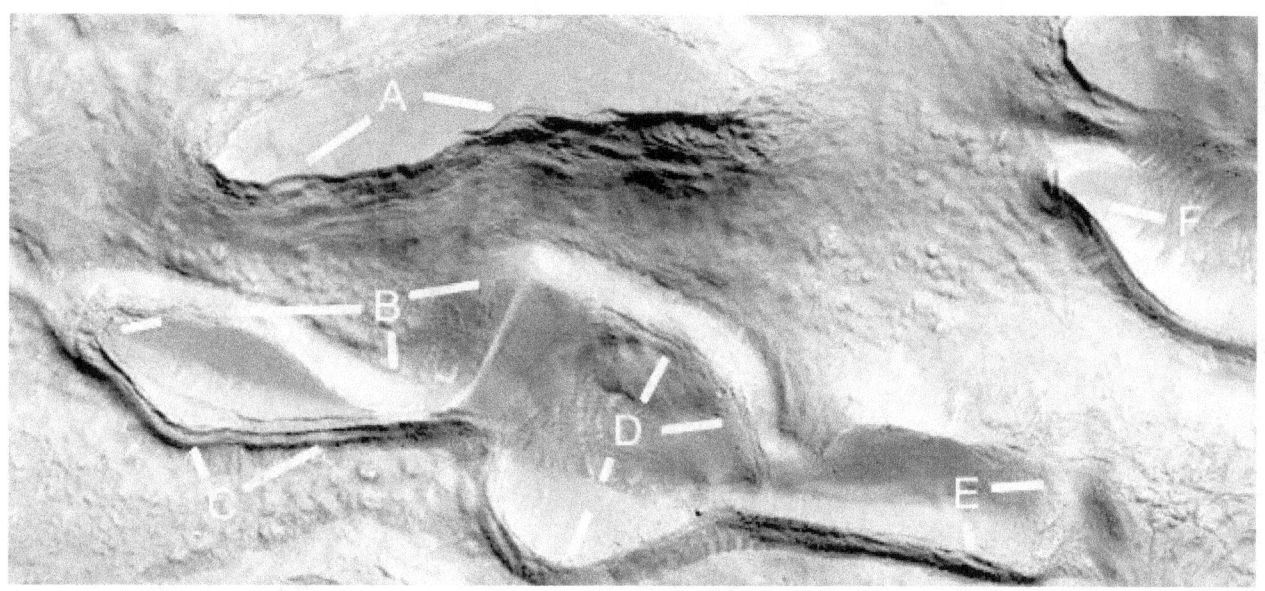

Cymd346i2

Hypothesis

A parabola is shown.

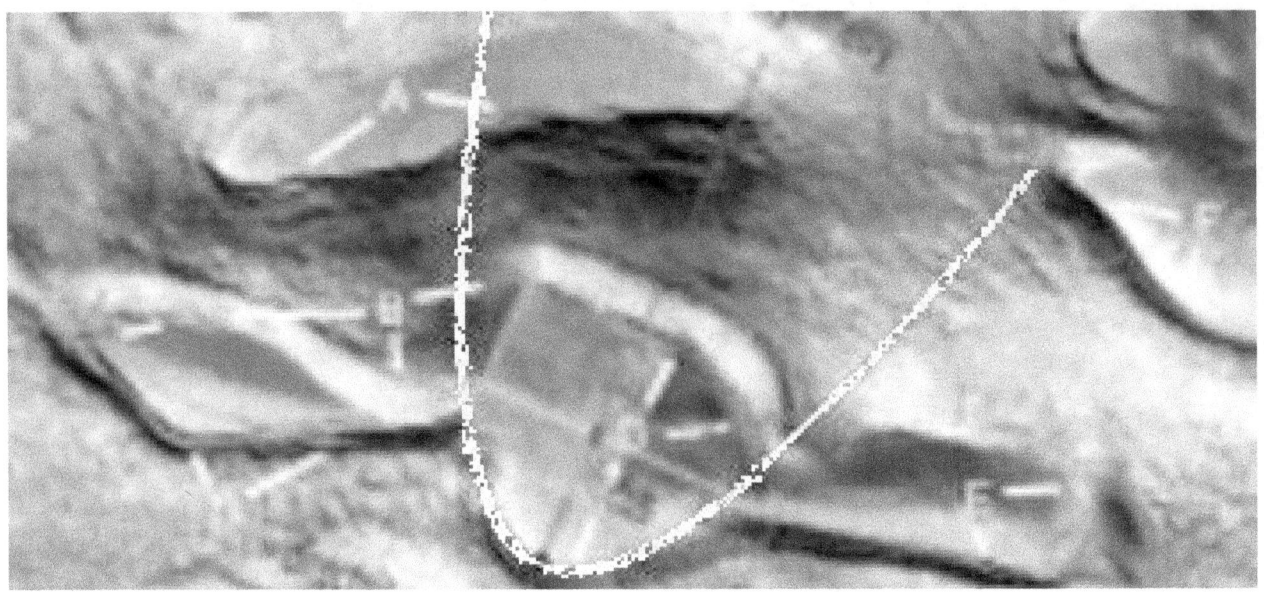

Cymd346j

Hypothesis

A, D, E, and F appear to be a cement retaining wall perhaps the crater was more unstable in its soil. B and G show two dams, E is a hollow under a dam wall.

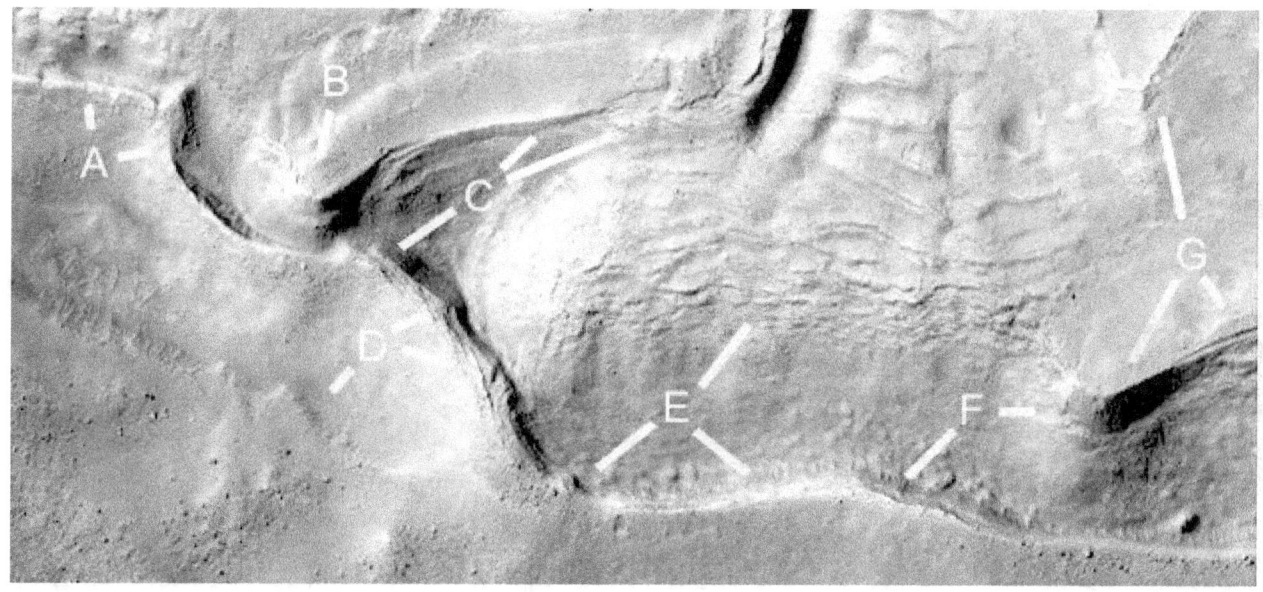

Cymd346j2

Hypothesis

Another two parabolas are shown here.

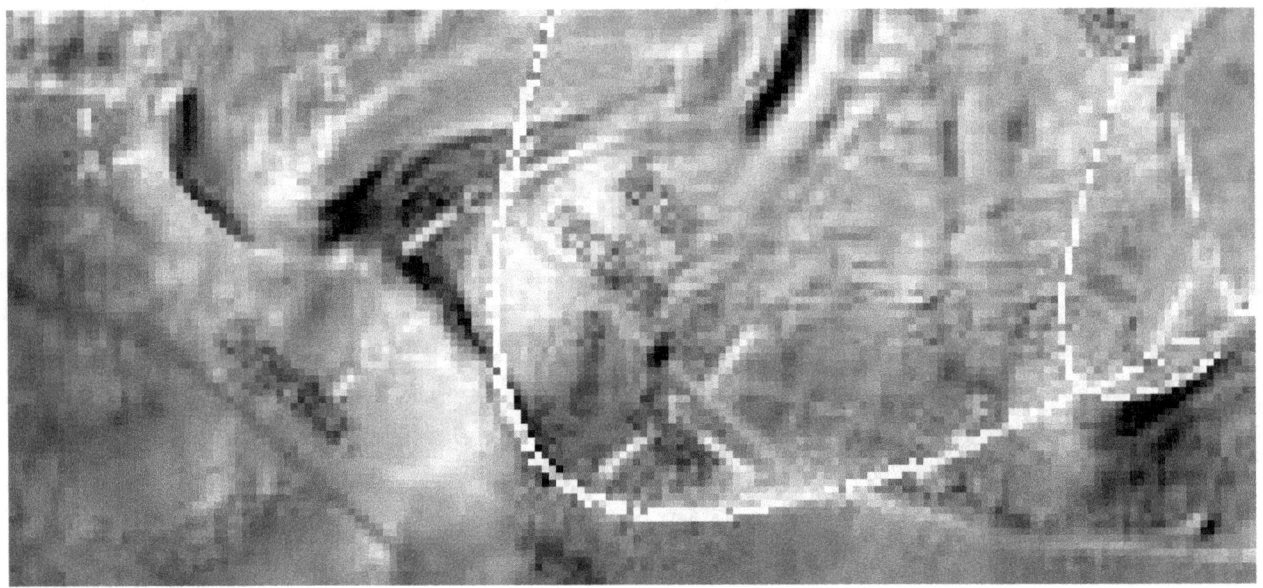

Cymd346l

Hypothesis

The lip of the excavation dam at A and B is thinner but has not cracked here. It might indicate the construction techniques took into account how the ground would move. Some cement skins were much thicker but still cracked over time, here it was estimated a thicker skin was not necessary. C shows the edge of the smooth dam floor. D and G may be cracks from creep. E shows the same smooth and thin skin. F shows a groove around the upper edge of the dam.

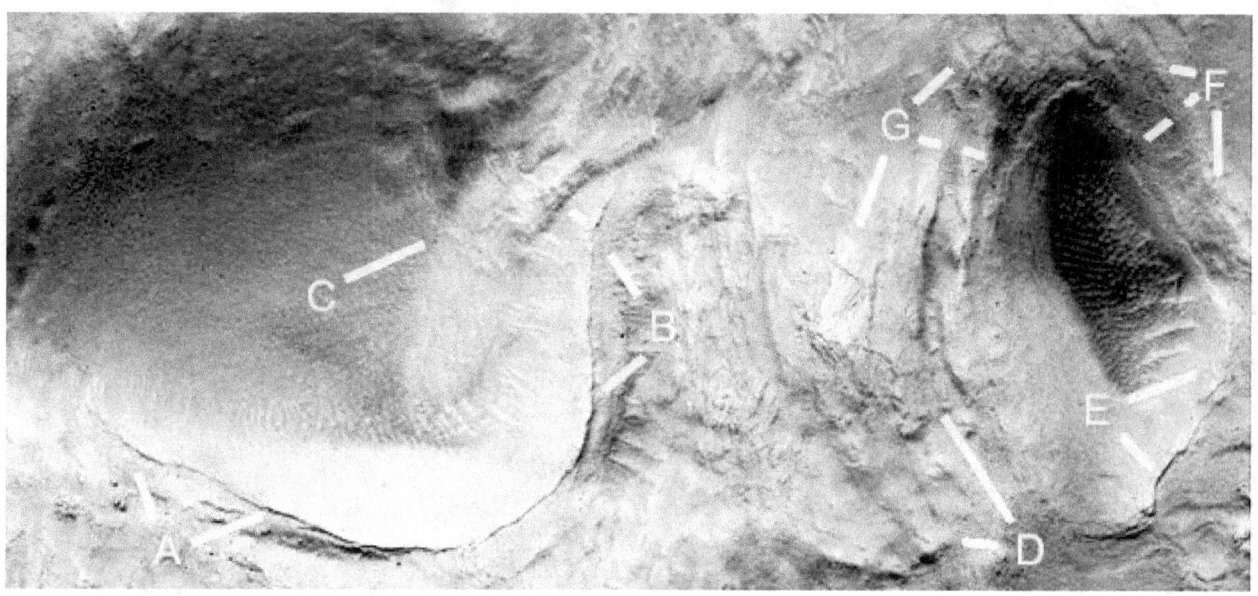

Cymd346l2

Hypothesis

This shows another parabola.

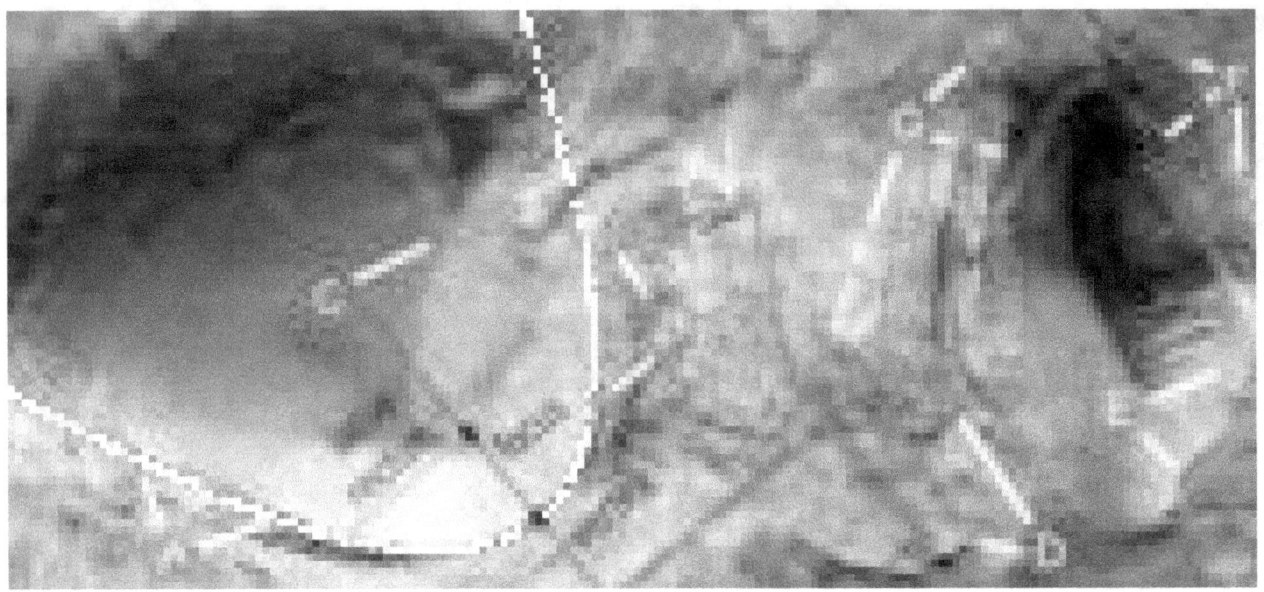

Cymd346m

Hypothesis

A and B show an artificial looking edge to this material, also how thick it is yet has still cracked over time. To the left of A is another water channel.

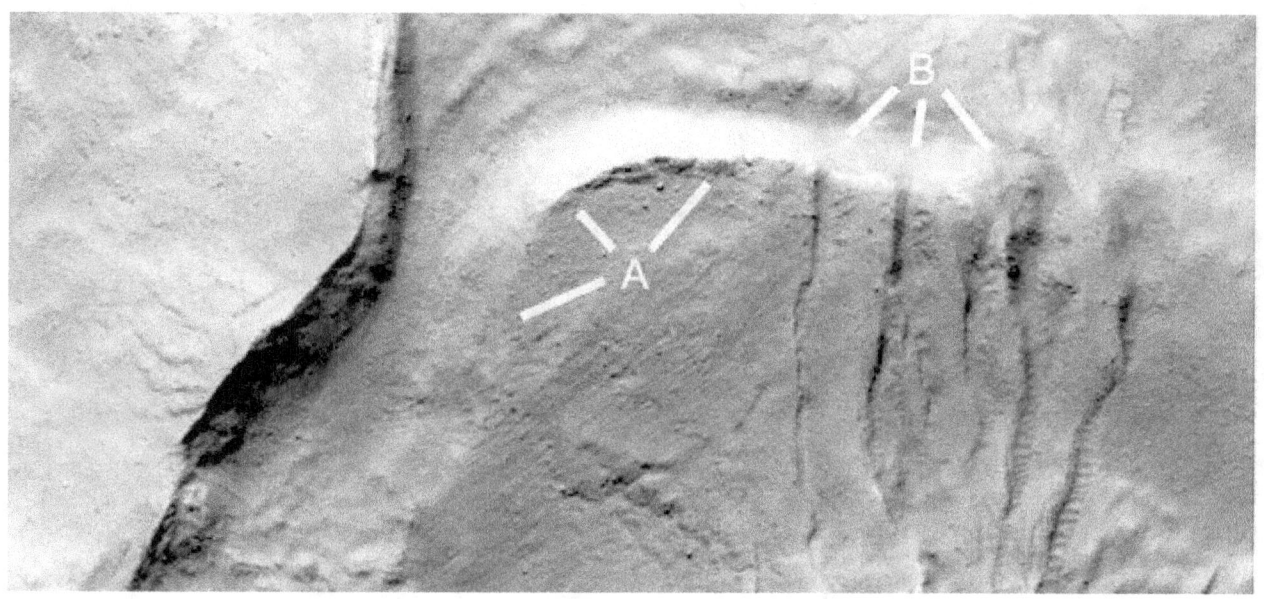

Cymd346m2

Hypothesis

A parabola is shown.

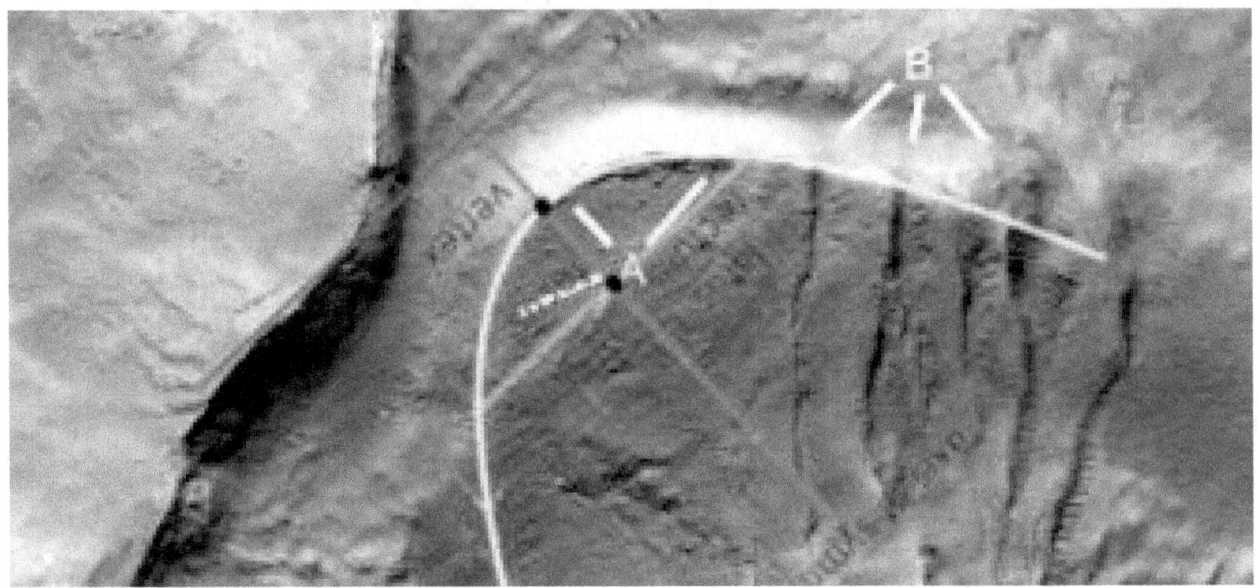

Cymd348

Hypothesis

This shows another small dam, A shows some erosion and B where the water seems to have damaged the wall as it overflowed. C shows the path of some of this water may have eroded the dam floor.

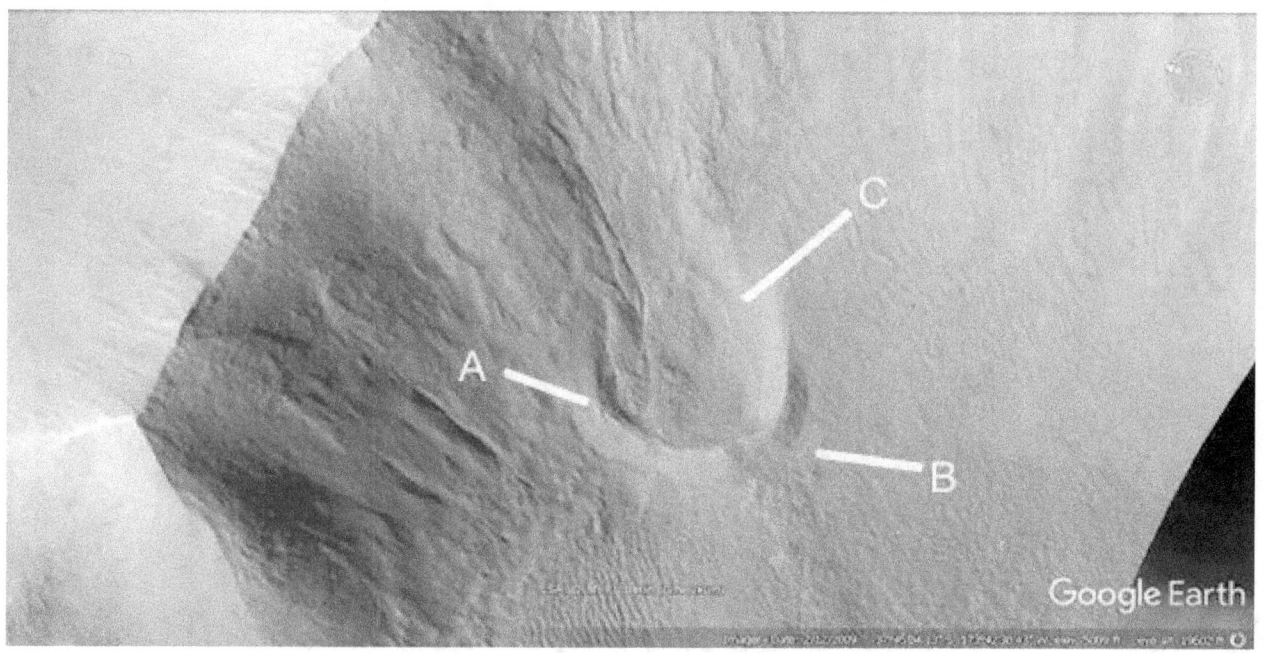

Cymd348a

Hypothesis

A parabola is shown.

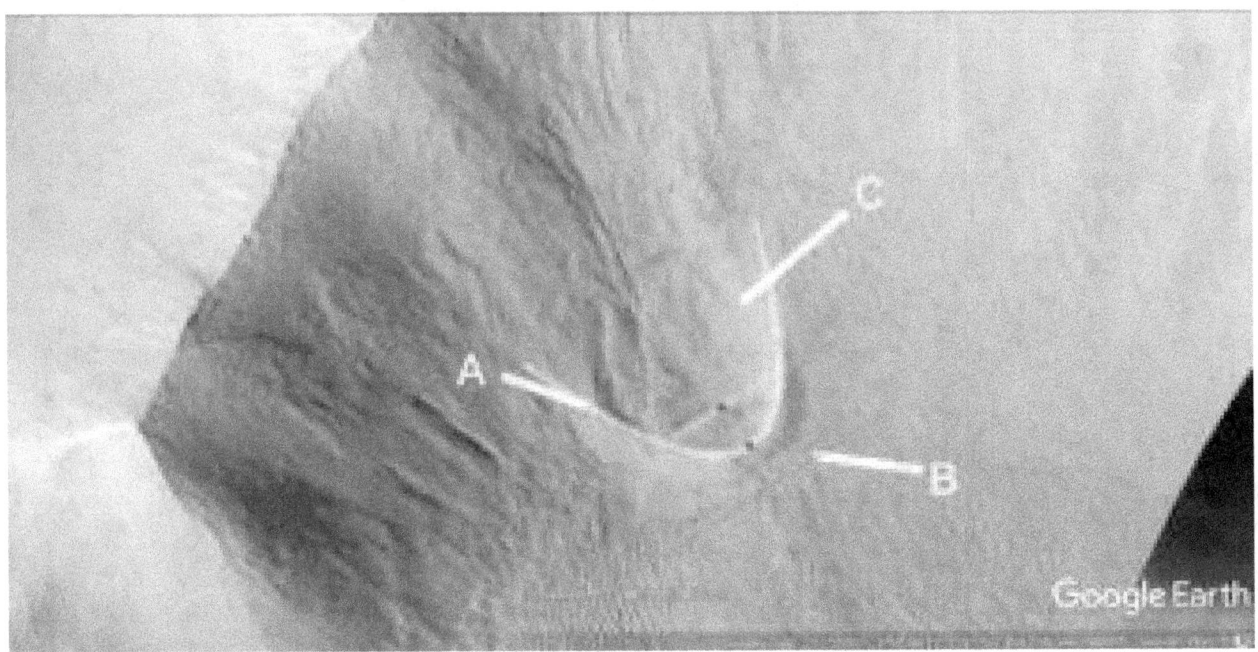

Cymhh355b

Hypothesis

At A the walls may have eroded down to the floor, alternatively they are sticking through an eroded ceiling material. B shows some clear walls, to their right C looks like similar walls that are either buried or under the ceiling material. D is more eroded than E, it is as if E will eventually disappear like at D. F probably has walls under it.

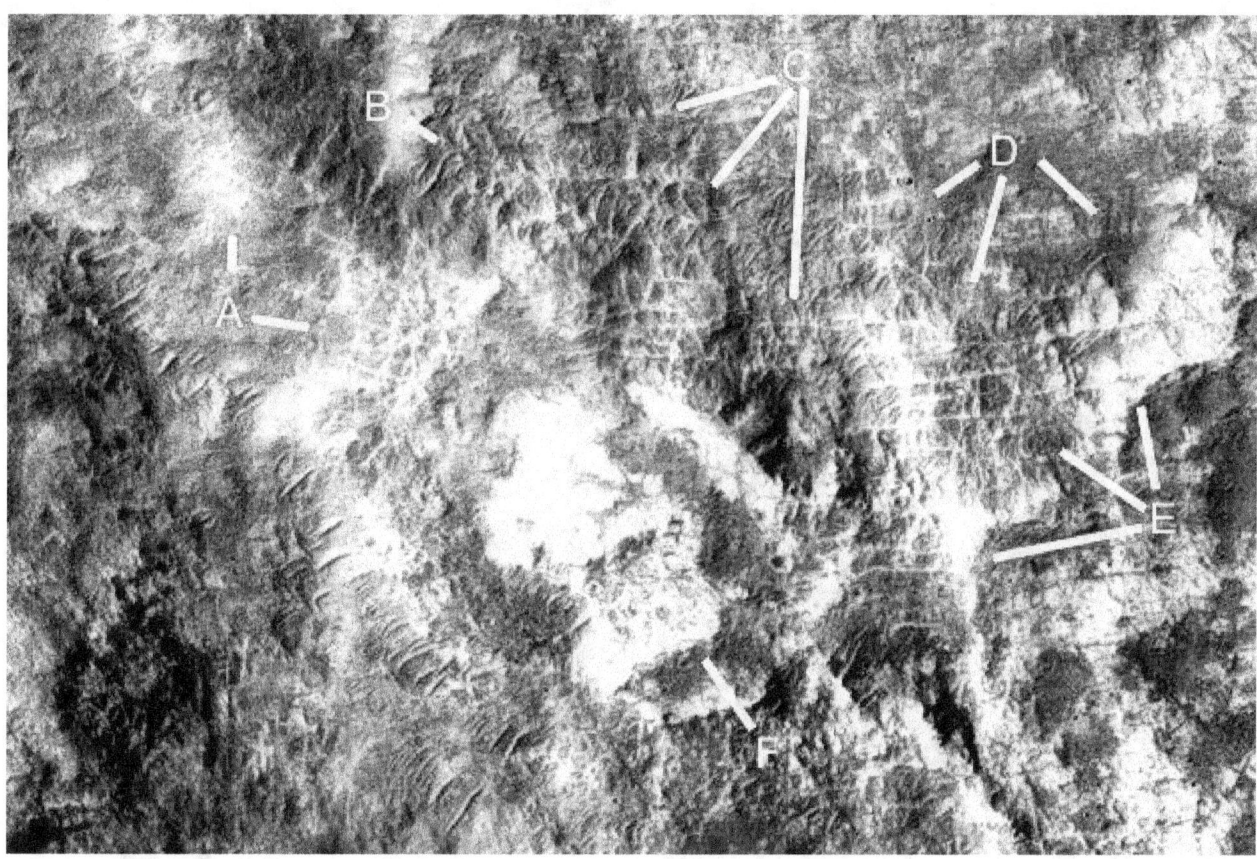

Cymhh355c

Hypothesis

A and B show an artificial looking shape, perhaps part of a pit dam. B at 10 o'clock shows how straight the edge is here.

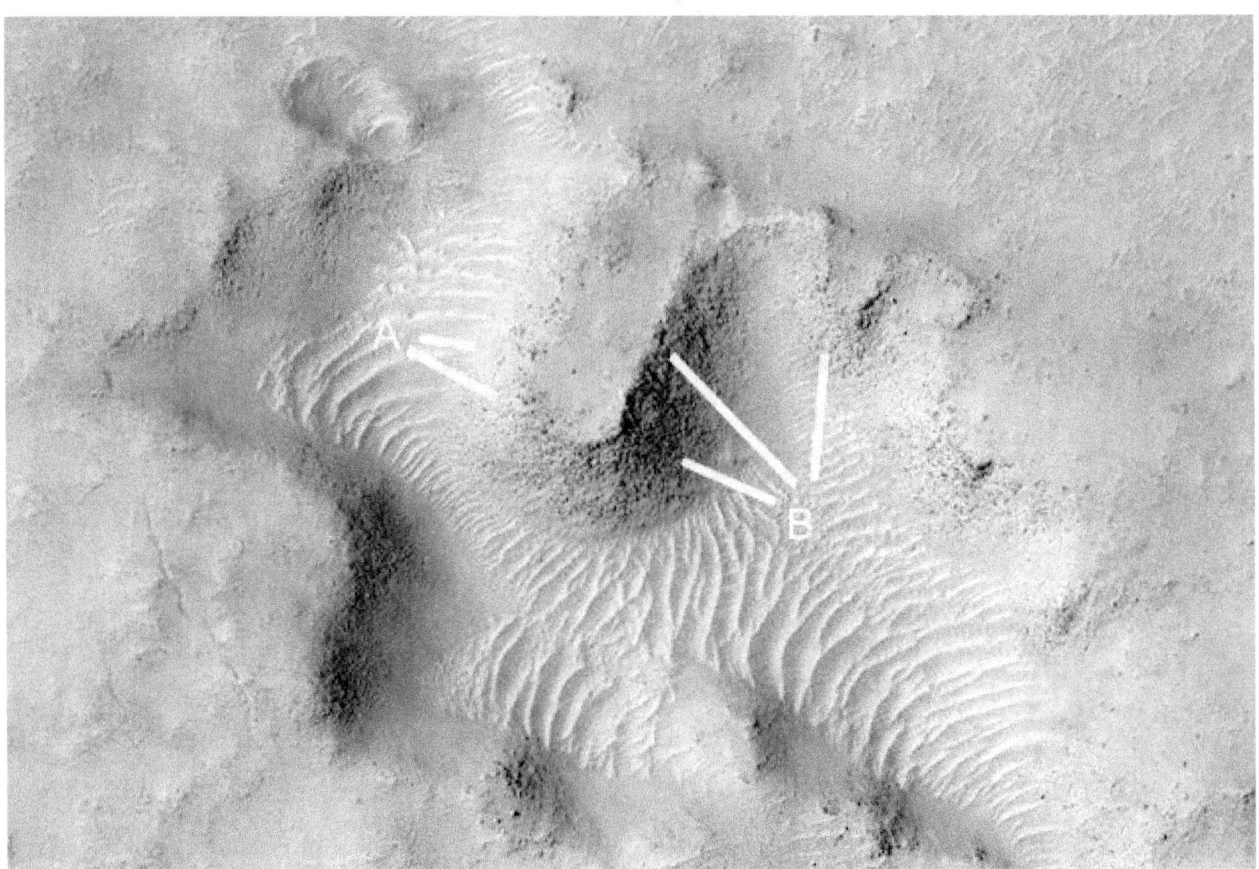

Cymhh355c2

Hypothesis

The bottom of the formation is shaped like a parabola.

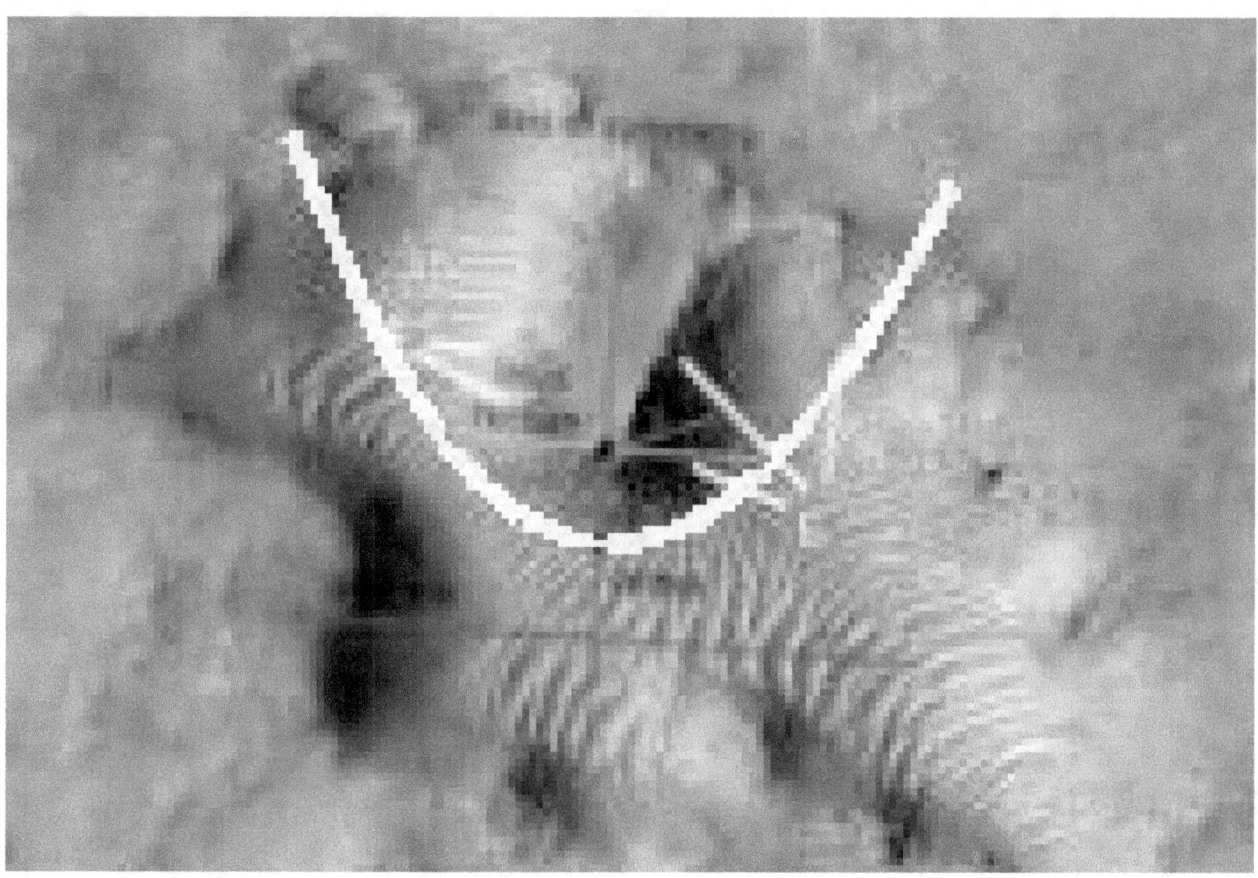

Cymhh361d

Hypothesis

A at 1 o'clock, B at 11 and particularly at 2 o'clock, and D at 6 o'clock shows one of these ridges is hollow like a tube. A at 2 and 4 o'clock shows more room shapes with dark borders, B at 3 o'clock shows more clearly defined rooms with a three dimensional shape. B at 7 o'clock shows the pit wall. C at 8 o'clock shows more rooms, at 11 o'clock the edge of the pit. D at 8 o'clock also shows more rooms and at 2 o'clock the pit wall.

Cymhh361f

Hypothesis

A at 12 o'clock shows a higher area with room like walls in it, the impression is of this eroding down to the other parts of A and B. C shows more rooms, D some ceiling material. Many of these rooms are likely to be full of dust from the collapsed roof.

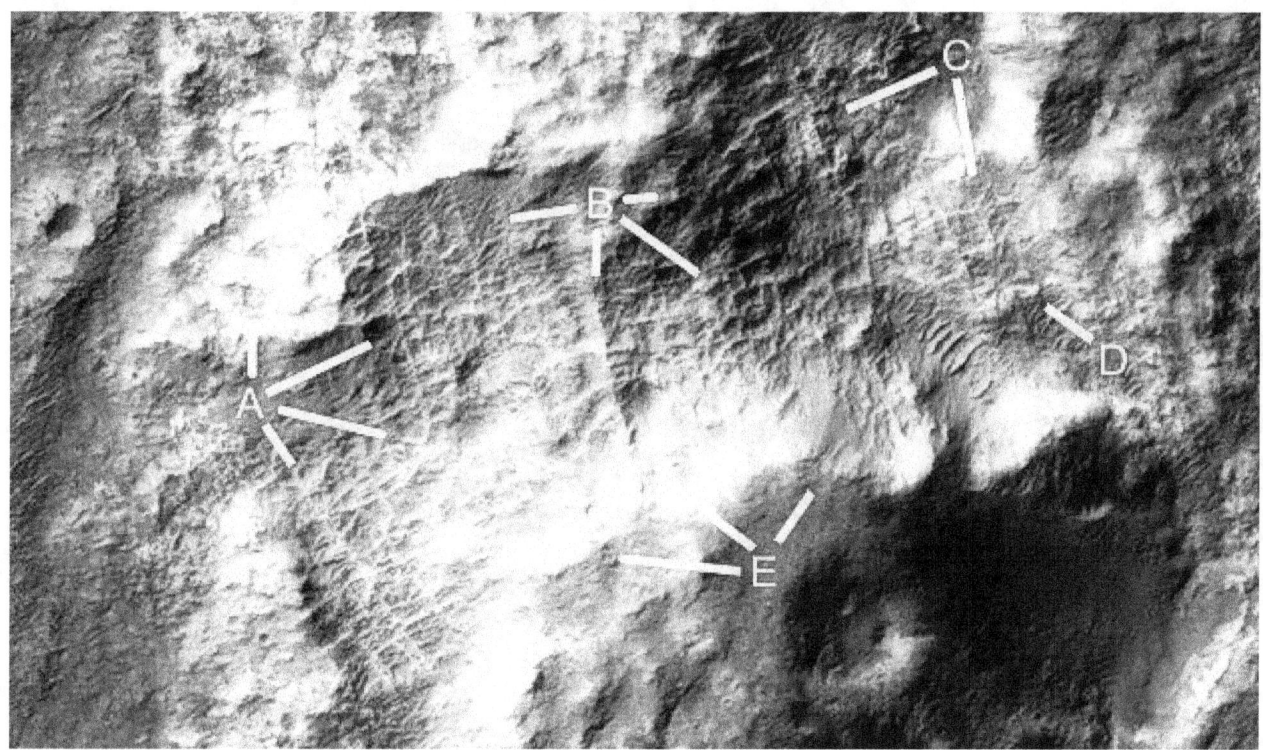

Cymhh361i

Hypothesis

The three dimensional impression is even stronger here, A shows rooms appearing under the smooth ceiling material. B may also be tubes or suspended roads as there is an impression of empty space under them. C at 9 o'clock shows rooms with no ceilings, at 4 o'clock there is still some ceiling or they are full of soil. D at 9 o'clock is like a hill of rooms, at 1 and 2 o'clock there is a road like formation that goes on to 12 and 2 o'clock. The letter E is in a depression surrounded by higher rooms like at 7 and 8 o'clock. F shows more variations in the elevations of the rooms from the shadow. G has many straight walls and may have right angles from directly above it. The rooms at H appear to be partially eroded.

Cymhh361I

Hypothesis

This pit is bare of rooms, A may be the normal terrain around these hollow hills. B shows the smooth ground inside the pit and the pit walls are well defined. C may have some remaining rooms. D may be a tube that continues on from 11 to 2 o'clock.

Cymd361l2

Hypothesis

The pit at B has a parabolic shape, the Latis Rectum is parallel to the tube at D.

Cymhh362e

Hypothesis

This shows a smaller hollow hill, A is where just an outline of the walls remains at 5 o'clock and the pit wall at 4 o'clock. From B across to E the hill has rooms throughout it. C may have been a larger room, D shows other walls. F at 10 o'clock has some clear rooms at 10 o'clock similar to Earth construction techniques probably with near equal sizes in right angles, at 4 o'clock is the pit wall.

Cymhh362e2

Hypothesis

A parabola is shown.

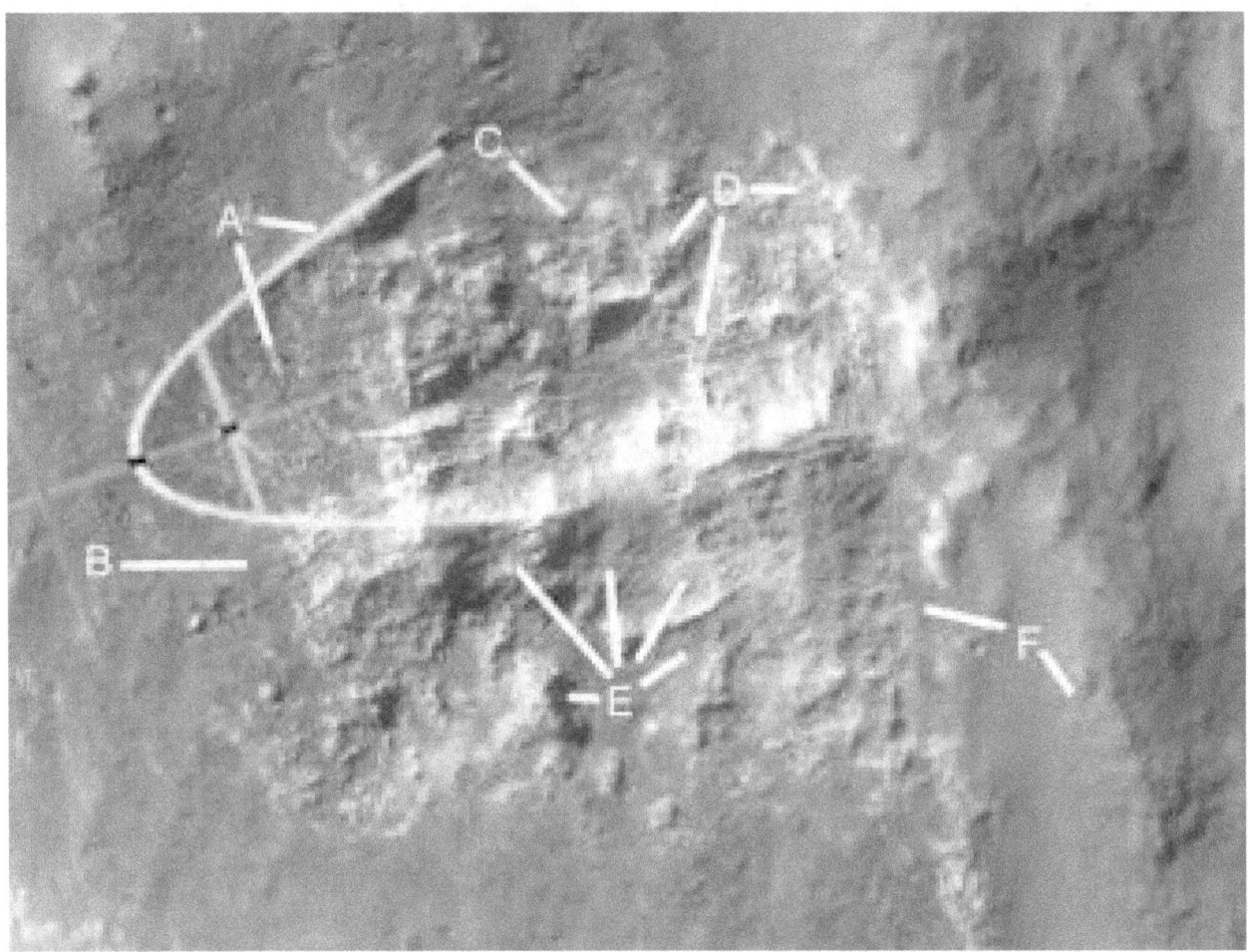

Cymhh362g

Hypothesis

A also looks like rooms under the ceiling material, B seems to be where they are being exposed as the ceiling recedes. C is more ceiling material, below this there is a hill shape with more rooms or perhaps a second floor of them. D shows tube like connections form from one hill of rooms to the main area above it. E shows progressive erosion of the rooms, at 4 and 7 o'clock would be the pit wall. F at 10 and 11 o'clock may show a steeper side as if there are surviving walls here, at 6 o'clock is the pit wall.

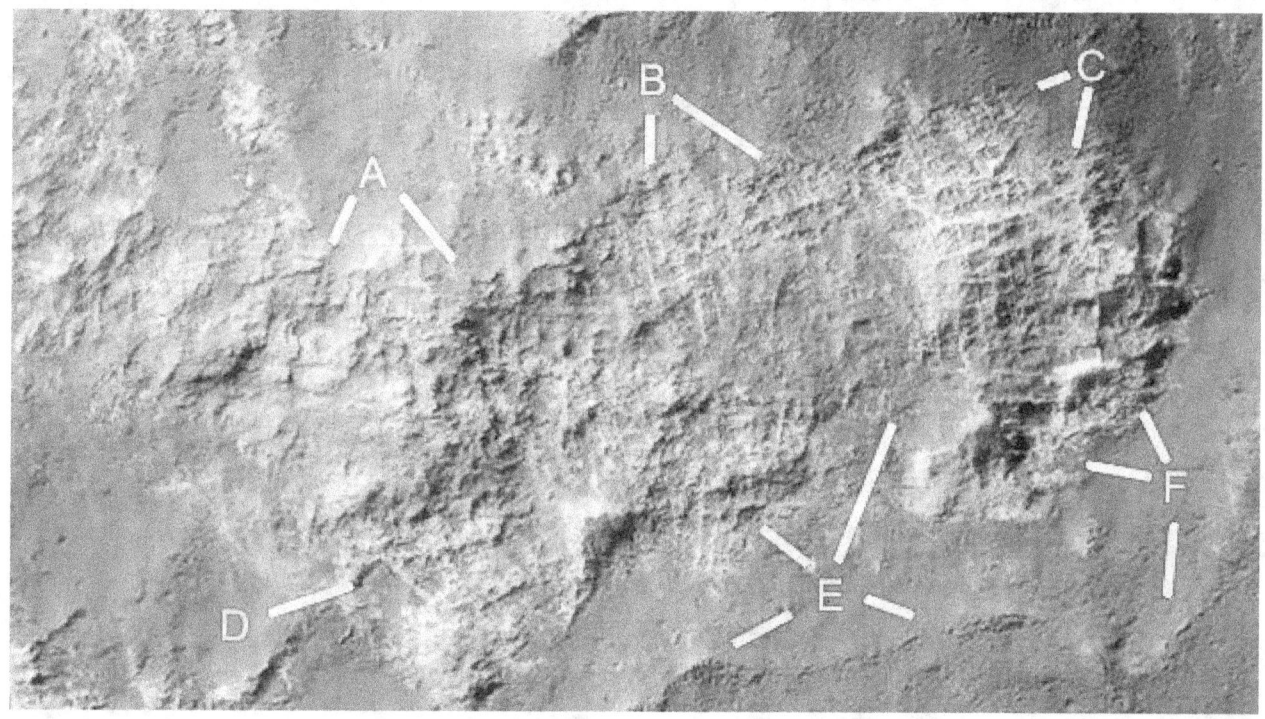

Cymhh363a

Hypothesis

These rooms have a much more radial pattern, A shows more rooms under the ceiling material at 7 o'clock. At 4 and 5 o'clock the walls are roughly parallel with some getting closer together towards the right. The rooms at B seems to go into the slope like a layer of them, many are triangular or irregular to give this radial shape. C may show rooms set into the slope like the layer of rooms broke off here, it may be possible to follow these deeper into the slope. The walls vary in height at 7 o'clock. D shows more rooms but to its right they have either eroded away or are still buried. E shows an edge of this radial array, below these there may be more rooms under the ground. At F the walls are more sparse perhaps with bigger rooms or they could be tubes and tunnels. G shows how these appear to have altered the crater shape on its lower side with a wavy ridge or interior support. H shows more rooms and I how these come to the edge of another crater. These walls would be fragile and unlikely to survive the fracturing in the ground from impacts of this size.

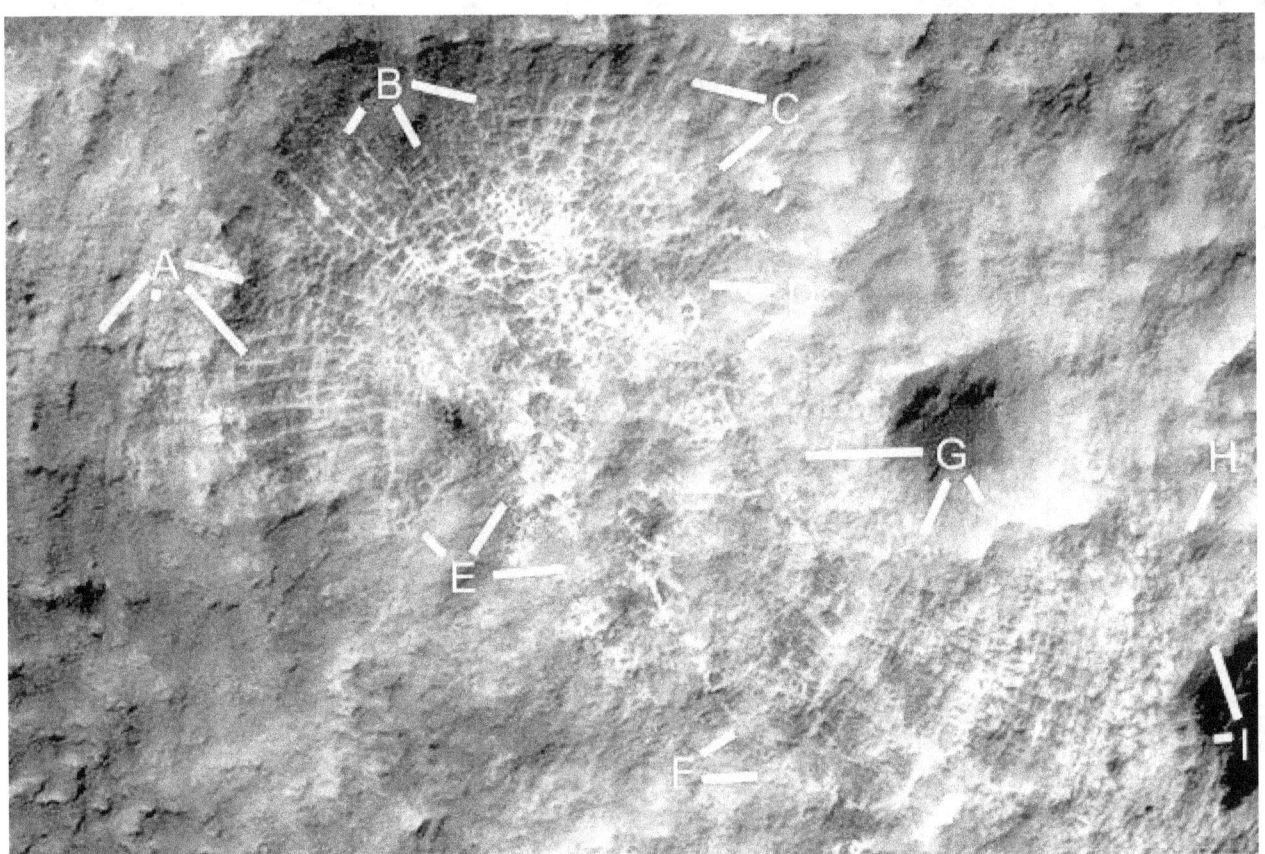

Cymhh363e

Hypothesis

The lower part of this image is much smoother as if the skin is still intact, A shows it eroding at 2 o'clock, at 7 and 8 o'clock there appear to be the edges of the skin. At 4 and 5 o'clock, and B at 10 o'clock is a smooth shape like cement. C and D also show where this skin is breaking up.

Cymhh363e2

Hypothesis

This cement like shape closely follows a double parabola as shown. The latis rectums are also parallel t each other, still less likely to occur by chance.

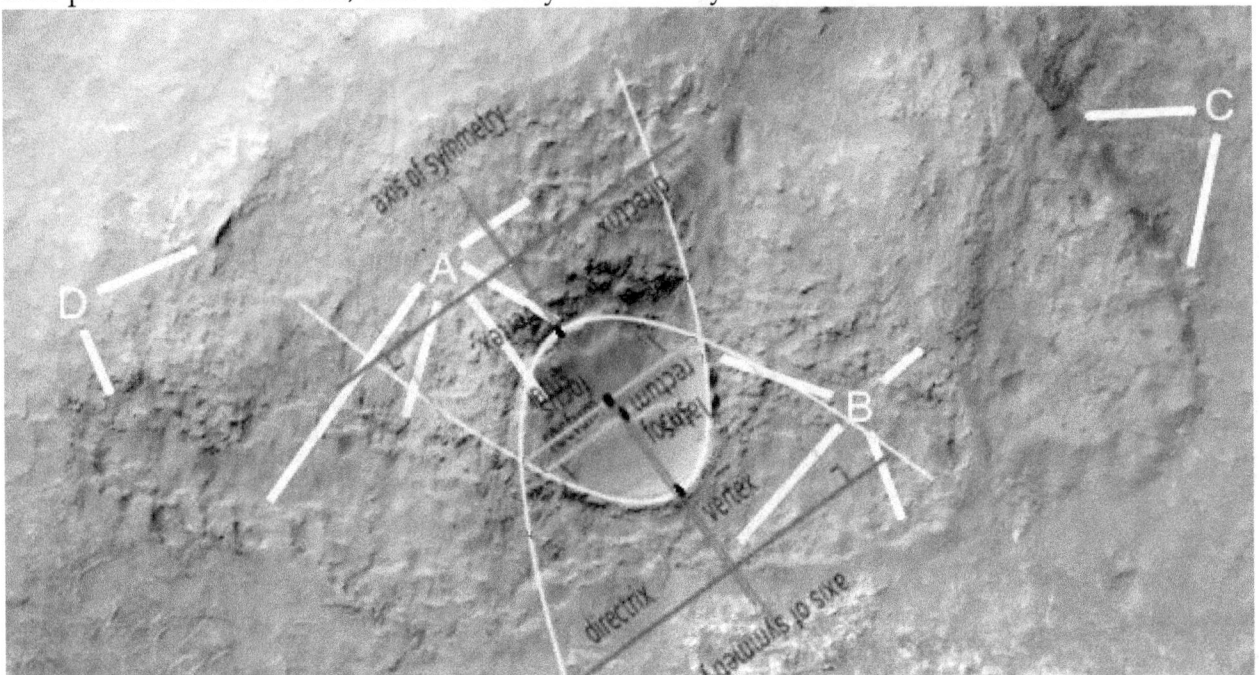

Cymhh363e3

Hypothesis

Four other shapes are like parabolas as shown.

Cymhh363f

Hypothesis

A shows more walls in good condition, many appear to be at right angles. B shows some smaller rooms at 10 o'clock and much clearer rooms at 10 o'clock. C shows more irregular walls perhaps from erosion at 9 o'clock, at 3 and 5 o'clock the rooms are much more regular. D also shows regular rooms at 6 o'clock perhaps more eroded at 4 o'clock. E shows some rooms covered by ceiling material in patches at 10 and 2 o'clock, much more covered at 3 and 9 o'clock.

Cymhh363j

Hypothesis

These rooms are particularly clear because of the shadows, A, B, C, and D show rooms being exposed under the ceiling material. E appears to be highly eroded rooms, F and G are very clear rooms perhaps with more walls seen inside them.

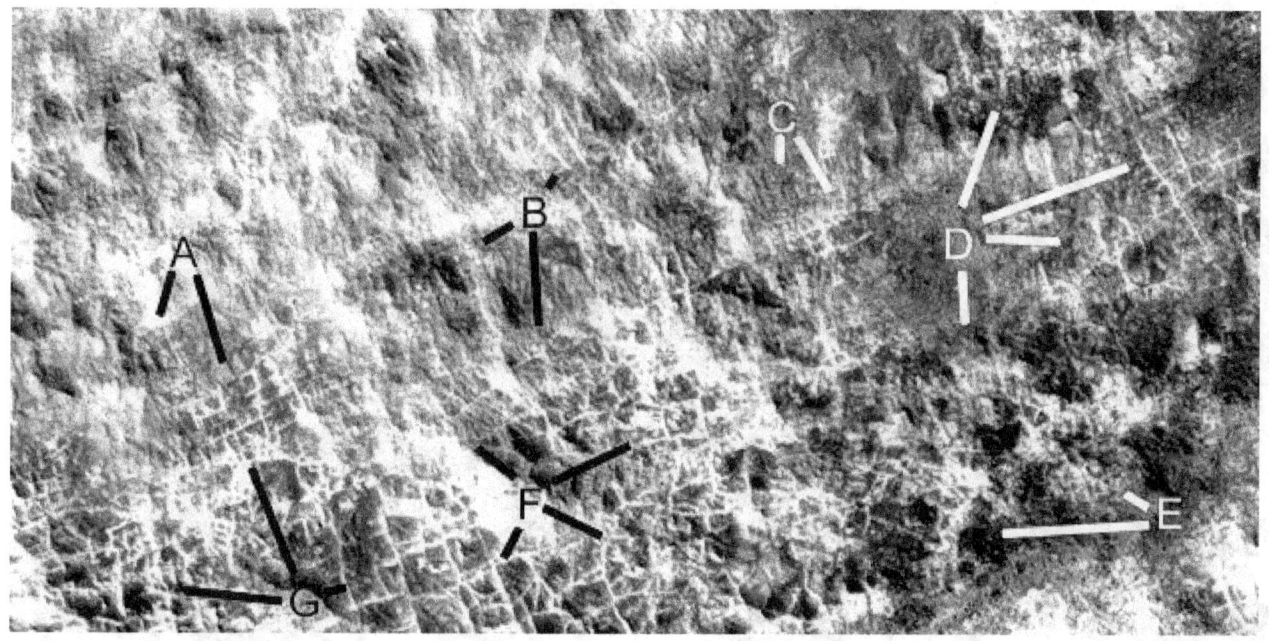

Cymhh363m

Hypothesis

From A to F there are many right angled rooms, at B these are more eroded. C also appears eroded as if the walls are partially collapsed. To the left of D it appears 3 dimensional with the shadows, E is also 3 dimensional but there is more ceiling material or debris between the walls.

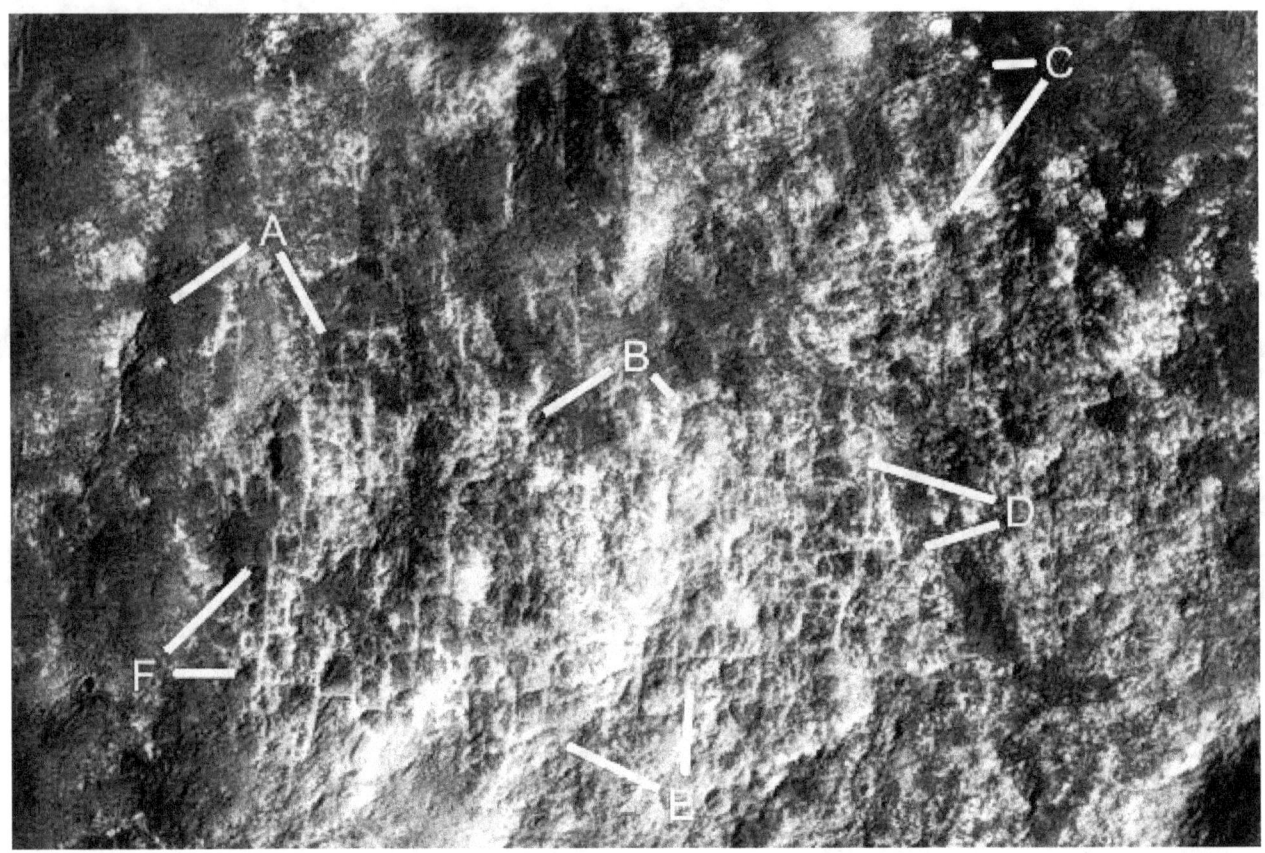

Cymhh363n

Hypothesis

A, B, and C show more highly eroded walls, at D and E these are much clearer with more right angles. F appears much lower than G, as if G has more rooms buried under it.

Cymhh368

Hypothesis

These darker areas have very sharp boundaries, they may have been farms. A may have been for drainage, B and C are very consistent in their shade and this has not blown across the boundary at D like with dark dunes. E has some unusual shapes, perhaps small hollow hills.

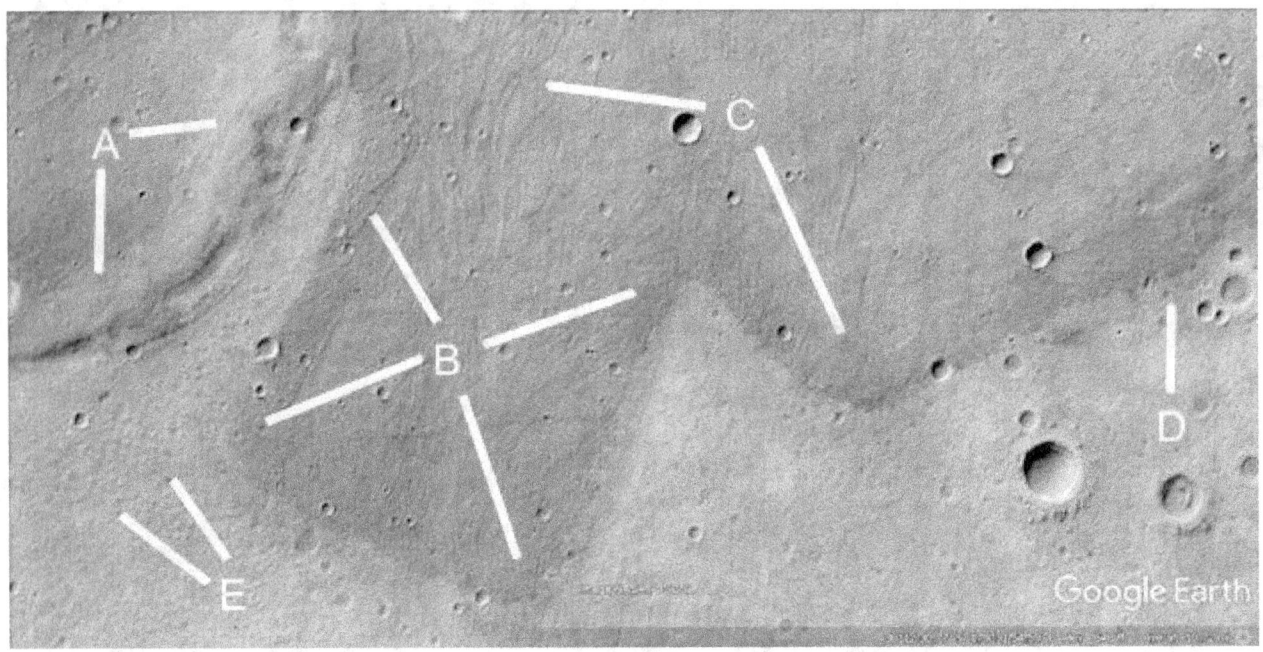

Cymhh368a

Hypothesis

These shows two parabolas in the image. B is also close to a rectangle in shape.

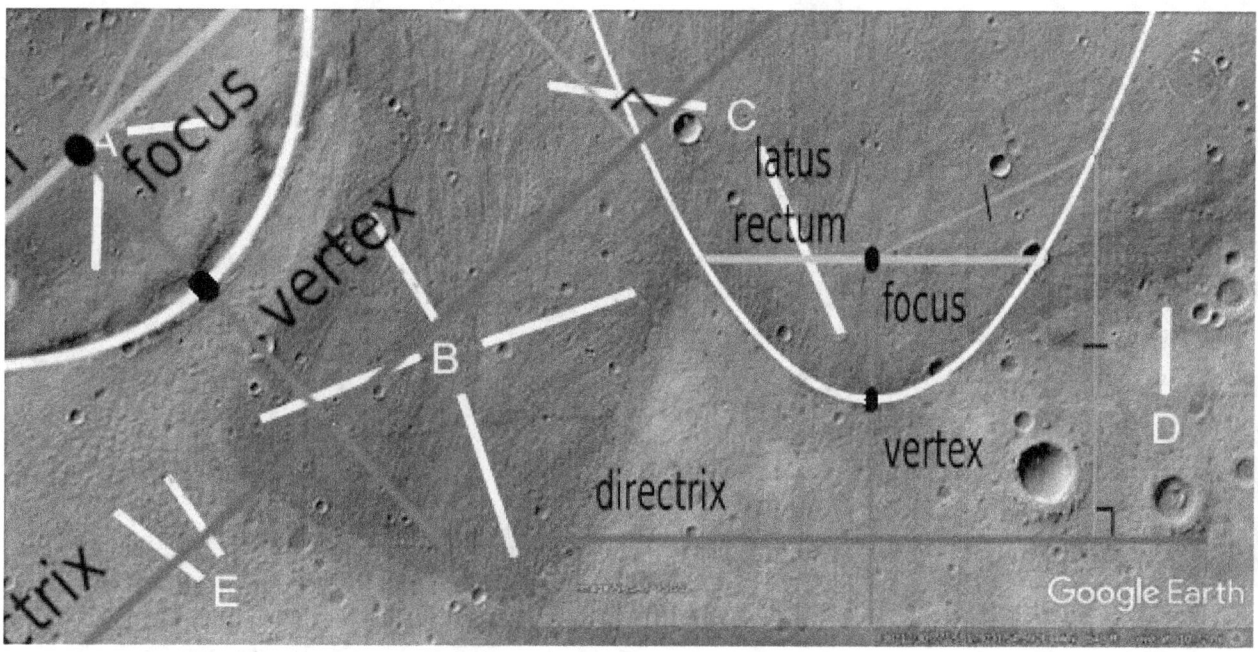

Cymd370a

Hypothesis

A shows a brick like consistency on the bottom of the dam wall, as if there are rows of material and columns in its construction. It is also regular in shape though it protrudes from the cliff.

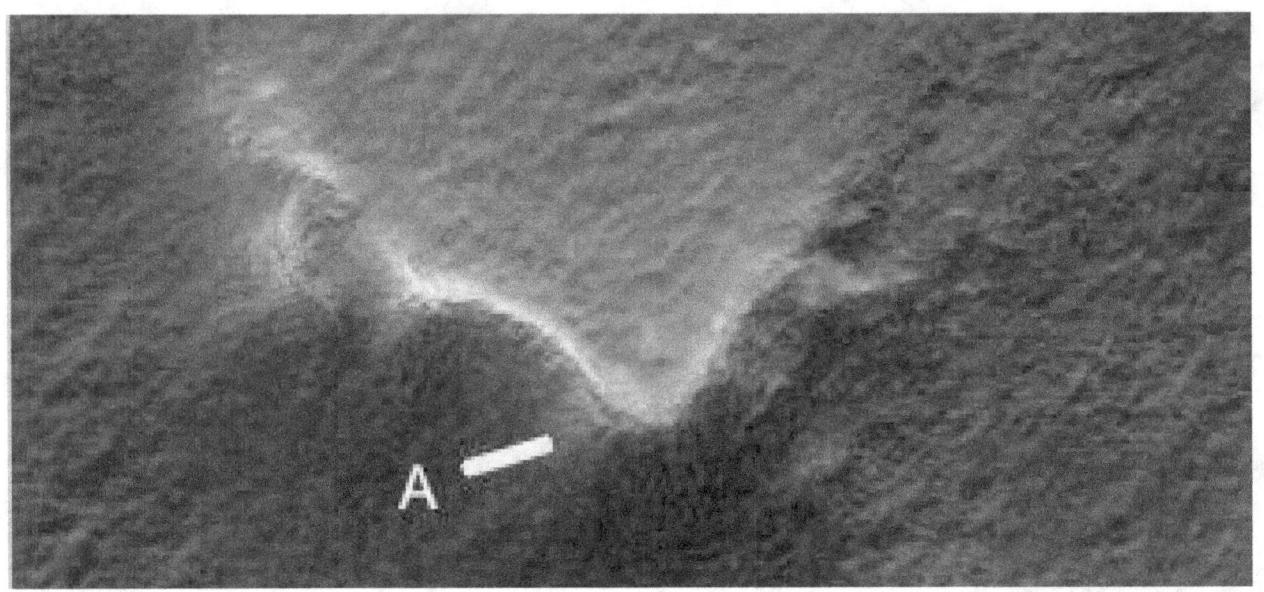

Cymd370a2

Hypothesis

The middle of the dam also forms a parabola.

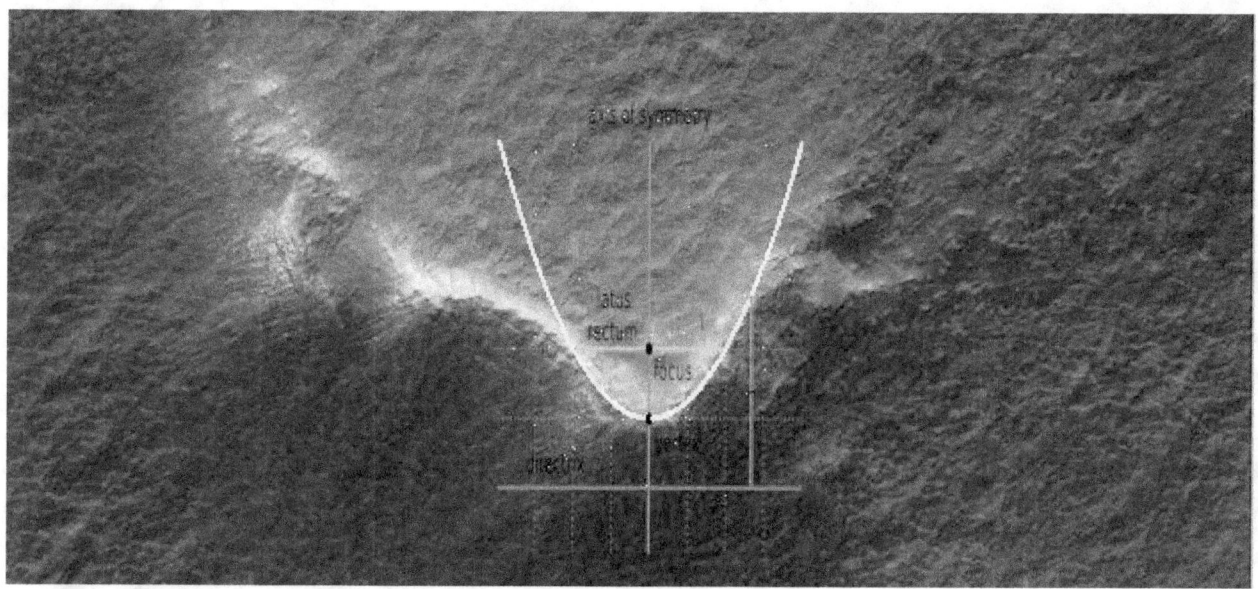

Cymd370b

Hypothesis

This shows how the dam would have caught the water from the top of the crater. It is the most unlikely place for this formation to occur as water would tend to erode this away if natural. Instead the rest of the crater wall around it is smooth.

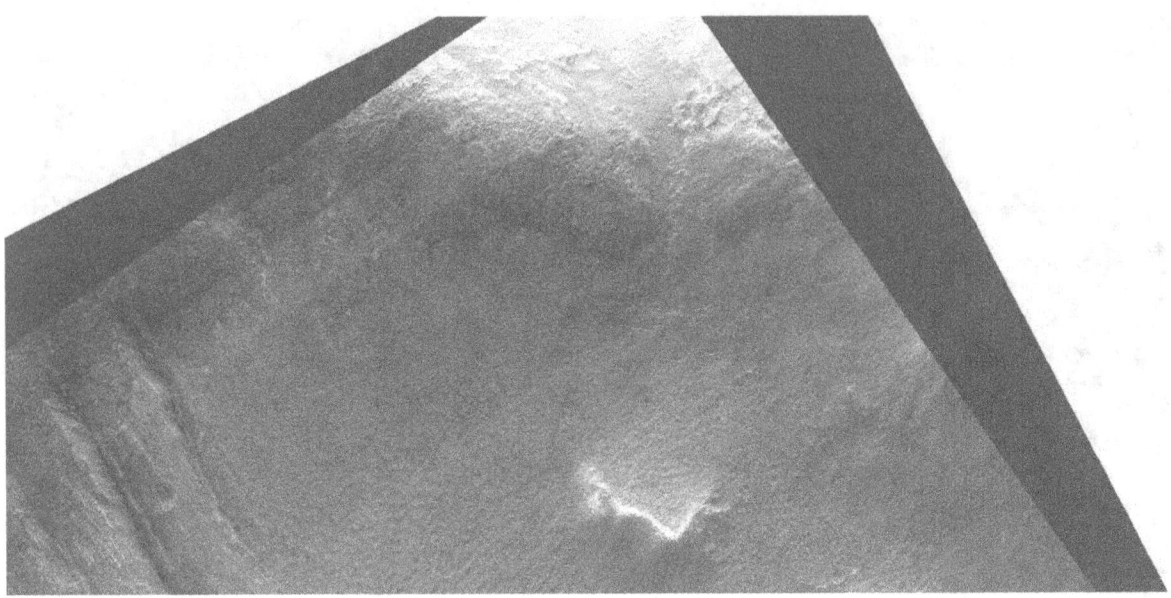

Cymd370c

Hypothesis

A and B show a former dam, it looks like the dam wall broken off leaving a parabolic groove. C shows another probable dam, the stones are like columns in the former dam wall. D shows another dam with the columns intact, it indicates the construction technique of the wall. It may have had the columns inserted first here and then the wall connected between them.

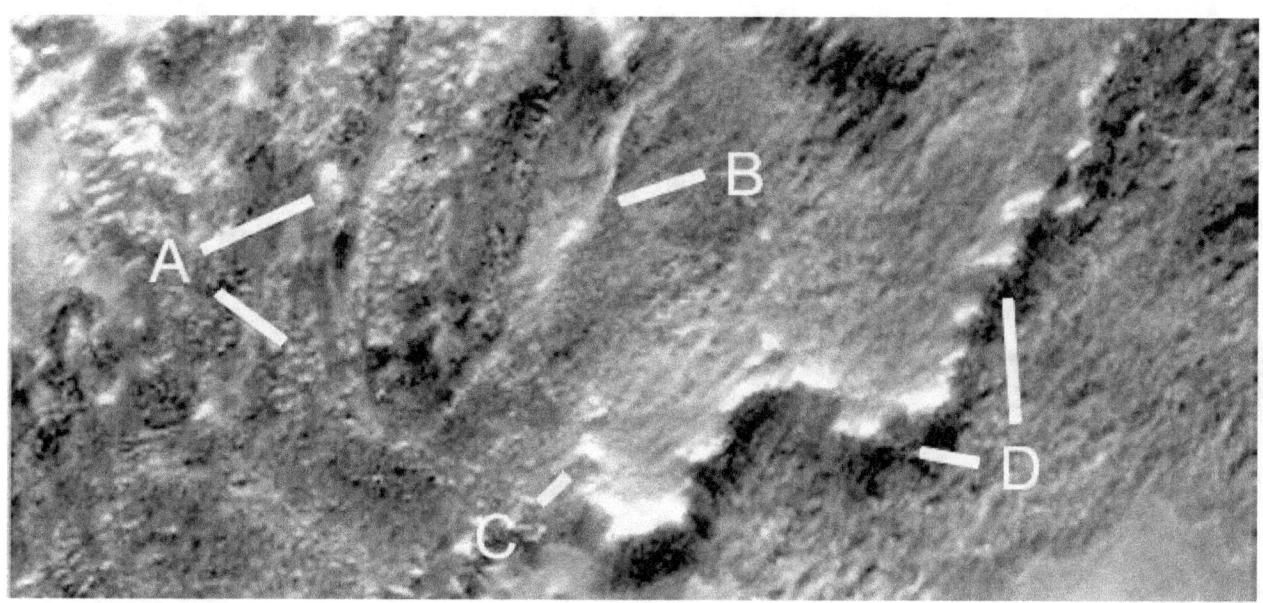

Cymd2370c2

Hypothesis

This shows how the former dam had a parabolic shape.

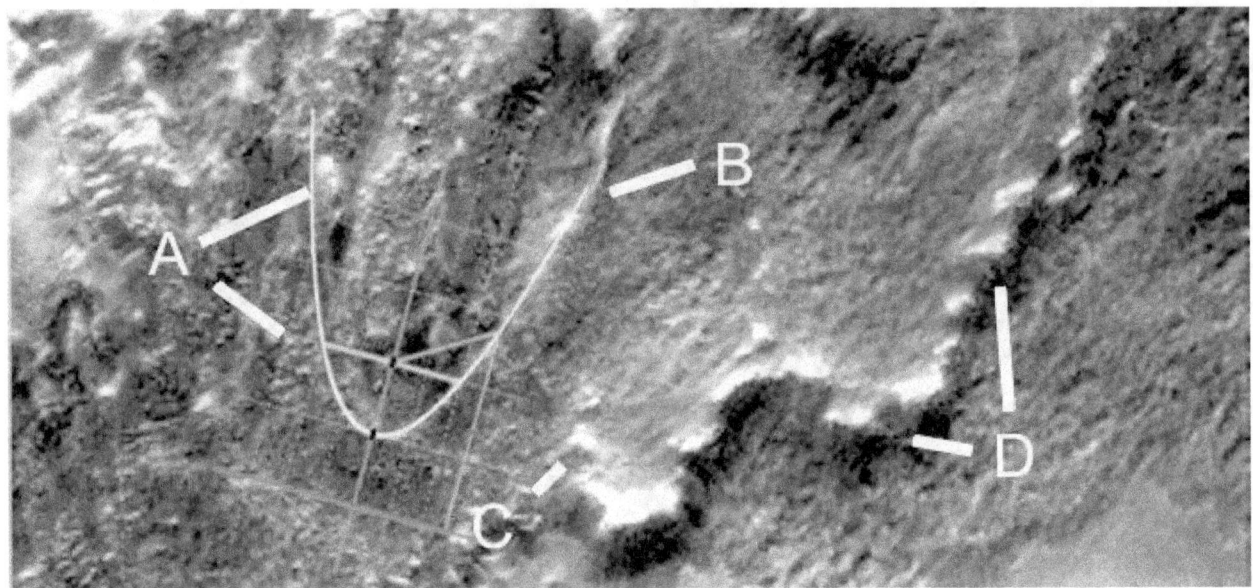

Cymd370d

Hypothesis

This may also have been some dams, above A is smooth ground like a cement floor of a dam. Below this is a small dip like the apex of a dam wall. B also appears to have columns like the dam wall collapsed. C is approximately a parabolic arch, the two often go together.

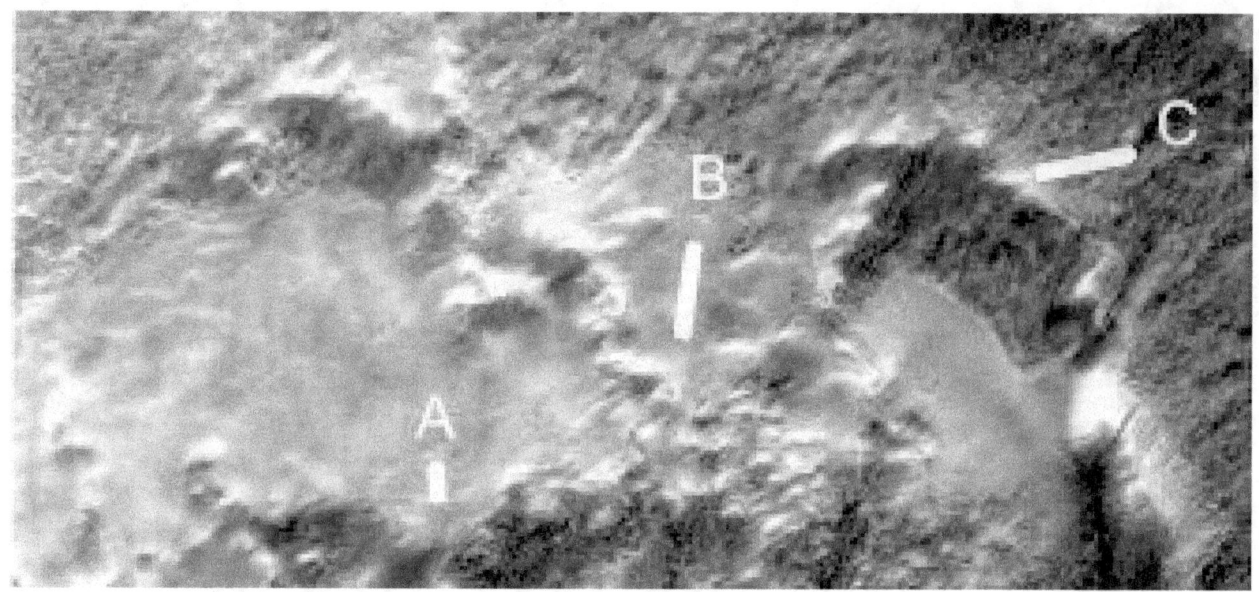

Cymd370d2

Hypothesis

While highly eroded, there may have been a parabolic dam and a parabolic arch here.

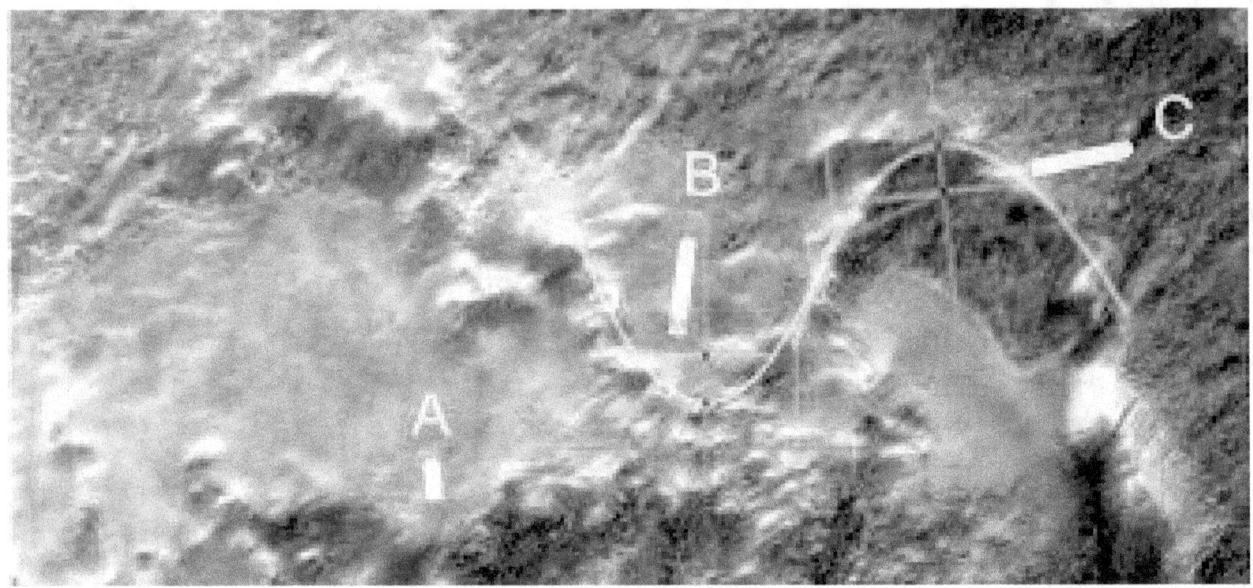

Cymd370e

Hypothesis

A shows how the dam floor is much smoother than the terrain under it, like cement. B at 5 o'clock shows the dam wall nearly eroded away. At 8 o'clock there is a break in the cement floor like it is breaking up.

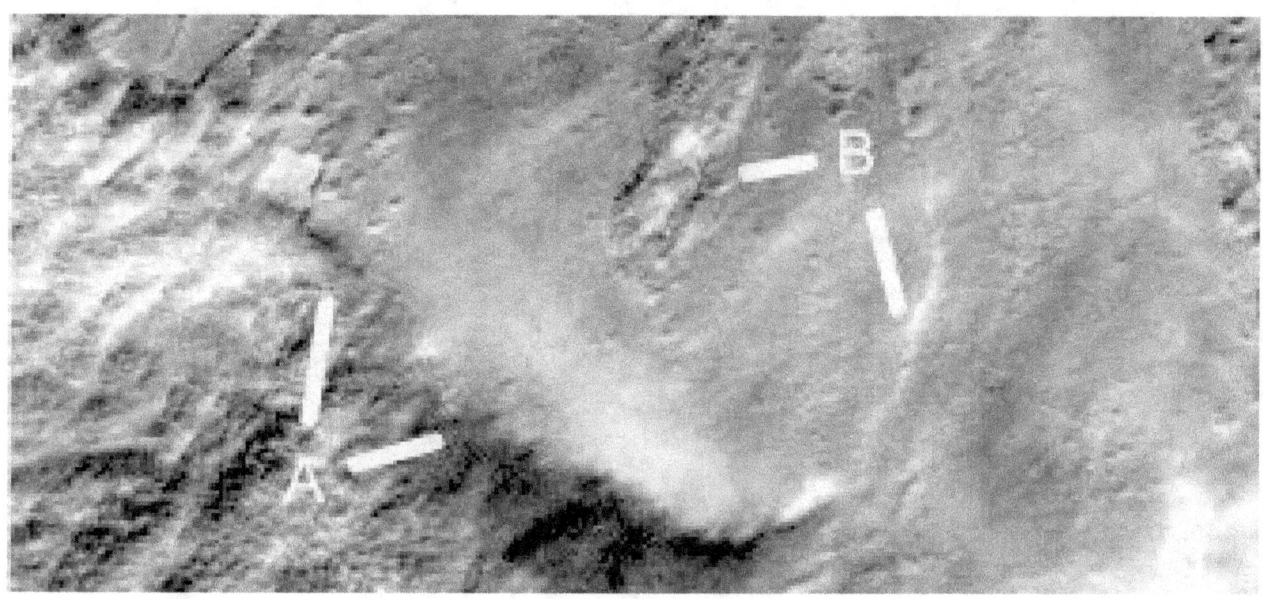

Cymd370e2

Hypothesis

This shows the parabolic shape.

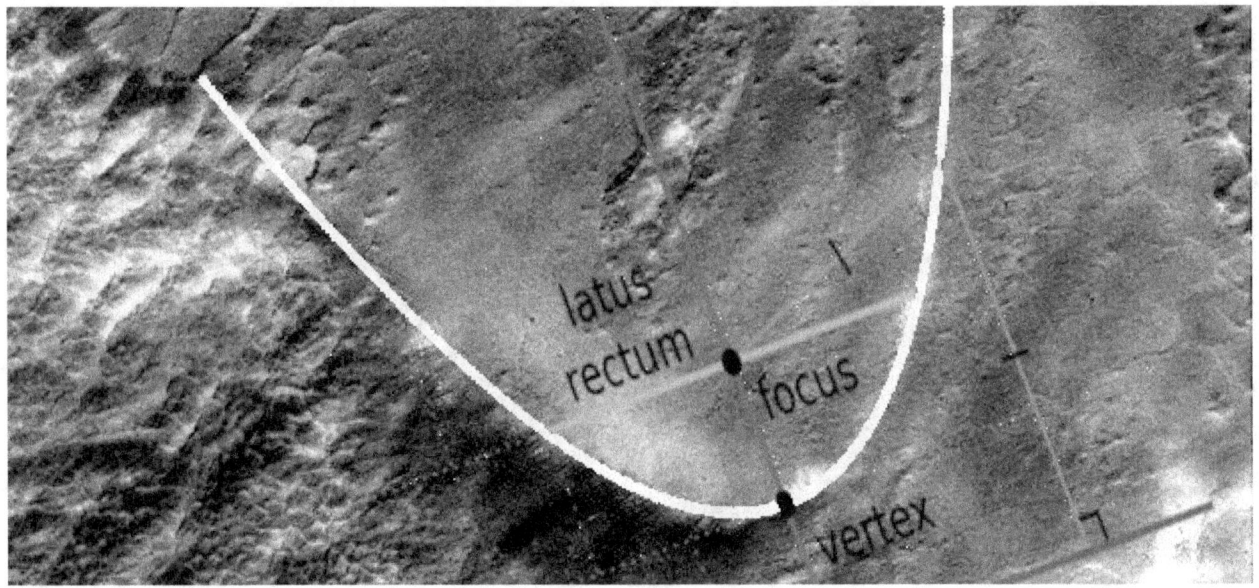

Cymd370f

Hypothesis

A may show a small dam where the columns remain. B shows a more complete wall at 10 and 12 o'clock, at 3 o'clock is probably the edge of the cement floor of another dam. C shows the cement edge above the dam.

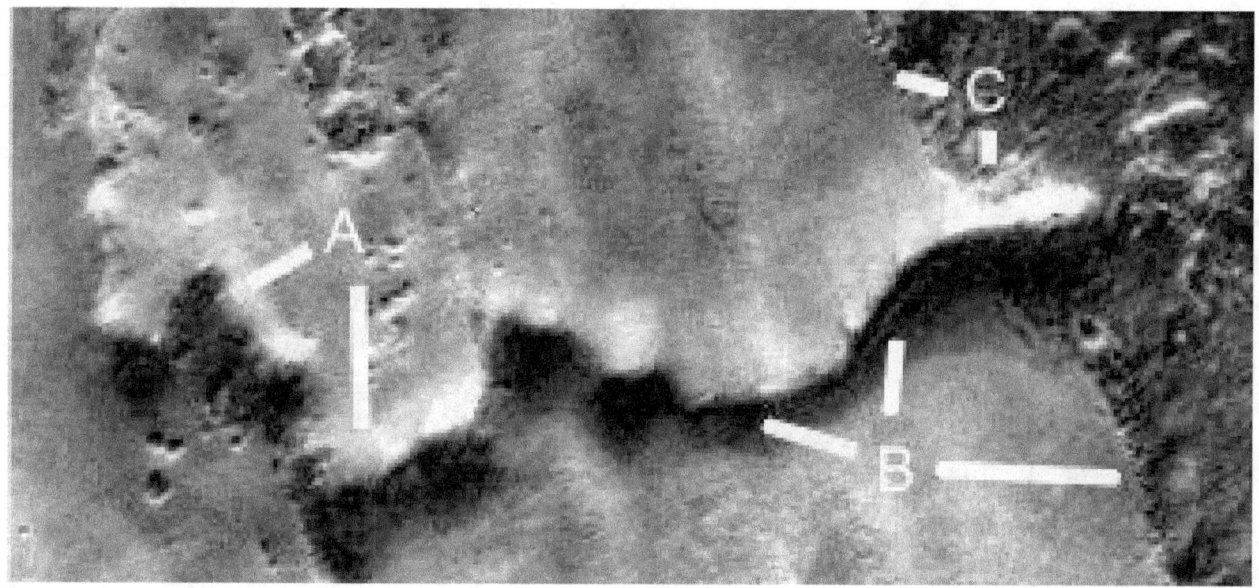

Cymd370f2

Hypothesis

This also has a parabolic shape.

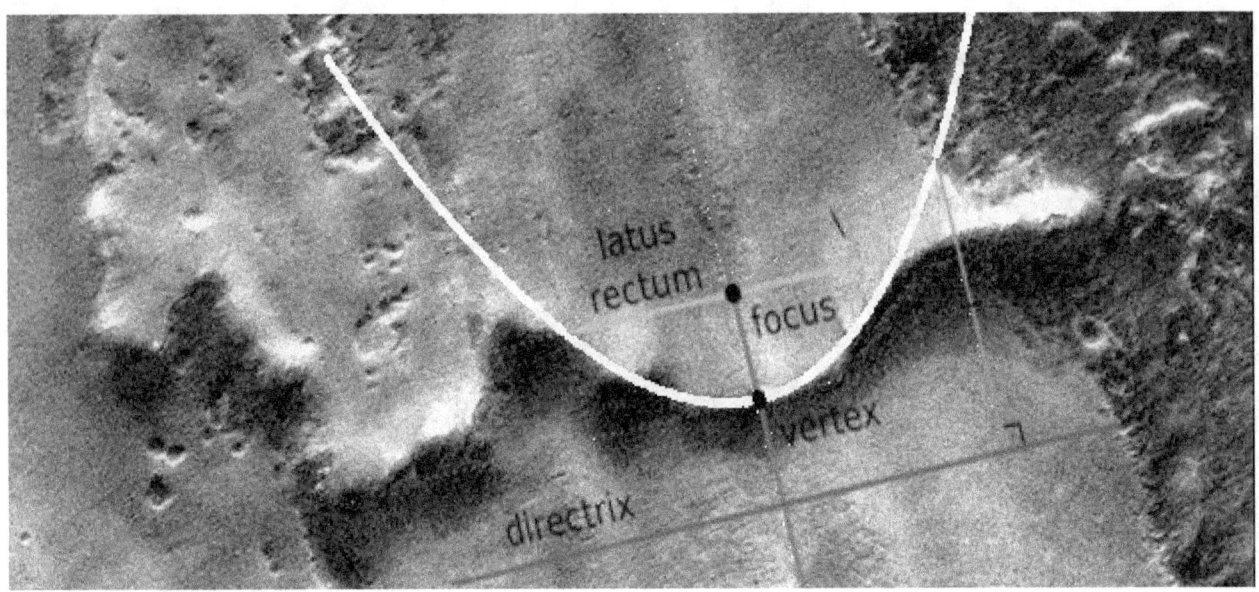

Cymd370h

Hypothesis

A shows the smooth floor of the dam, different to the rougher terrain on the left. B shows creep or cold flow in the cliff material, but the dam floor shows no sign of creep. It implies it is a different material. C shows the different dam walls.

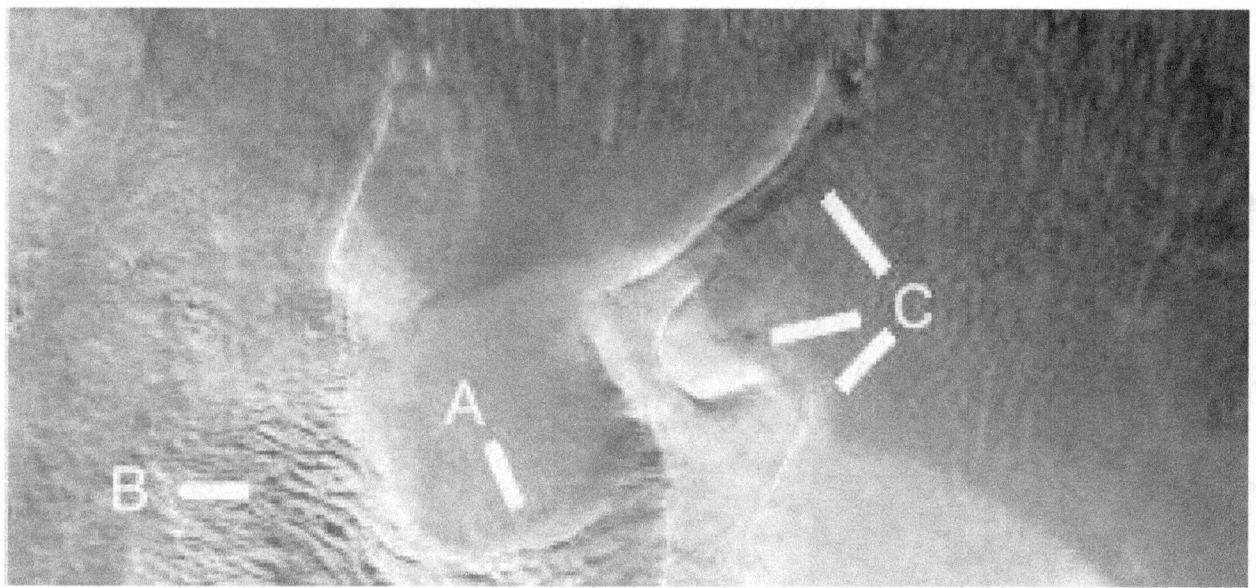

Cymd370h2

Hypothesis

This shows the dam is a parabola, also the arch under the dam on the right is a parabola.

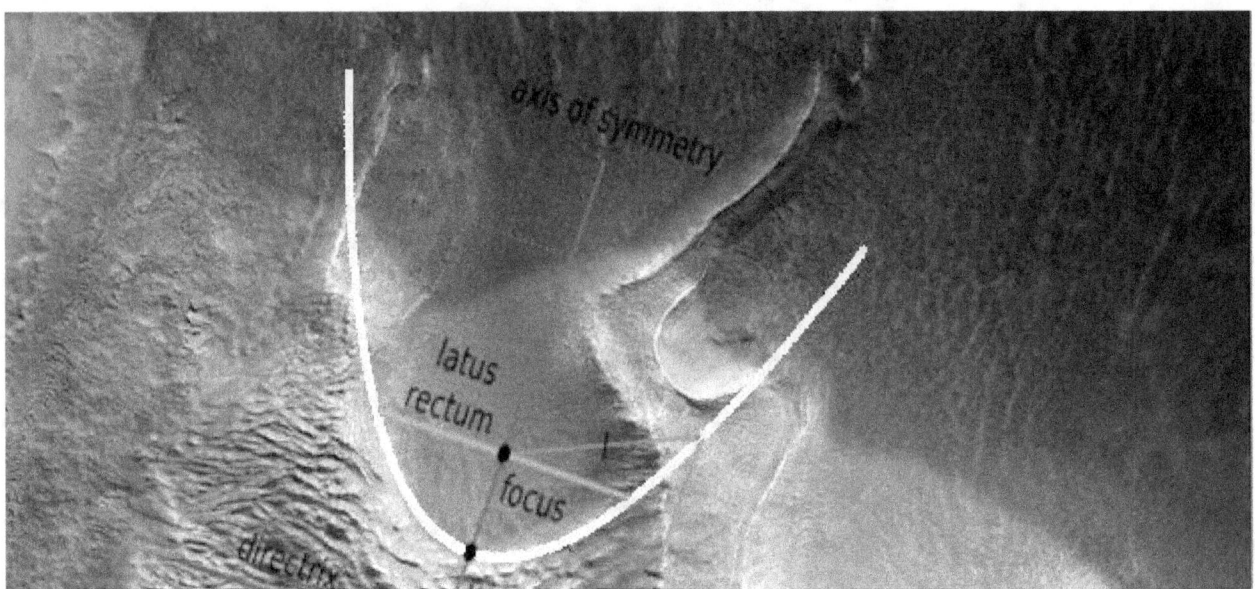

Cymd370k

Hypothesis

A is like the support for the dam, it extends with a constant thickness over to D at 2 o'clock and E at 5 o'clock. It also has a smooth curve in it like an arch. B would show the degraded dam floor, C would be either creep or is designed to slow the water going down the slope. D at 9 o'clock appears to be where this cement like exterior is degrading. E is another dam with a small parabolic shape at 6 o'clock.

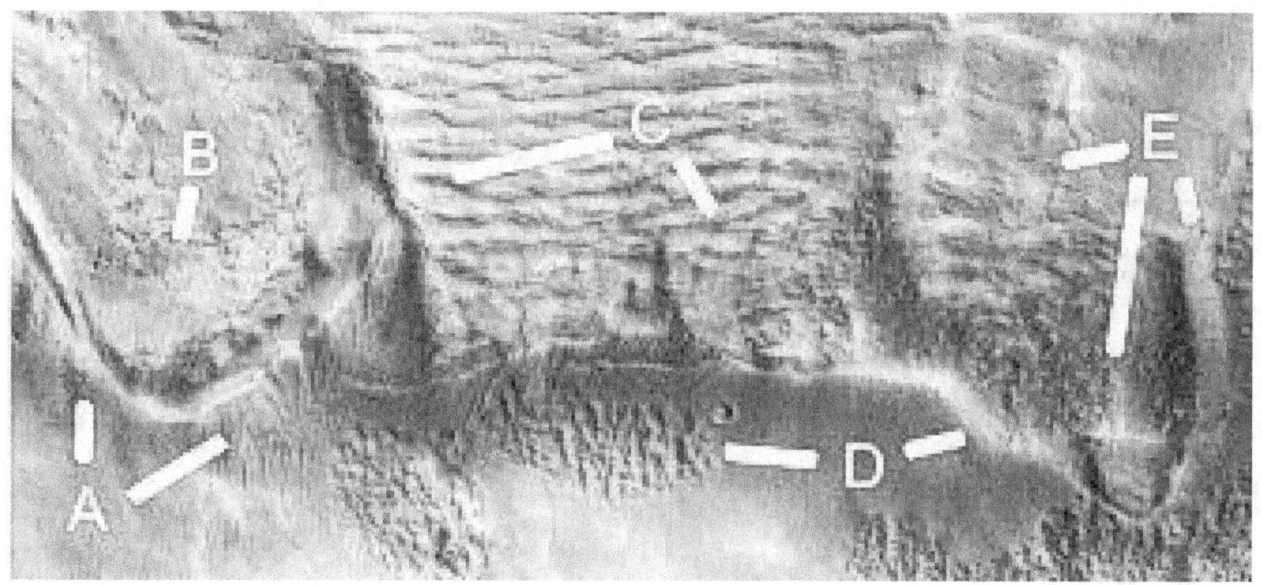

Cymd370k2

Hypothesis

This shows two parabolic dams.

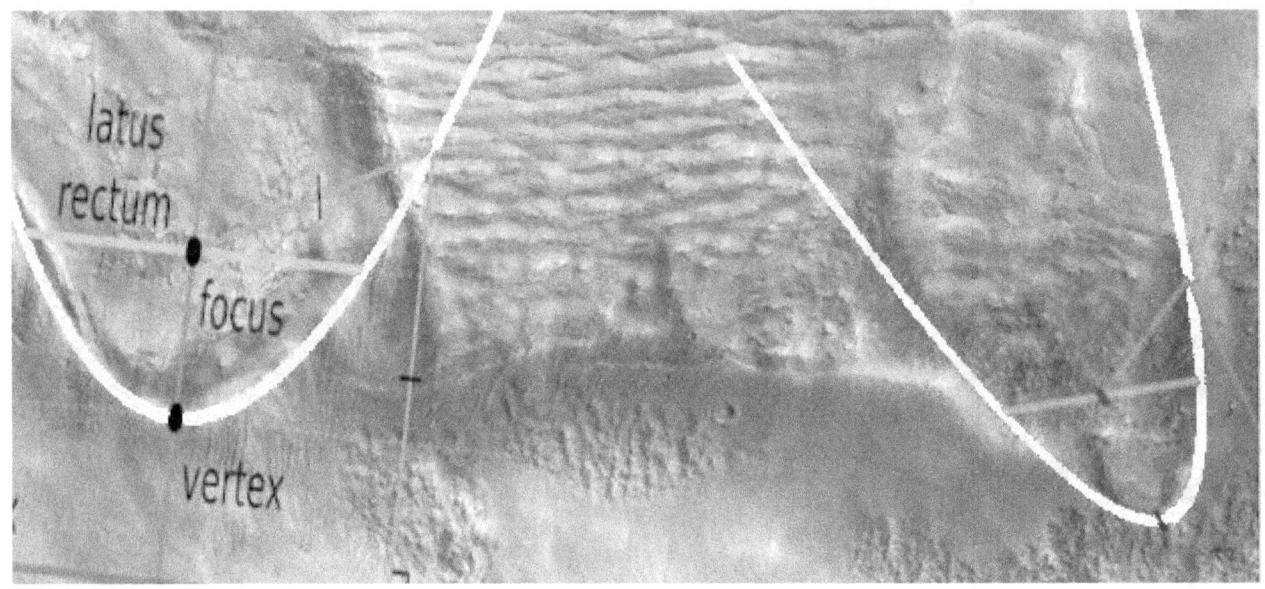

Cymd396a

Hypothesis

A shows a dam where the wall appears to have broken off, the shape remains approximately parabolic. The dam floor is also smoother than the surrounding terrain. B shows another dam which is deeper, C looks like a water flow come down the side of the dam.

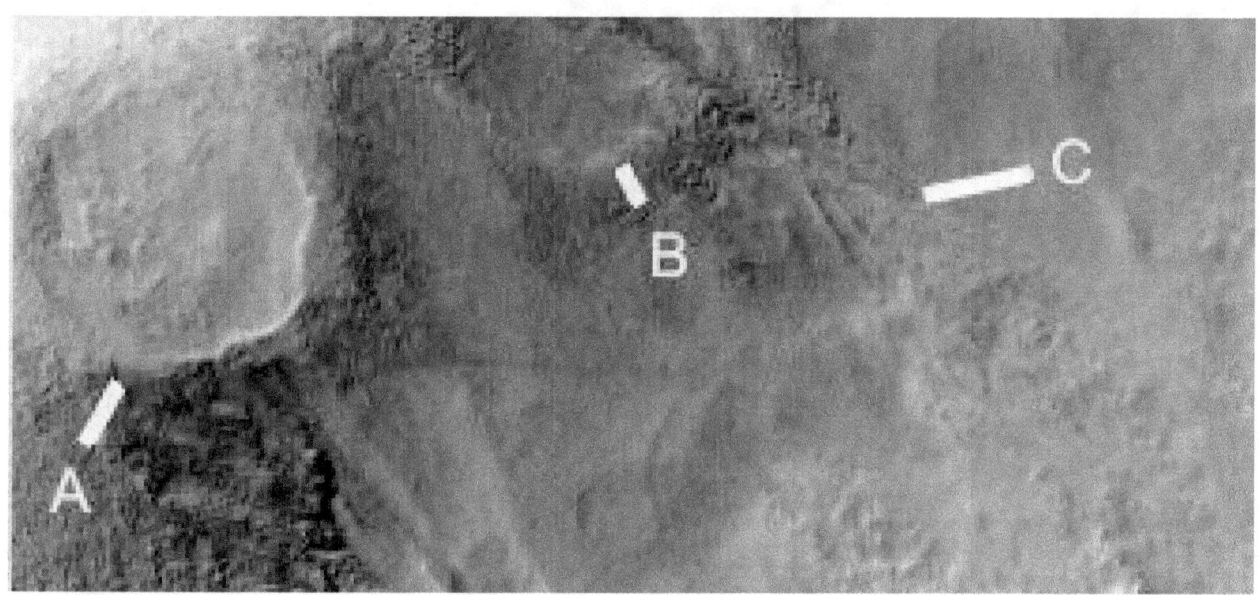

Cymd396a2

Hypothesis

This shows two parabolic dams.

Cymd396b

Hypothesis

A and B are probably two highly eroded dams.

Cymd396b2

Hypothesis

A has a parabolic shape.

Cymd396c

Hypothesis

A shows a parabolic dam, the dam floor is highly degraded as if cement is flaking off. The wall under it is smoother with a rounded shape like an arch for support. B is shaped like a parabolic arch, C shows the edge of another parabolic dam with some damage to the wall.

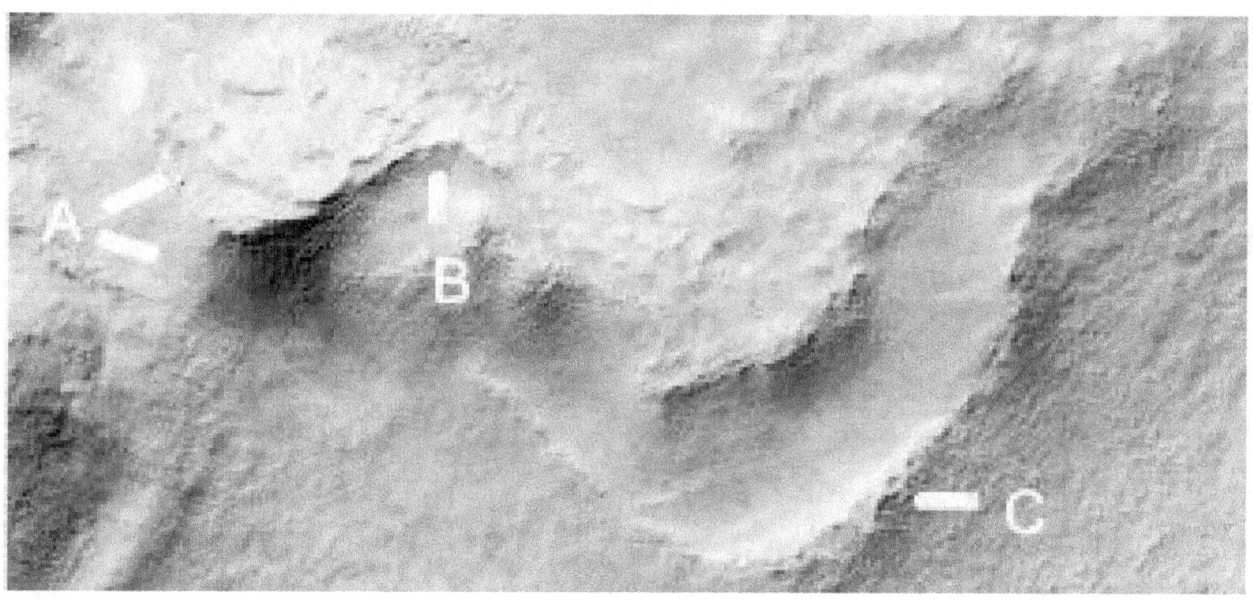

Cymd396d2

Hypothesis

This shows one of the parabolic dams.

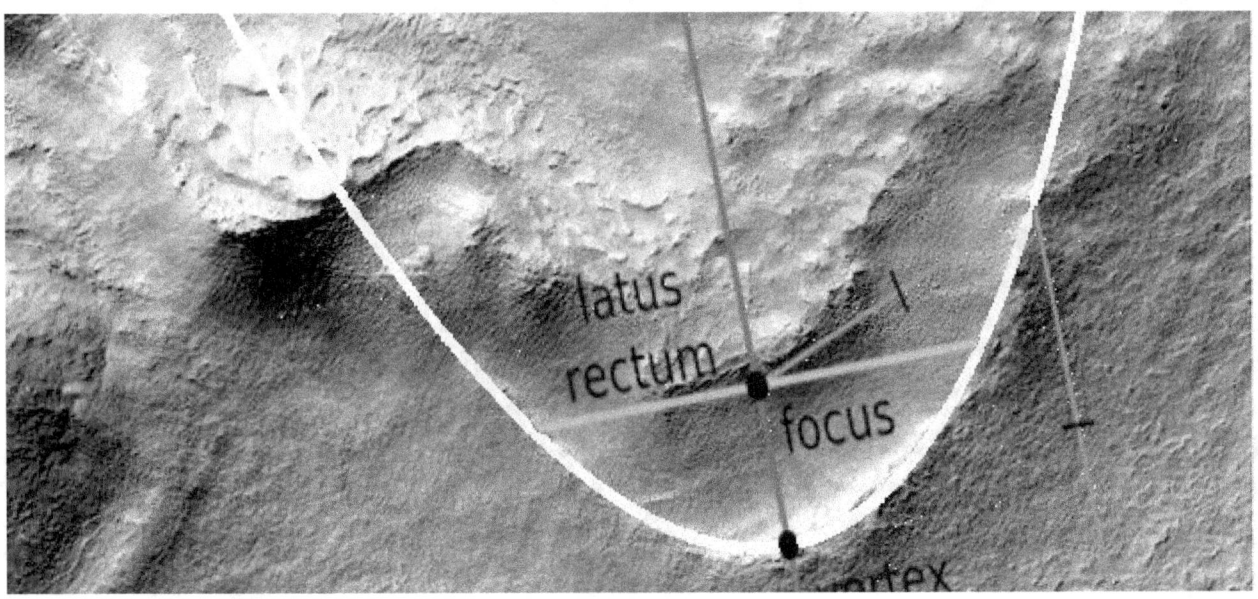

Cymd396c3

Hypothesis

There is a parabolic dam on the left, on the right there are signs of a parabolic arch and a large parabolic dam.

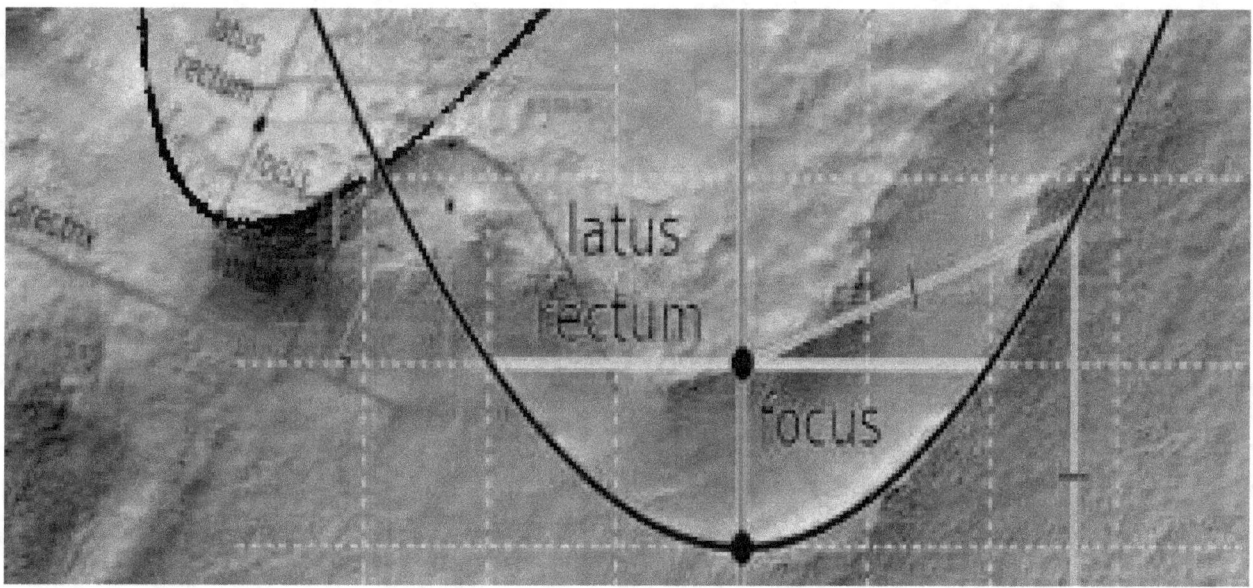

Cymd396d

Hypothesis

From A to C is a parabolic arch, the ground is flat with a similar albedo above it and below it is smoother than the surrounding terrain. B shows where the parabolic dam is cracking on its wall.

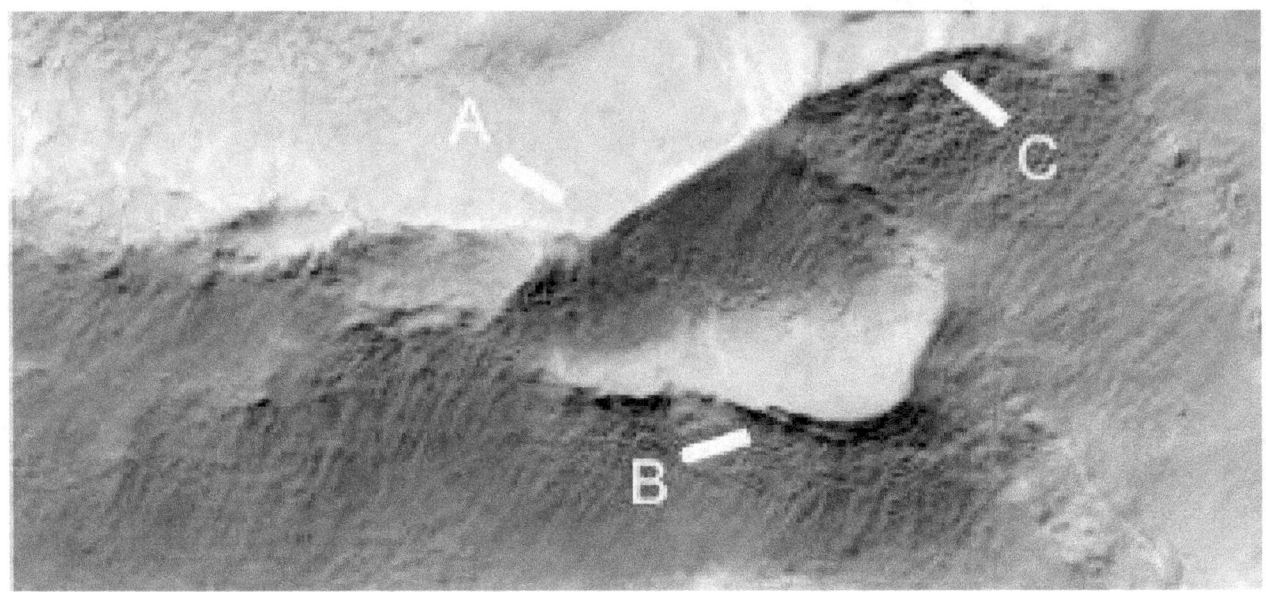

Cymd396d2

Hypothesis

This shows the parabolic dam.

Cymd396d3

Hypothesis

This shows the parabolic arch.

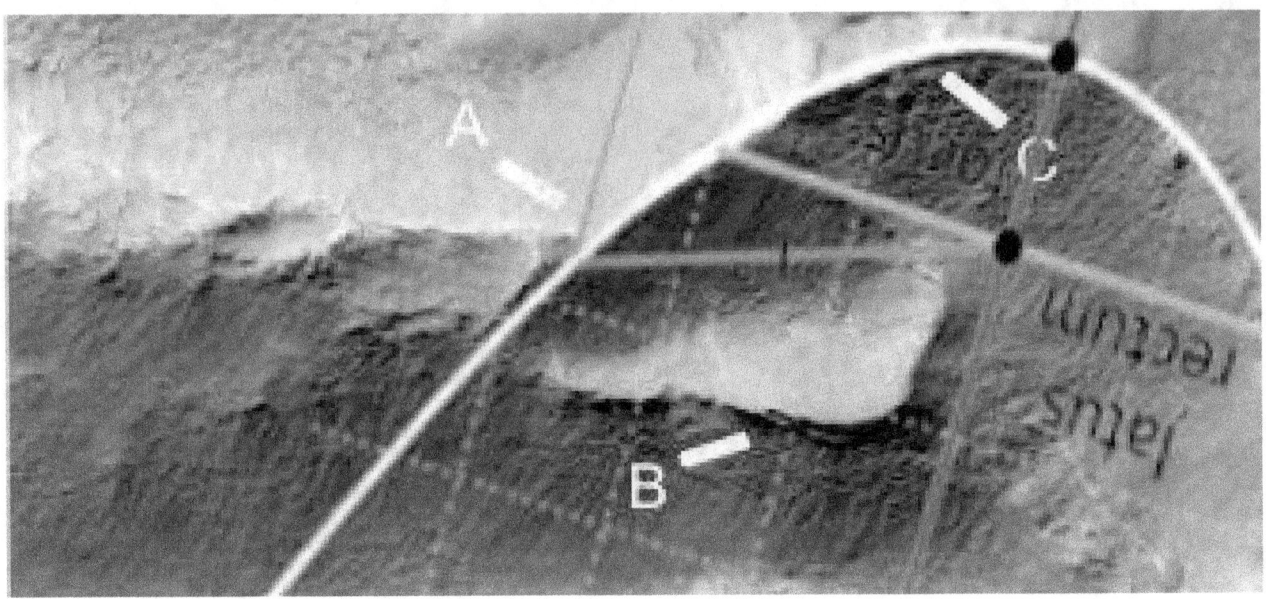

Cymd408a

Hypothesis

An unusual shape pointing up the crater wall, A is one dam, B shows some creep in the dam at C. D at 7 o'clock shows the smooth dam floor compared to the ground above it. At 2 o'clock the wall is eroded or breaking.

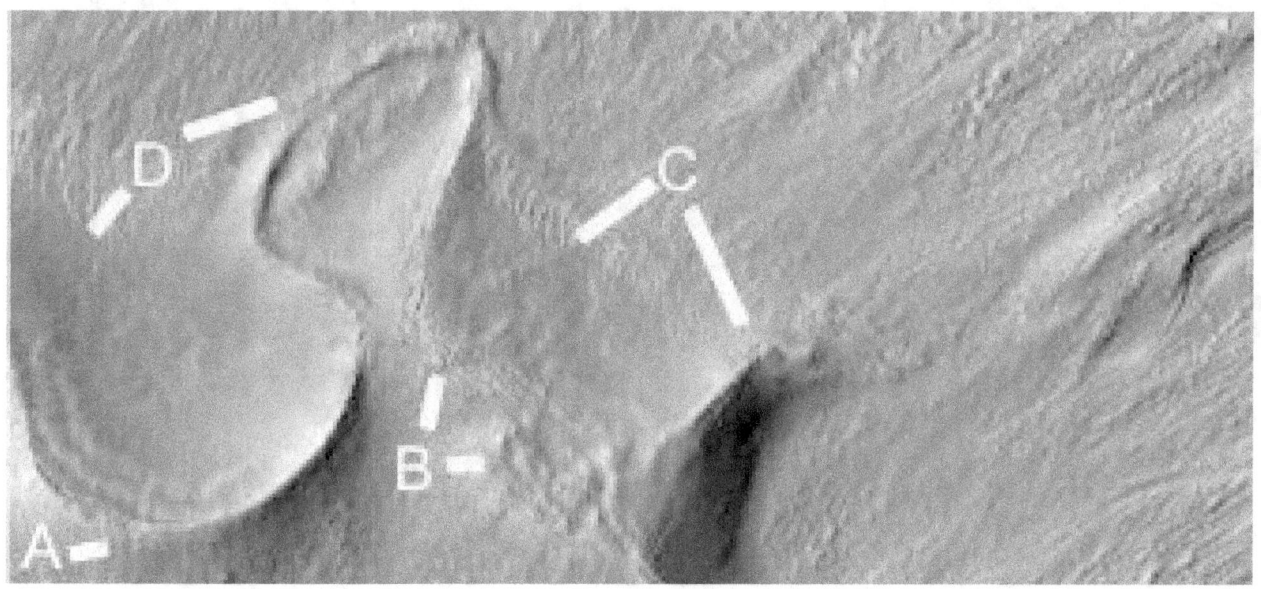

Cymd408a2

Hypothesis

This shows 4 parabolas making up the formation.

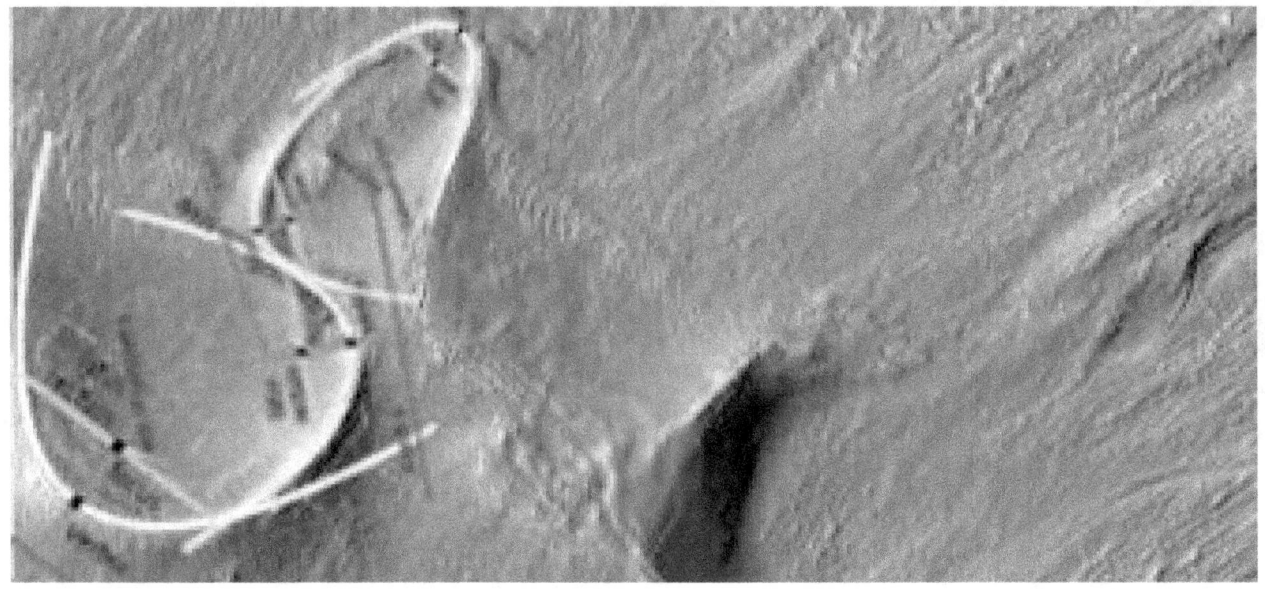

Cymd409b

Hypothesis

A shows a wall associated with a parabolic dam between it and B. C and D are also parabolic, E shows a dam wall with the smooth dam floor above it.

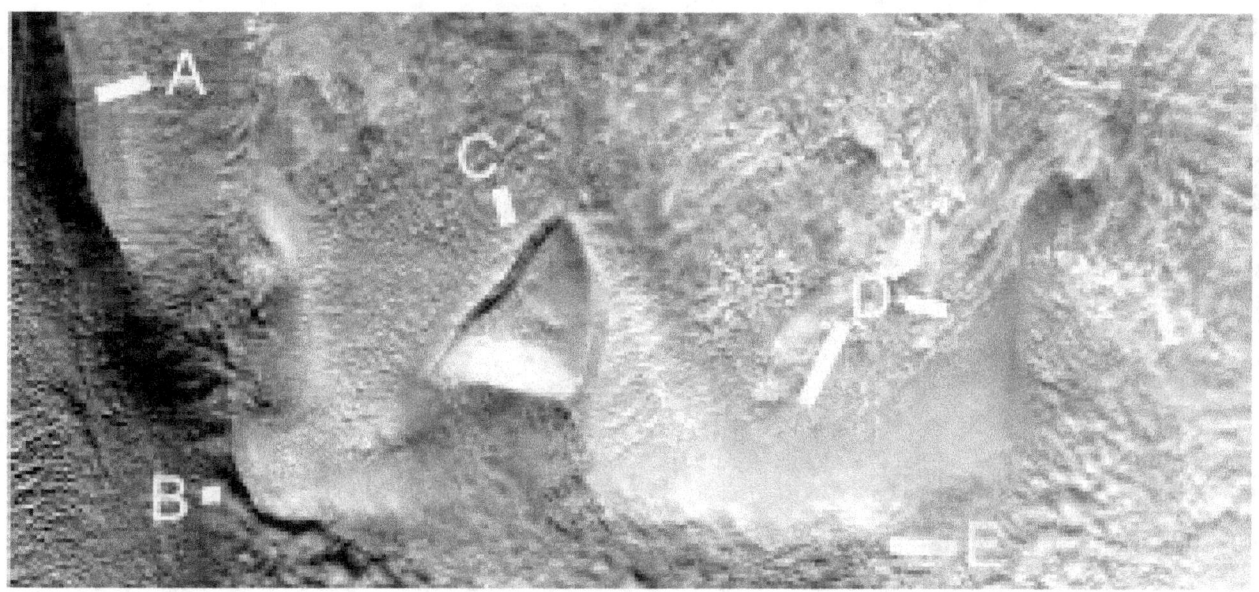

Cymd409b2

Hypothesis

Four parabolas are shown.

Cymd412a

Hypothesis

A shows a dam that is more squarish, B shows how smooth and constant in shape the dam wall is. C and D may show another dam.

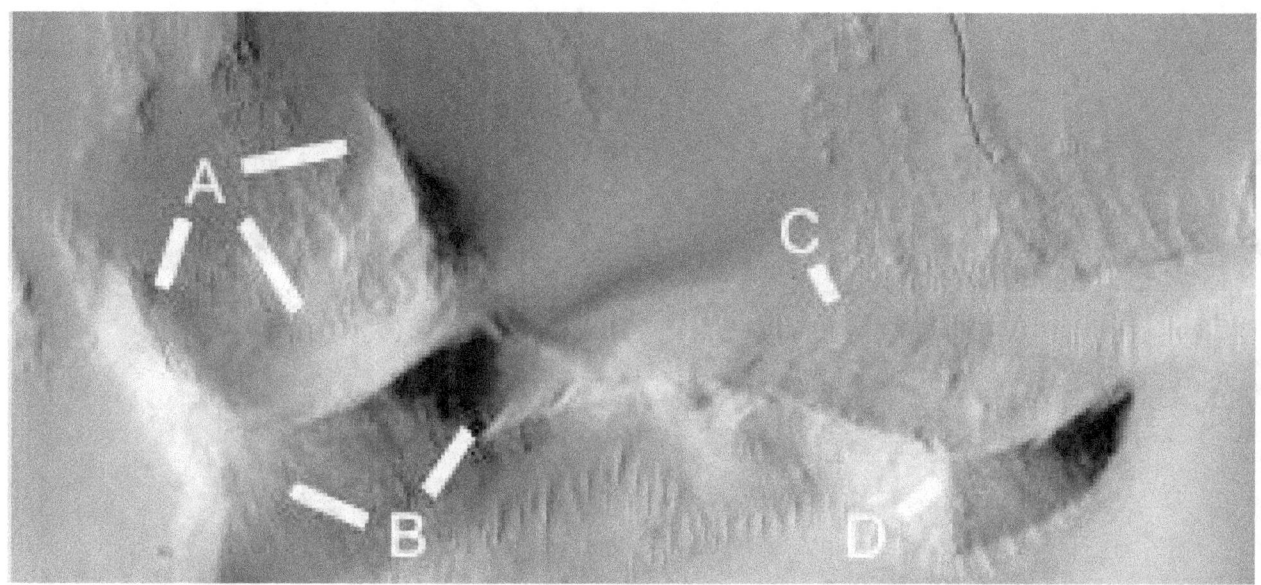

Cymd412a2

Hypothesis

A parabola is shown in the side of the dam at A.

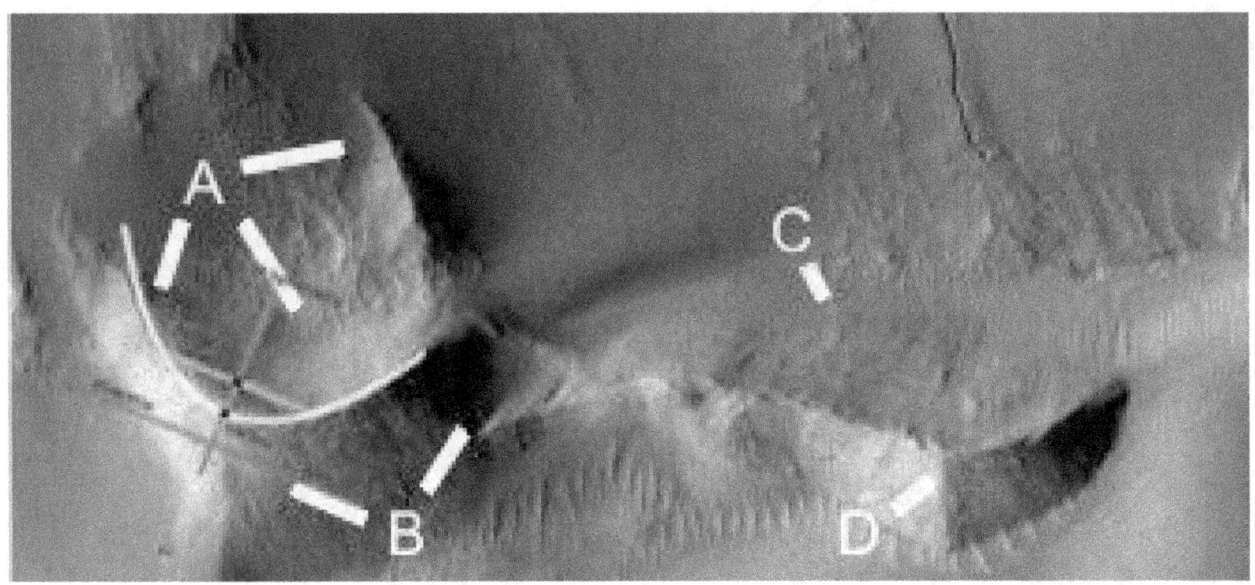

Cymd412b

Hypothesis

A shows the smooth wall of a dam, B the flatter dam floor in it. C may be a parabolic dam filled with silt.

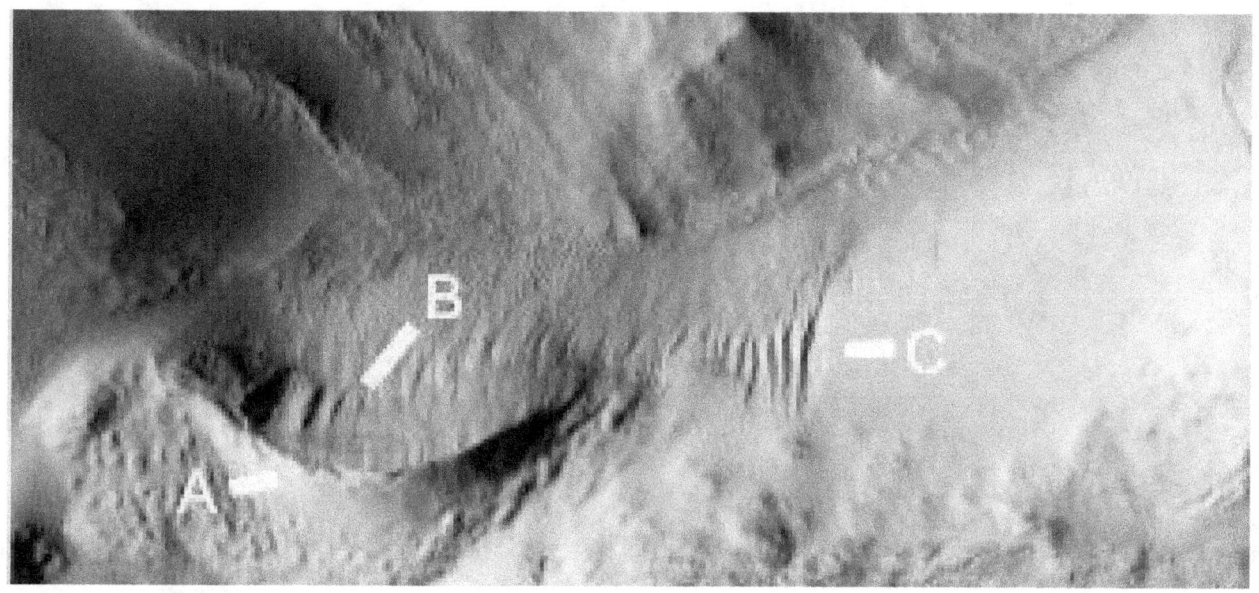

Cymd412b2

Hypothesis

A parabola is shown.

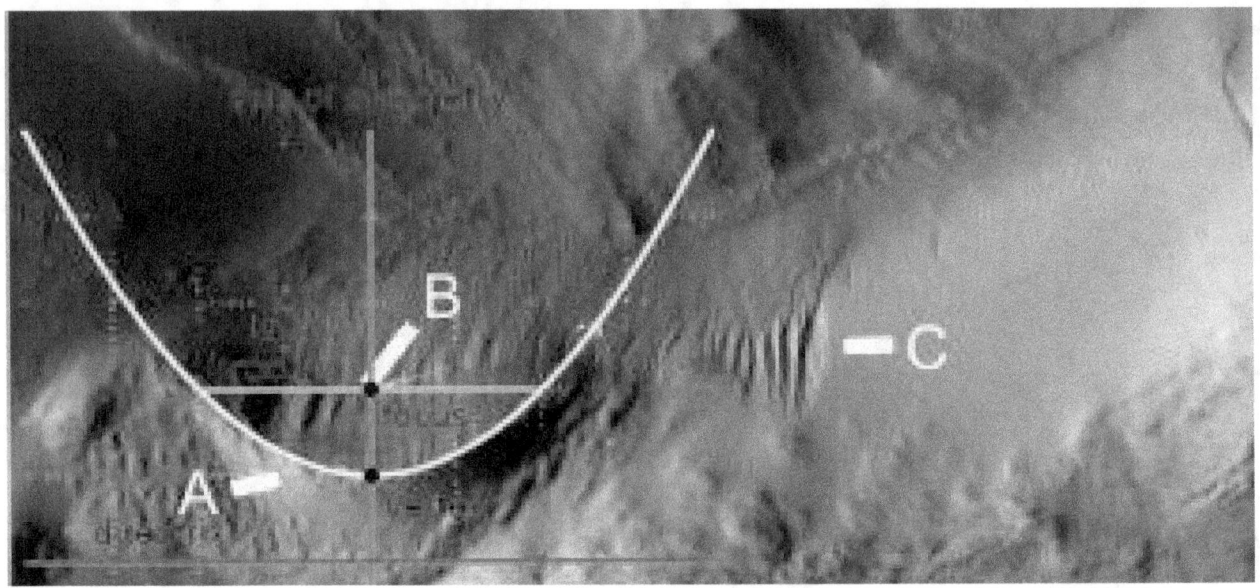

Cymd412d

Hypothesis

A shows a pit dam, where the water would collect in the crater. B shows another pit dam.

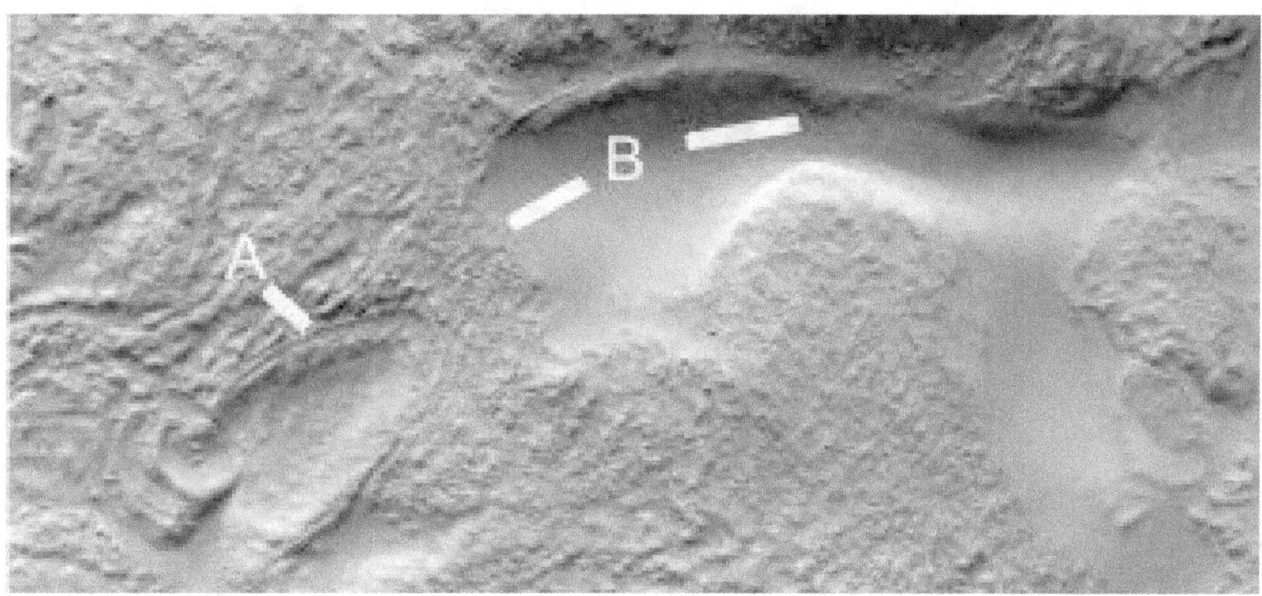

Cymd412d2

Hypothesis

Two parabolas are shown here.

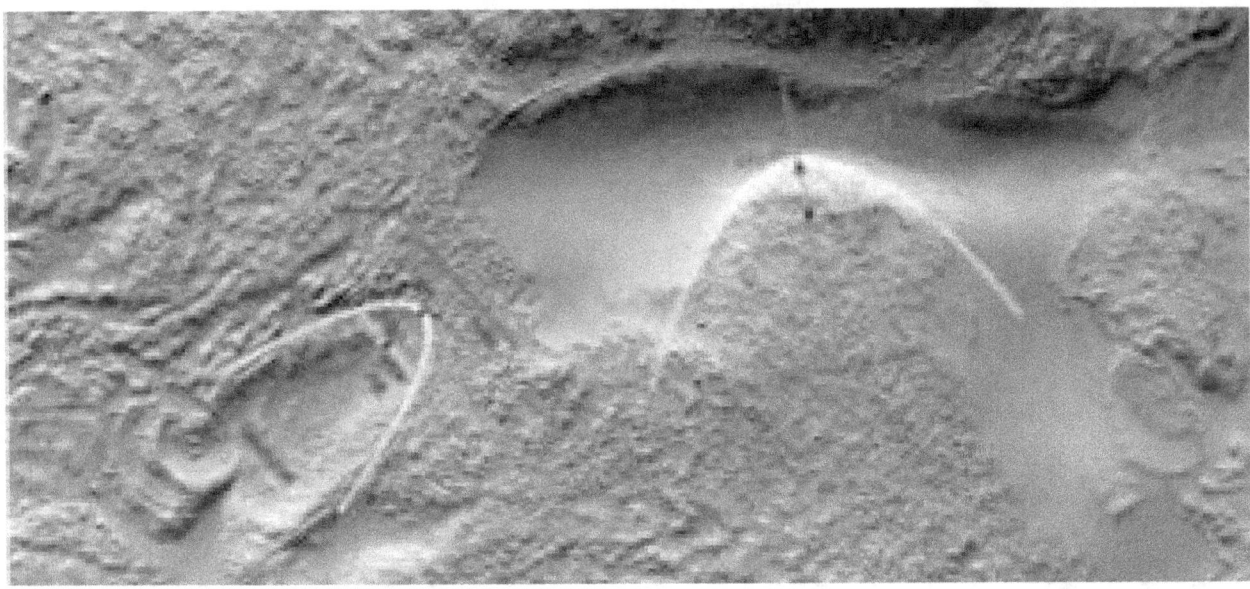

Cymd438a

Hypothesis

A and B are close to the shape of three parabolic dams, C may show a parabola at 2 o'clock along with 1 at 10 o'clock. D shows a parabolic dam at 10 o'clock and close to a parabolic arch at 1 o'clock. The formations to the left of E appears to be parabolic, with the upper and lower parts of the pit to the right both seem to be parabolas.

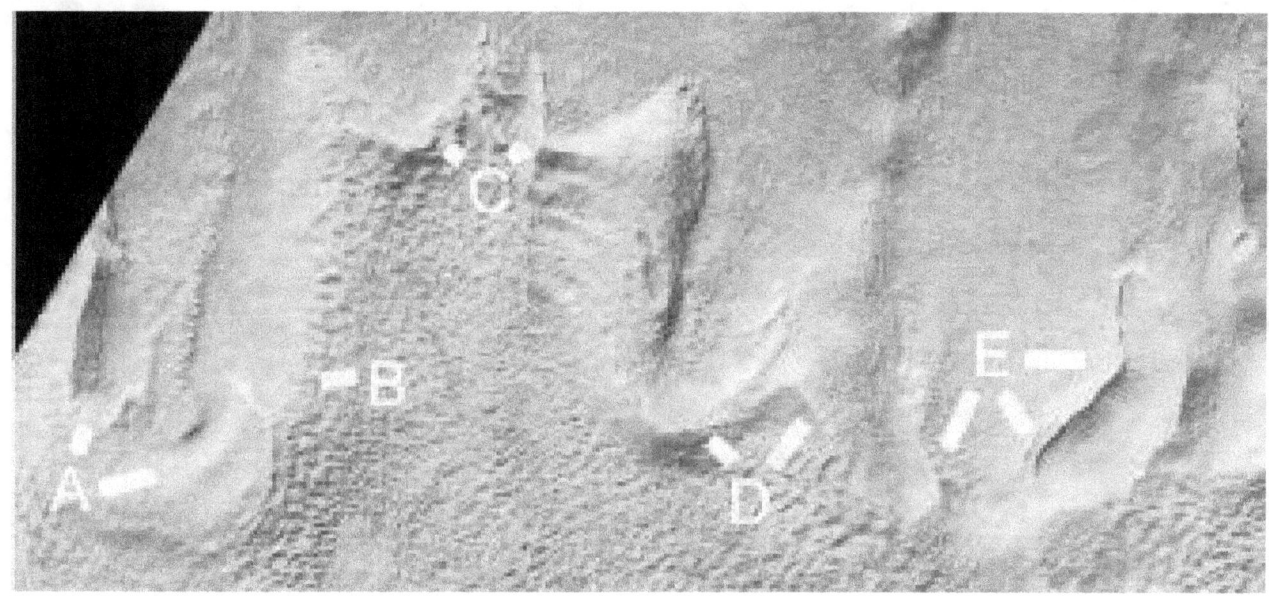

Cymd438a2

Hypothesis

Two parabolas are shown here, the others are too small to confirm.

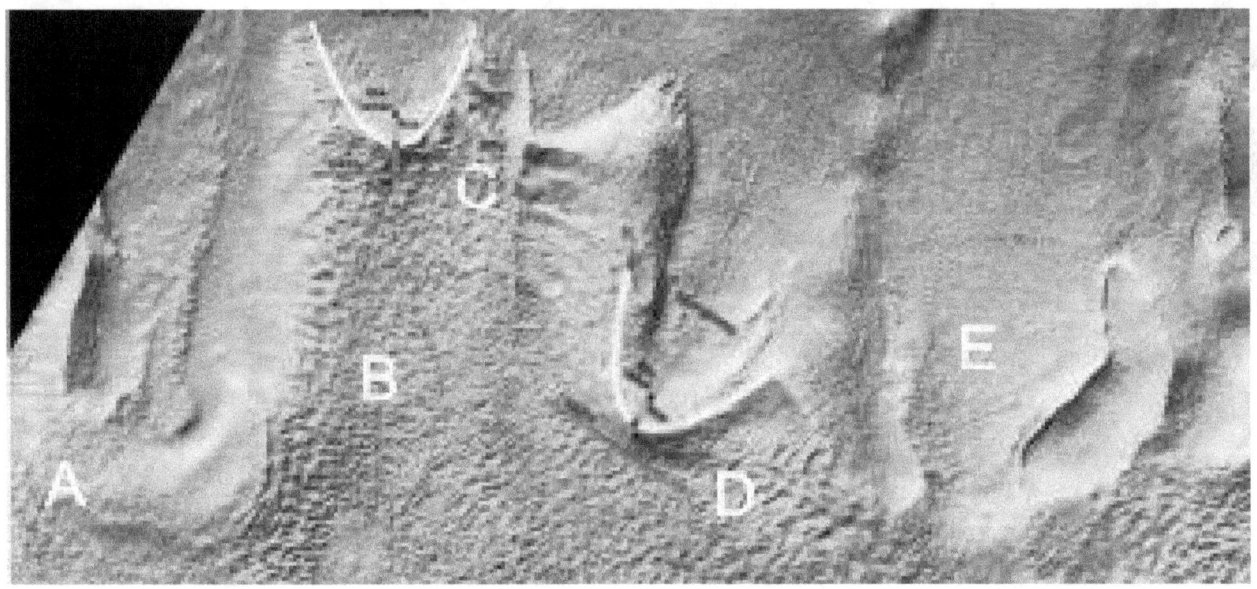

Cymd438c

Hypothesis

A shows a dam, the regular grooves on the dam wall may be pillars to give it strength. B shows where the dam wall is cracking. C shows a smooth dam floor at 5 o'clock, also at 8 o'clock the edges of this shape are smooth. The dam wall comes out of this shape up the image. D shows a dam wall and a smooth dam floor.

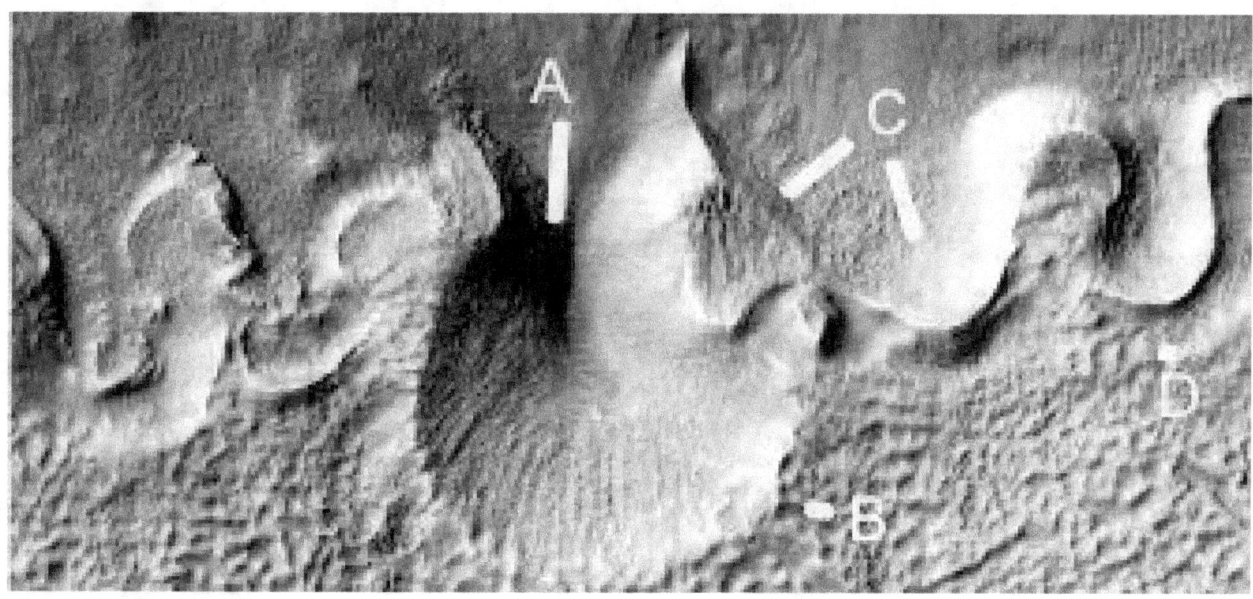

Cymd438c2

Hypothesis

Four parabolas are shown.

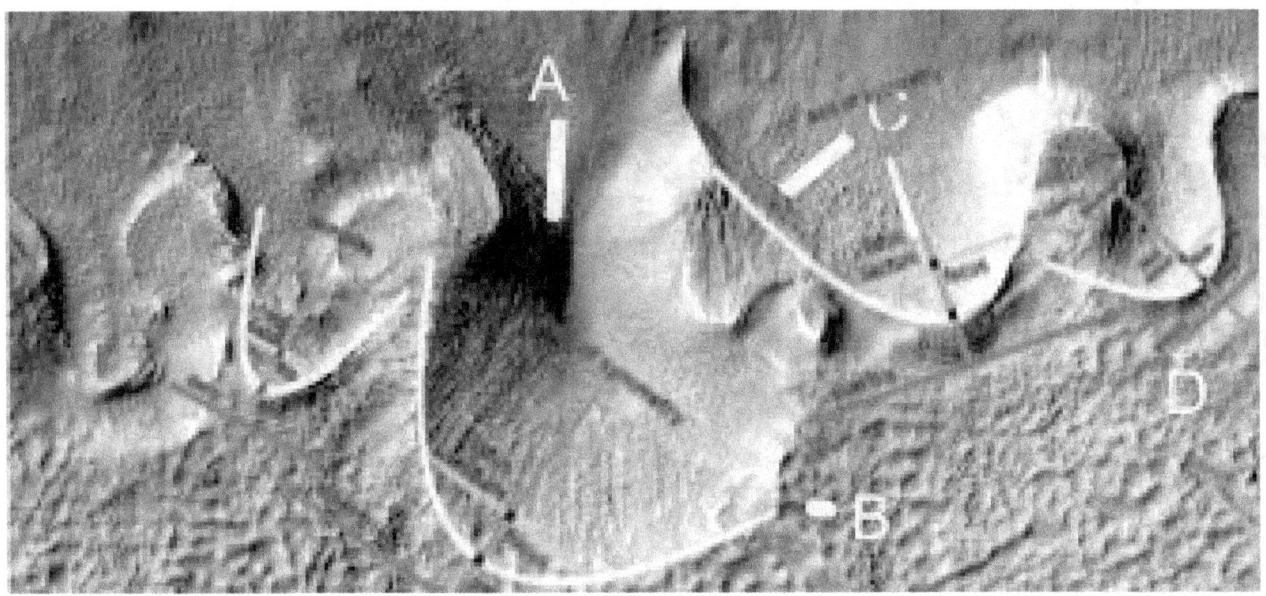

Cymt440b

Hypothesis

Water probably ran into the bottom of the crater, covering the floor with cement allowed for a larger dam. The center remains rough like the original dam floor.

Cymt440b2

Hypothesis

Six parabolas are shown.

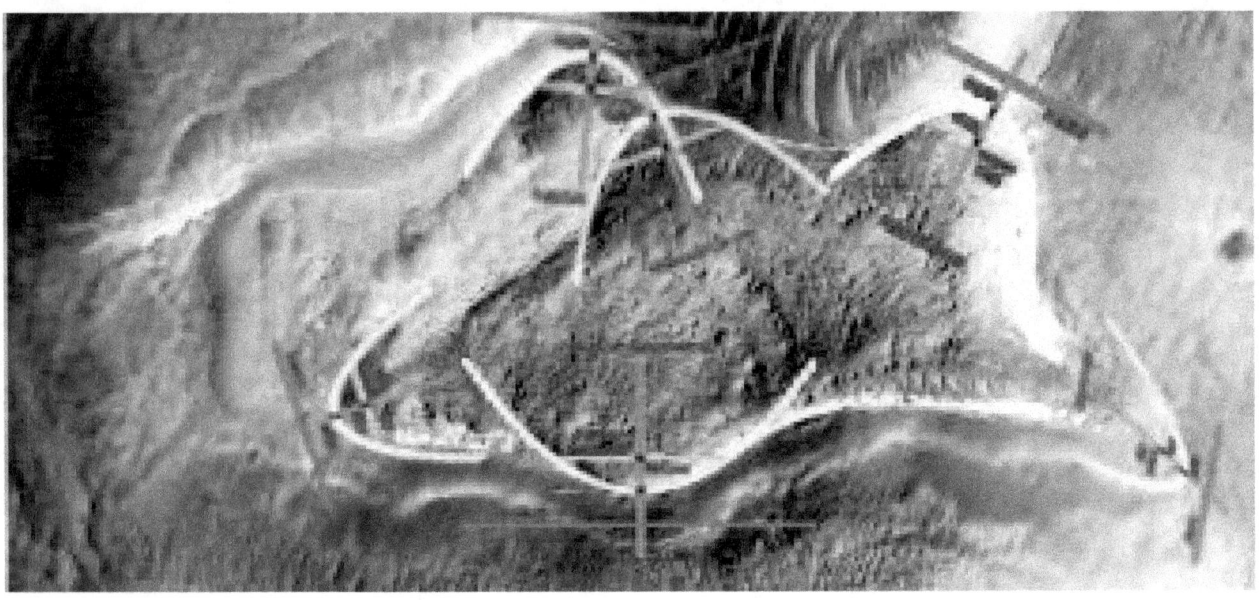

Cymd454c

Hypothesis

A shows a parabolic arch, the remains of smooth cement on it. This goes down to a small dam at 7 o'clock. B shows more of this cement running into a dam at 7 o'clock.

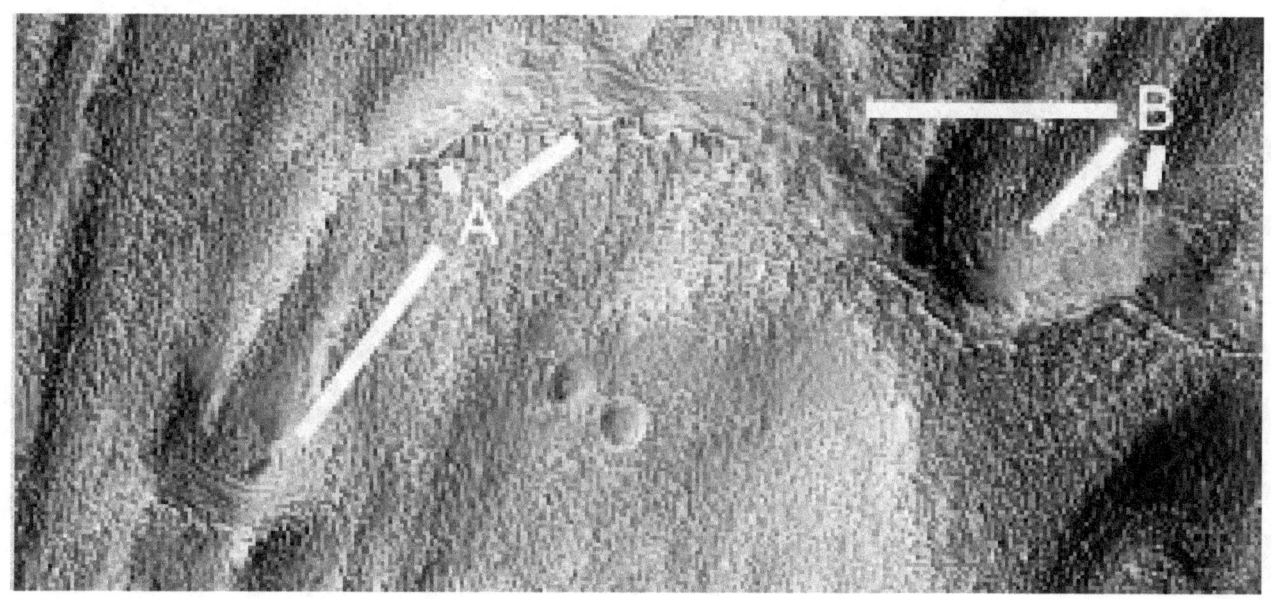

Cymd454c2

Hypothesis

A parabola is shown.

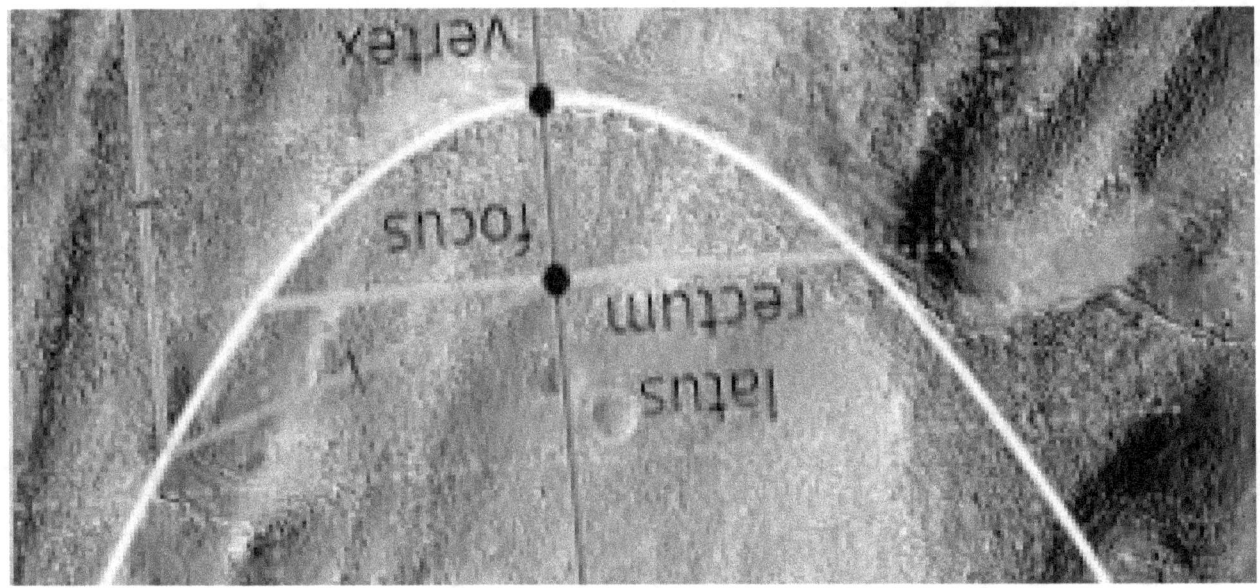

Cymd454g

The dam is in good condition, there are regular stripes along the dam floor perhaps as struts making it stronger. Also there are small vertical grooves on the dam wall perhaps pillars in it.

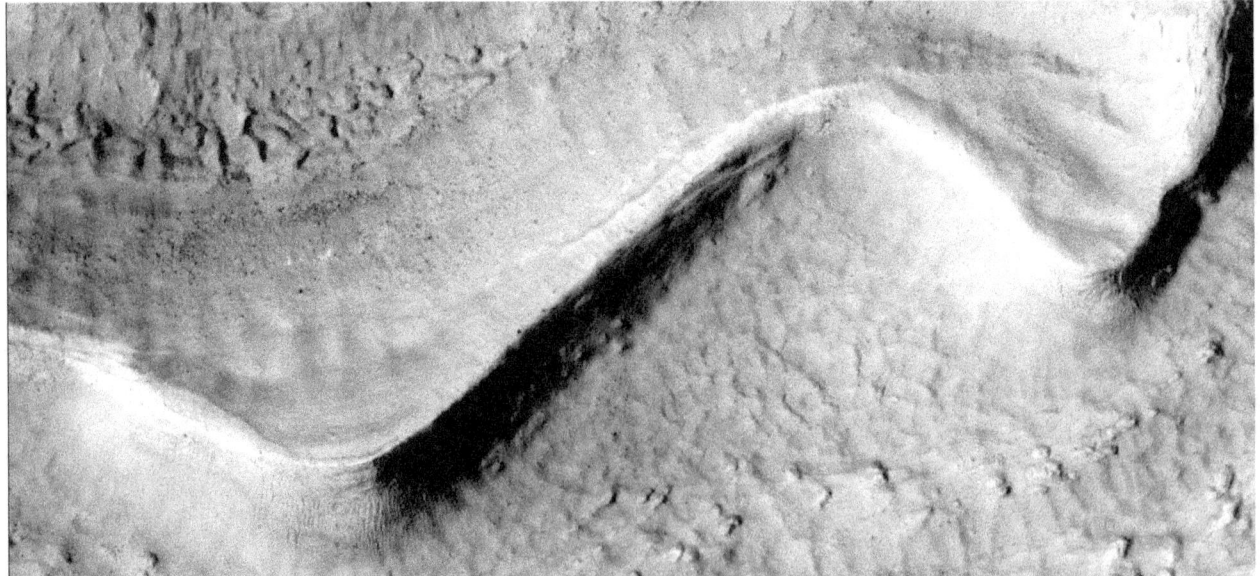

Cymd454g2

Hypothesis

Three parabolas are shown.

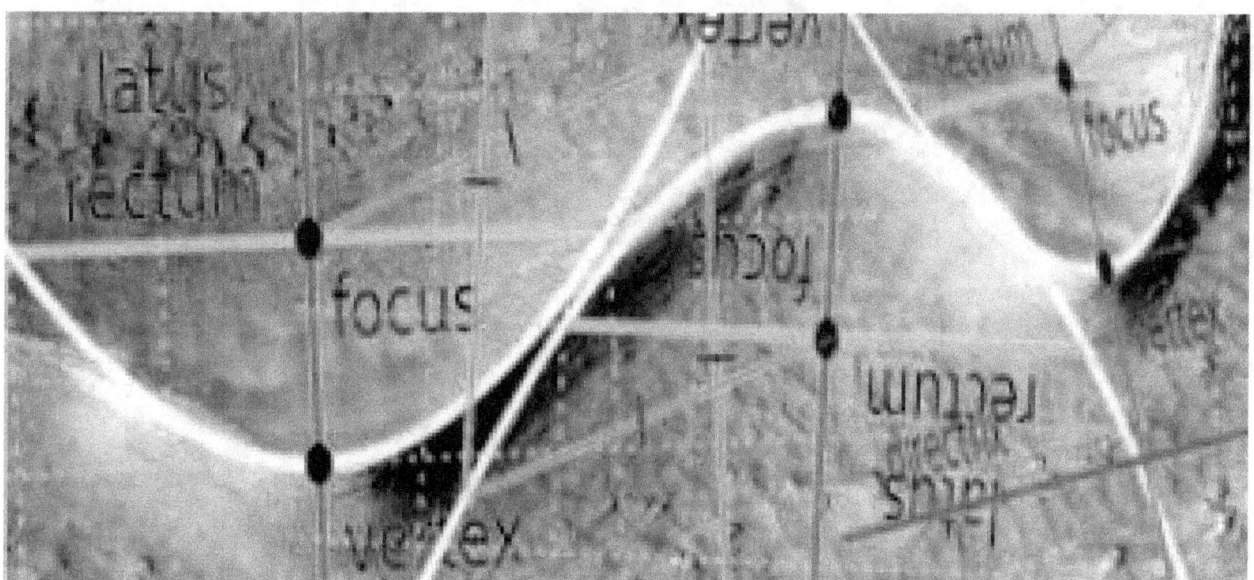

Cymd454h

Hypothesis

A and B show the sides of a water channel, water would have flowed across this at C to another dam. The shape appears so artificial that a natural explanation is hard to sustain.

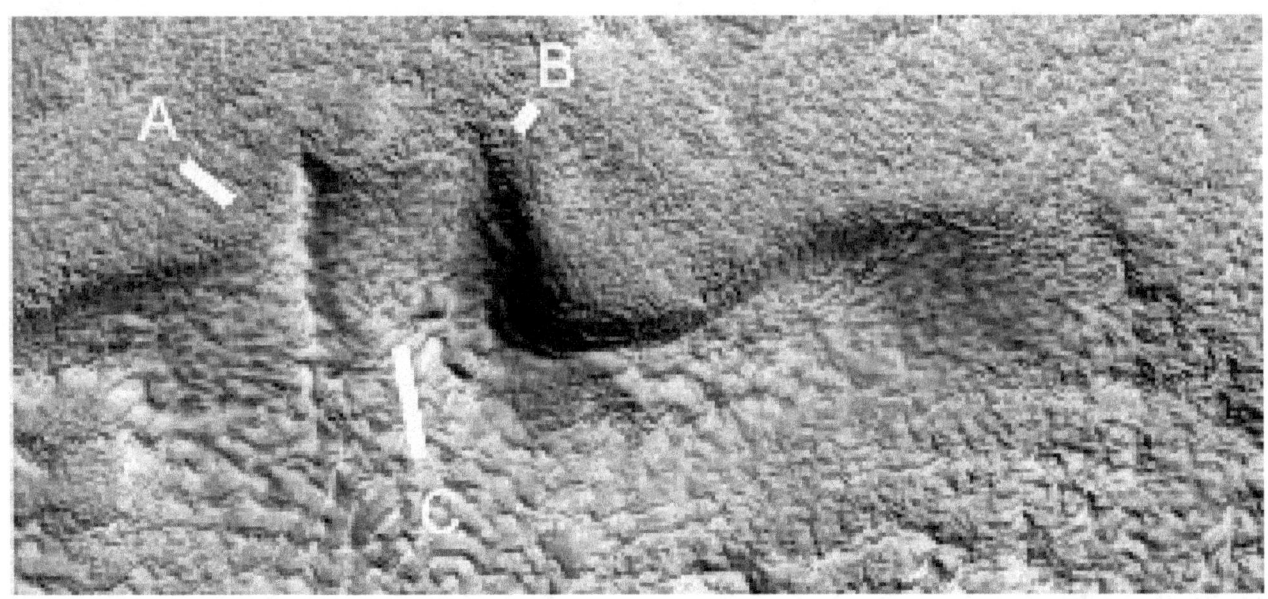

Cymd454h2

A parabola is shown.

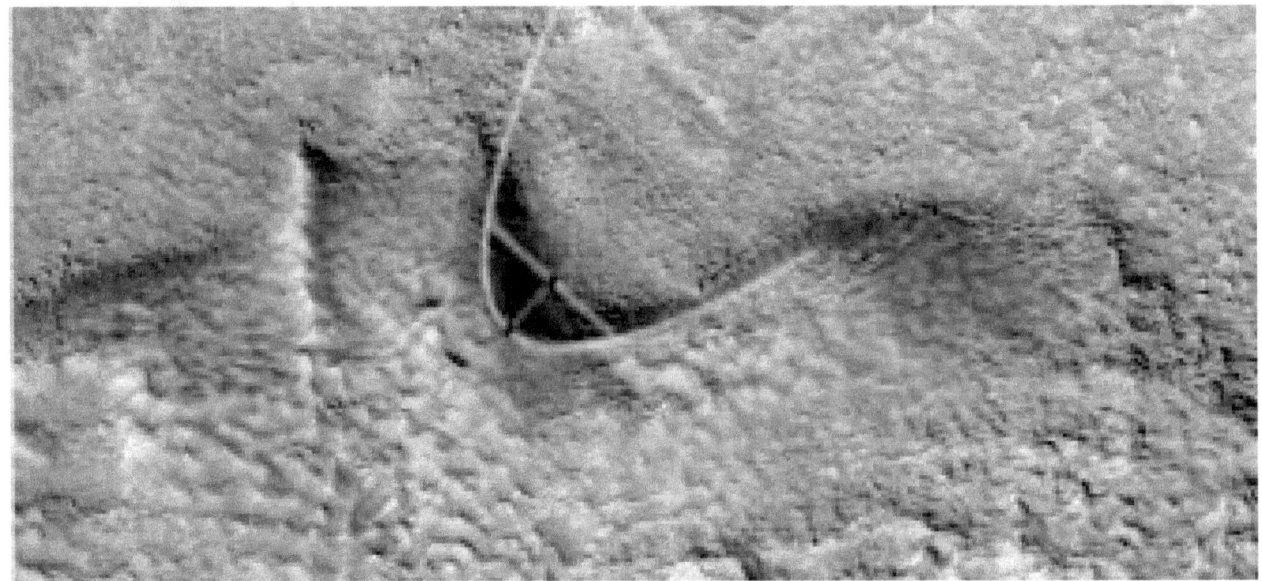

Index

wider tube, 71

---

[i] https://en.wikipedia.org/wiki/Parabolic_arch
[ii] OPTIMUM DESIGN OF ARCH DAMS FOR FREQUENCY LIMITATIONS, INTERNATIONAL JOURNAL OF OPTIMIZATION IN CIVIL ENGINEERING Int. J. Optim. Civil Eng., 2011; 1:1-14